Strategies for Nutrient Use Efficiency Improvement in Plants

Strategies for Nutrient Use Efficiency Improvement in Plants

Guest Editor

Hongmei Cai

MDPI

Basel • Beijing • Wuhan • Barcelona • Belgrade • Novi Sad • Cluj • Manchester

Guest Editor
Hongmei Cai
College of Resources and
Environment
Huazhong Agricultural
University
Wuhan
China

Editorial Office
MDPI AG
Grosspeteranlage 5
4052 Basel, Switzerland

This is a reprint of the Special Issue, published open access by the journal *Plants* (ISSN 2223-7747), freely accessible at: https://www.mdpi.com/journal/plants/special_issues/J562G86VN9.

For citation purposes, cite each article independently as indicated on the article page online and as indicated below:

Lastname, A.A.; Lastname, B.B. Article Title. *Journal Name* **Year**, *Volume Number*, Page Range.

ISBN 978-3-7258-4885-0 (Hbk)
ISBN 978-3-7258-4886-7 (PDF)
https://doi.org/10.3390/books978-3-7258-4886-7

Contents

About the Editor

Hongmei Cai

Hongmei Cai is currently an Associate Professor of the Department of Soil and Plant Nutrition at the College of Resources and Environment of Huazhong Agricultural University in China. She received her MSc and PhD in Biochemistry and Molecular Biology from Huazhong Agricultural University in China, and has more than 20 years of experience in plant nutrition research. As a member of the Society of Plant Nutrition and Fertilizer in China, her research interests include gene function and the regulatory pathway of plant nutrition, the physiological and molecular response in plant nutrition, nutrient interactions, high nutrient-use-efficient variety selection, and optimal fertilization management.

Preface

Nutrient elements are both essential for plant growth and development and the limiting factors for grain yield and quality in crops. Today, intensive high-yield agriculture is highly dependent on the addition of fertilizers such as nitrogen, phosphorus, and potassium. However, further increases in fertilizer application are unlikely to be as effective at increasing yields, as efficiency declines at higher levels of addition. Therefore, understanding how plants respond to environmental nutrient levels to identify appropriate approaches and strategies to improve nutrient use efficiency is of significance for crop yield and quality. This reprint covers the effects of different fertilizer management strategies on plant growth, yield production, physiological response, and nutrient transfer and utilization, as well as the evaluation of nutrient use efficiency in different plant species and varieties.

Hongmei Cai
Guest Editor

plants

MDPI

Article

Productivity of Three Pea (*Pisum sativum* L.) Varieties as Influenced by Nutrient Supply and Meteorological Conditions in Boreal Environmental Zone

Daiva Janusauskaite

Department of Plant Nutrition and Agroecology, Institute of Agriculture, Lithuanian Research Centre for Agriculture and Forestry, Instituto al. 1, LT-58344 Akademija, Kėdainiai District, Lithuania; daiva.janusauskaite@lammc.lt

Abstract: In order to grow crops that reduce the negative impact on the environment, as well as meet the nutritional needs of the increasing human population, it is necessary to include new and more sustainable production strategies into current agricultural systems. The aim of our study was to evaluate the optimal nutritional conditions of semi-leafless pea productivity and ascertain the influence of meteorological factors on the productivity of these plants under boreal environmental conditions. The test involved three semi-leafless pea varieties, one of which was a new variety, and eight N fertilization treatments were used: (1) without fertilizers (N_0), (2) without N fertilizers (N_0), (3) N_{15}, (4) N_{30}, (5) N_{45}, (6) N_{15+15}, (7) N_{60}, and (8) N_{60}. Plots of the second–seventh treatment received a base application of $P_{40}K_{80}$; the eighth treatment received $P_{80}K_{160}$. Fertilizer efficiency depended on the meteorological conditions. Based on their productivity, the pea varieties were arranged in the following descending order: Ieva DS > Respect > Simona. Compared with unfertilized peas, NPK fertilizers enhanced the seed yield by 10.6–12.9% on average. Splitting the N_{30} rate and applying N_{60}, under a background of $P_{40}K_{80}$, was not efficient. The optimal rate of $N_{15–45}P_{40}K_{80}$ fertilizers for peas was determined. Meteorological factors significantly influenced seed yield by 75.2%, 44.1%, and 79.9% for varieties Ieva DS, Simona, and Respect, respectively.

Keywords: NPK fertilization; seed yield; seed quality; semi-leafless pea; yield components

1. Introduction

Peas are cultivated in 84 countries around the world. The high prevalence of peas is due to their great yield potential, remarkable biological properties, and value for human and livestock nutrition. Pea seeds are high in nutritional value, rich in protein, carbohydrates, phosphorus, vitamins A and B, iron, and calcium, and are easily digested [1,2].

The goal of the "European Green Deal" is to identify ways to reduce the overabundance of nutrients in the environment that are a basic source of soil, water, and air pollution, and thus adversely affect biodiversity and climate. The target of the agricultural policy is to reduce fertilizer consumption by at least 20% by the year 2030 while ensuring retention of soil fertility [3]. One of the strategies to achieve these goals is the introduction of legumes, including peas, into crop rotation. The most essential profit of pea growing is the enrichment of soil with nitrogen collected during the nitrogen fixation process, which is also used by the succeeding crops [4,5]. Expanded pea cultivation would reduce the use of nitrogen from mineral fertilizers and maximize the use of biological nitrogen [6,7]. Residues of legumes contribute to the accumulation of organic matter in soil [8,9]. Like all legumes, peas activate the microbial functions of soil which are also related to the increase in organic C [8,10]. Aboveground legume biomass and belowground biomass directly contribute to soil organic carbon accumulation [10–12]. Peas are an important crop for a sustainable future and better food systems [6,13]. Peas are suitable as break crops in wheat-based rotations because they are not susceptible to the same diseases and pests as the cereal [11].

New pea types, the semi-leafless varieties, are characterized by improved disease tolerance, easier harvesting, and higher yield potential [14]. There is a lack of knowledge about the nutrient requirements of semi-leafless peas as they differ from leafy varieties. There are different opinions about pea fertilization. Little data are available on the nutrient requirements of the new semi-leafless cultivars and their response to NPK fertilization. Research is needed on the nutrient requirements of Lithuania's newest pea cultivar Ieva DS and to compare the productivity with already cultivated varieties. The importance of nitrogen for pea is unquestionably high, as in all plants [4,15]. There is an opinion that it is enough to fertilize pea only with "starting" nitrogen rates [16]. However, other studies have shown that in soil with a high nitrate–nitrogen content, starter N application had a detrimental effect on pea germination, and nodulation and seed yield [17]. The efficiency of fertilizers is governed by meteorological conditions and the nutrient content in the soil [18,19]; therefore, recommended NPK rates can vary widely [20,21]. The phosphorus and potassium requirements of peas vary depending on the growing conditions [20]. Potassium stimulates the transport of nitrogen from root nodules to the aboveground parts and affects protein synthesis [22].

The response of different pea varieties to fertilization mostly varies [23–25].

When including peas in the crop sequence, seed inoculation with a specific Rhizobium strain is an important practice for optimizing N nutrition in legumes [1,5]. In previous studies, it was found that the inoculation of pea seed had a strong positive impact on pea productivity, especially in the fields where they were not previously cultivated [6]; however, it did not affect the yield of peas in the fields where they were previously grown [17]. We did not use inoculation because peas are included in our fields' crop rotation.

This study aimed (I) to evaluate the productivity potential of semi-leafless peas and to determine the optimal NPK rates, and (II) to establish the relationship between pea productivity and its components and meteorological factors under boreal environmental conditions.

2. Results

2.1. Pea Seed Yield

Weather conditions differed in the amount of rainfall and its distribution between the experimental seasons (Table 1); therefore, the results were discussed separately for all experimental years. A two-way ANOVA showed that pea seed yield was influenced by variety (factor A) ($p \leq 0.01$), fertilization (factor B) ($p \leq 0.01$), and their interaction (A $\times$ B) ($p \leq 0.05$ and $p \leq 0.01$) (Table 1). There was one exception, when fertilization and A $\times$ B interaction did not have significant influence on seed yield in the 2015 year. This year was distinguished from the others by low rainfall and low HTC, which corresponded to the lowest limit of optimal irrigation. The lack of moisture negatively affected the pea yield and the efficiency of the fertilizers.

Variety was the main factor determined, at 29.7–56.9% of the total variability of the seed yield. Fertilization was responsible for 18.3–34.0% of the differences between the treatments. The interaction A $\times$ B explained the least part of the differences (9.1–8.9%).

The variety Ieva DS showed the highest yield (3.88–5.16 t·ha^{-1}), or 5.9–13.8% higher than the trial mean, regardless of the different weather conditions in the experimental years. The variety Respect exceeded the trial mean by 7.8% and 3.2% only under sufficiently wet weather conditions (HTC 1.6 at 2014 and 1.4 at 2017, respectively). The seed yield of the variety Simona was 9.3–13.8% lower than the trial mean, with one exception, when it exceeded the trial mean by 3.9% under the dry weather conditions in 2015.

Table 1. Influence of variety and fertilization on seed yield (t·ha^{-1}) of pea.

Variety (Factor A)	Fertilization (Factor B)	2014	2015	2016	2017
	The effect of variety				
Ieva DS		5.16 a§	3.88 a§	4.70 a§	4.69 a§
Simona		4.19 c	3.77 a	3.68 c	4.02 c
Respect		5.24 a	3.25 c	4.02 c	4.57 a
Trial mean		4.86	3.63	4.13	4.43
	The effect of fertilization				
	1. NPK 0:0:0	4.29 b†	3.53 b†	3.58 b†	3.78 b†
	2. 0:40:80	4.72 a	3.44 b	3.85 a	4.35 a
	3. 15:40:80	4.88 a	3.61 b	3.96 a	4.37 a
	4. 30:40:80	4.95 a	3.64 b	4.16 a	4.66 a
	5. 45:40:80	5.10 a	3.75 b	4.40 a	4.64 a
	6. 15 + 15:40:80	4.76 a	3.60 b	4.10 a	4.27 a
	7. 60:40:80	4.99 a	3.73 b	4.39 a	4.68 a
	8. 60:80:160	5.24 a	3.75 b	4.62 a	4.67 a
Mean of NPK fertilized treatments		4.98	3.65	4.21	4.52
Differences fertilized NPK (3–8) vs. unfertilized (1)		0.69	0.11	0.63	0.74
Contribution (%) of sum squares) of variety (factor A), fertilization (factor B), and their interaction					
A		56.9 **	29.7 **	56.2 **	34.0 **
B		18.3 **	4.3	30.8 **	34.0 **
A × B		8.9 *	1.8	6.1 **	8.5 *

Different letters in column denote a statistically significant difference (at $p \leq 0.05$ according to LSD): §—between treatments and trial mean; †—between treatments; * $p \leq 0.05$; ** $p \leq 0.01$.

The mean of the fertilized treatments varied from 3.65 to 4.98 t·ha^{-1}. Depending on the year's weather conditions, the efficiency of the fertilizers was not the same during different years. Under the effect of NPK fertilization, seed yield increased by 0.11–0.74 t·ha^{-1}.

A three-way ANOVA showed that the weather conditions of the experimental year (factor A) was the main factor responsible for 39.4% of the total variability of the averaged data of the seed yield (Table 2). Variety (factor B) explained 15.8%, fertilization (factor C) governed 11.2%, and A × B and A × C interactions explained 12.3% and 2.4%, respectively, of the yields' averaged data variations. In all cases, the influence of the mentioned factors was significant at $p \leq 0.01$. No significant effect of interactions B × C and A × B× C on seed yield was established.

The most productive pea variety was Ieva DS, which produced the average seed yield of 4.61 t·ha^{-1} and, compared with the control, exceeded the trial mean by 0.35 t·ha^{-1}, or 8.2%. According to the average yield, varieties were arranged in the following descending order: Ieva DS > Respect > Simona.

NPK fertilizer increased the seed yield by 0.55 t·ha^{-1} (or 15.6%) on average, compared with unfertilized pea. The influence of fertilizers on seed yield was significant ($p \leq 0.05$) in all cases. Under a background of $P_{40}K_{80}$, the application of N_{15}, N_{30}, and N_{45} rates increased seed yield by 0.42 t·ha^{-1}, 0.57 t·ha^{-1}, and 0.69 t·ha^{-1} (or 11.1%, 15.1%, and 18.3%, respectively), compared with the control ($N_0P_0K_0$) (Figure 1). Splitting the N_{30} rate (N_{15} + N_{15}) (sixth treatment) did not exceed its single application and yielded the same efficiency as N_{15} (third treatment). Seed yield did not increase with increasing the nitrogen fertilizer rate above N_{45}. The most abundant fertilization ($P_{80}K_{160}$ + N_{60}) resulted in the highest (4.57 t·ha^{-1}) yield, or 20.9% higher, compared with unfertilized pea.

Table 2. Influence of year, variety, and fertilization on seed yield and yield components of pea (averaged 2014–2017).

Year (Factor A)	Variety (Factor B)	Fertilization (Factor C)	Number of Pods Per Plant	Number of Seeds Per Plant	Weight of Seeds Per Plant (g)	Number of Pods Per m^2	Seed Yield ($t \cdot ha^{-1}$)
		The effect of years					
2014			8.1 a§	29.9 a§	6.4 c§	839 a§	4.86 a§
2015			4.9 c	19.6 c	5.3 c	450 c	3.63 c
2016			5.1 c	18.6 c	4.9 c	568 c	4.13 c
2017			5.7 c	22.1 b	11.6 a	574 c	4.43 a
		The effect of variety					
	Ieva DS		6.6 a§	22.6 b§	9.9 a§	655 a§	4.61 a§
	Simona		6.0 b	23.8 a	5.9 c	604 b	3.91 c
	Respect		5.2 c	21.2 c	5.3 c	564 c	4.27 b
	Trial mean		6.0	22.5	7.0	608	4.26
		The effect of fertilization					
		1. NPK 0:0:0	5.4 b†	20.7 b†	6.5 b†	552 b†	3.78 b†
		2. 0:40:80	5.7 b	21.7 b	6.7 b	562 b	4.09 a
		3. 15:40:80	5.8 a	22.1 a	6.9 b	593 a	4.20 a
		4. 30:40:80	6.1 a	22.5 a	7.3 a	607 a	4.35 a
		5. 45:40:80	6.1 a	23.1 a	7.3 a	614 a	4.47 a
		6. 15 + 15:40:80	6.2 a	23.3 a	7.3 a	629 a	4.18 a
		7. 60:40:80	6.1 a	23.0 a	7.1 a	644 a	4.45 a
		8. 60:80:160	6.2 a	23.8 a	7.4 a	662 a	4.57 a
Mean of NPK fertilized treatments (3–8)			6.1	23,0	7.2	625	4.37
Differences fertilized NPK (3–8) vs. unfertilized (1)			0.7	2.1	0.7	73	0.55
Contribution (%) of sum squares) of year (factor A), variety (factor B), fertilization (factor C), and their interaction							
	A		59.9 **	57.0 **	29.1 **	62.1 **	39.4 **
	B		11.4 **	3.1 **	16.8 **	4.3 **	15.8 **
	C		2.6 **	2.5 **	0.34 **	3.9 **	11.2 **
	A × B		7.2 **	13.3 **	48.7 **	4.5 **	12.3 **
	A × C		0.8	1.6	0.4	3.8 **	2.4 **
	B × C		1.0	1.1	0.2	1.1	1.0
	A × B × C		1.7	2.4	0.7	1.9	2.25

Different letters in column denote a statistically significant difference (at $p \leq 0.05$ according to LSD): §—between treatments and trial mean; †—between treatments; ** $p \leq 0.01$.

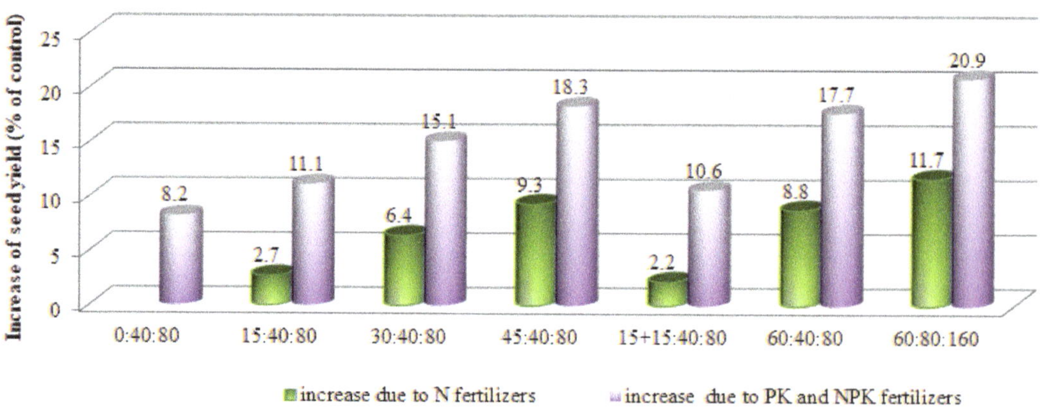

Figure 1. The influence of fertilization on increase in seed yield (% of control) (averaged across varieties).

The seed yield of pea varieties was found to be significantly correlated with the N rate (Figure 2). The relationship was strong and ranged from 0.789 * to 0.956 *.

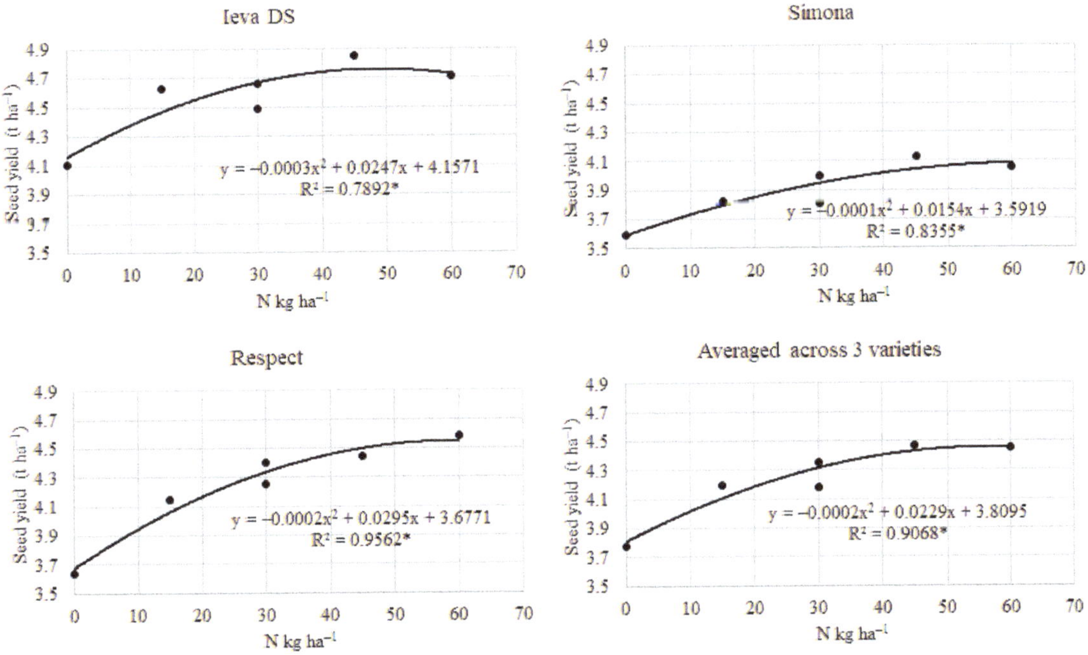

Figure 2. The relationship between N rate and pea seed yield (averaged 2014–2017). * significant at $p \leq 0.05$.

2.2. Seed Yield Components

The analysis of variance revealed that all seed yield components were significantly ($p \leq 0.01$) influenced by year (factor A), variety (factor B), fertilization (factor C), and A × B interaction (Table 2). Year was the main factor determining 59.9% of the number of pods per plant (NPP), 57.0% of the number of seeds per plant (NPS), 29.1% of the weight of seeds per plant (WSP), and 62.1% of the number of pods per m^2 (NPm^2) total variability. During the wet year, 2014 (HTC = 1.6), the conditions were most favourable for the formation of seed yield components. Compared with the trial mean, NPP, NSP, and NPm^2 were found to be significantly higher by 35.0%, 32.9%, and 38.0%, in 2014. Meanwhile, during other research years, yield components were less than the trial mean.

Variety explained 11.4%, 3.1%, 16.8%, and 4.3% of the total variability of NPP, NSP, WSP, and NPm^2, respectively. The variety Ieva DS was distinguished from other varieties by its values of the yield components: NPP, WSP, and NPm^2 were higher by 10.0%, 41.4%, and 7.7%, respectively, compared with the trial mean. Meanwhile NPP, NSP, WSP, and NPm^2 for the variety Respect were lower by 13.3%, 5.6%, 24.3%, and 7.2% than the trial mean.

Fertilization resulted in the lowest part of the yield components variation and explained only 2.6%, 2.5%, 0.34%, and 3.9% of NPP, NSP, WSP, and NPm^2, respectively, of their total variability. Fertilization resulted in 11.5%, 9.2%, 8.3%, and 10.1% higher NPP, NSP, WSP, and NPm^2, respectively, compared with unfertilized peas.

2.3. Seed Quality

The thousand seed weight (TSW) was significantly ($p \leq 0.01$) influenced by year (factor A), variety (factor B), fertilization (factor C), and A × B and A × C interactions (Table 3). Year was the major factor responsible for most (48.9%) of the TSW total variability. A × B interaction governed 36.2% of the differences between treatments. The influence of variety, fertilization, and A × C interaction was small and explained only 1.2–3.7% of the TSW variation. In 2014, the TSW was 9.5% lower than the trial mean; meanwhile, the TSW exceeded the trial mean by 4.3% and 4.7% in 2016 and 2017, respectively. NPK fertilization significantly decreased TSW by 2.7–3.5%.

Table 3. Influence of year, variety, and fertilization on thousand seed weight (TSW), protein content of pea, and root nodules per plant (averaged 2014–2017).

Year (Factor A)	Variety (Factor B)	Fertilization (Factor C)	TSW (g)	Protein (%)	Root Nodules (Number Per Plant)
The effect of years					
2014			229 c§	24.4 a§	8.4 b§
2015			253 b	21.7 c	17.0 a
2016			264 a	21.8 c	24.8 a
2017			265 a	22.0 c	42.8 a
The effect of variety					
	Ieva DS		258 a§	22.7 a§	23.9 b§
	Simona		248 a	23.4 a	24.6 b
	Respect		252 c	21.5 c	21.2 b
	Trial mean		253	22.5	23.2
The effect of fertilization					
		1. NPK 0:0:0	259 b†	23.0 b†	25.4 b†
		2. 0:40:80	252 c	22.1 c	24.9 b
		3. 15:40:80	251 c	22.4 c	24.4 b
		4. 30:40:80	252 c	22.5 c	23.7 b
		5. 45:40:80	253 c	22.5 c	20.8 b
		6. 15 + 15:40:80	251 c	22.6 c	23.2 b
		7. 60:40:80	253 c	22.5 c	20.4 b
		8. 60:80:160	250 c	22.3 c	23.1 b
Mean of NPK fertilized treatments (3–8)			252	22.4	22.6
Differences fertilized NPK (3–8) vs. unfertilized (1)			−8	−0.6	−2.8
Contribution (%) of sum squares) of year (A), variety (B), fertilization (C), and their interaction					
	A		48.9 **	44.7 **	50.7 **
	B		3.7 **	22.5 **	0.7
	C		1.8 **	2.2 **	0.9
	A × B		36.2 **	20.3 **	4.4 **
	A ×C		1.2 **	1.1 *	1.9
	B × C		0.5	0.5	2.7
	A × B × C		0.7	0.8	5.3

Different letters in column denote a statistically significant difference (at $p \leq 0.05$ according to LSD): §—between treatments and trial mean; †—between treatments; * $p \leq 0.05$; ** $p \leq 0.01$.

The seed protein content was significantly influenced ($p \leq 0.05$ and $p \leq 0.01$) by year (factor A), variety (factor B), fertilization (factor C), and A × B and A × C interactions (Table 3). The year explained a large part (44.7%) of the total variability and variety and A × B interactions were responsible for half as much variation (22.5% and 20.3%, respectively). Meanwhile, fertilization and A × C interactions determined only 2.2% and 1.1%, respectively, of the total variance of the protein content. Under the wettest conditions, in 2014, the protein content was 8.4% higher than the trial mean. The variety Simona significantly accumulated the highest amount of protein (by +4.0 percentage point (pp)), whereas the lowest amount of protein was accumulated by the variety Respect (by −4.4 pp), in comparison with the trial mean.

2.4. Root Nodules

The effect of year and A × B interaction on root nodules was significant ($p \leq 0.01$) (Table 3). All other factors had no significant effect on this indicator. Under the normal moisture regime in 2016 and 2017, the number of root nodules was the highest, and exceeded the trial mean by 6.7% and 84.5%, respectively.

2.5. Relationship between Seed Yield, Yield Components, Seed Quality Indices, and Meteorological Indices

We ascertained a correlation between seed yield and yield components (Table 4). The data showed that seed yield for the variety Ieva DS significantly ($p \leq 0.05$ and $p \leq 0.01$) correlated with all yield components, the number of root nodules (NRN), the insertion height of the first pod (IFP), and seed quality indices. The correlation of seed yield for the variety Simona with the mentioned indices was not significant in all cases: NRN did not correlate with seed yield. No correlation was found between seed yield for the variety Respect and WSP, NRN, and protein content. Data, averaged across varieties, revealed that this correlation ranged from weak to moderate and was significant in most cases.

Table 4. Correlation analysis among seed yield, yield components, and seed quality indices.

Index	Ieva DS	Simona	Respect	Mean across Varieties
PH	0.726 **	0.663 **	0.759 **	0.598 **
NPP	0.838 **	0.699 **	0.672 **	0.570 **
NSP	0.767 **	0.712 **	0.548 **	0.423 **
WSP	0.417 *	0.683 **	0.560	0.380 **
NPm	0.840 **	0.629 **	0.880 **	0.696 **
IFP	0.592 **	0.660 **	0.802 **	0.620 **
NRN	−0.369 *	−0.116	−0.035	−0.167
PC	0.690 **	0.509 **	−0.120	0.138
TSW	−0.813 **	−0.385 **	−0.529 **	−0.437 **

PH—plant height, NPP—number of pods per plant, NSP—number of seeds per plant, WSP—weight of seeds per plant, NPm—number of pods per m^2, IFP—insertion height of the first pod, NRN—number of root nodules, PC—protein content, TSW—thousand seed weight; * $p \leq 0.05$; ** $p \leq 0.01$.

A simple correlation analysis confirmed the existence of a correlation between seed yield, the morphological traits of peas, and meteorological indices throughout the growing season (Table 5). However, differences among the varieties were noticed in the strength of the relationship among the variables as shown by the values of the correlation coefficients. Averaged across varieties, seed yield positively corelated with precipitation ($p \leq 0.01$) and HTC ($p \leq 0.01$), and negatively correlated with sunshine duration ($p \leq 0.01$). The accumulated growing degree days (AGDD) > 5 °C, AGDD > 10 °C, and HTC positively correlated with morphological traits and PC, except when AGDD > 5 °C and AGDD > 10 °C correlated negatively with NRN and TSW. Sunshine duration and relative air humidity negatively correlated with all indices in most cases.

The multiple linear regression model showed that seed yield, morphological traits, and quality indices were intensely influenced by weather conditions all through the growing season (Table 6). It was found that the interaction of meteorological factors influenced seed yield by 75.2%, 44.1%, and 79.9% for varieties Ieva DS, Simona, and Respect, respectively. Averaged across varieties, meteorological conditions caused from 50.6 to 87.4% of morphological traits and 76.2% root nodule data variation. Protein content and TSW were influenced by weather conditions by 48.6% and 52,6%, respectively.

Table 5. Correlation analysis among the yield components and meteorological indices.

Variety	Indices	Precipitation	AGDD > 5 °C	AGDD > 10 °C	Sunshine Duration	Relative Air Humidity	HTC
Ieva DS	SY	0.501 **	0.290	0.200	−0.715 **	0.103	0.659 **
	PH	−0.007	0.357 *	0.426 *	−0.480 **	−0.455 **	0.600 **
	NPP	0.127	0.501 **	0.502 **	−0.496 **	−0.166	0.525 **
	NSP	0.228	0.264	0.297	−0.533 **	−0.332	0.615 **
	WSP	0.012	−0.241	−0.151	−0.332	−0.530 **	0.486 **
	NPm	0.328	0.365 *	0.320	−0.615 **	−0.033	0.601 **
	IFP	−0.086	0.093	0.214	−0.437 *	−0.671 **	0.627 **
	NRN	0.334	−0.857 **	−0.863 **	−0.230	−0.414	0.278
	PC	0.118	0.392 *	0.421 *	−0.555 **	−0.313	0.628 **
	TSW	−0.145	−0.709 **	−0.657 **	0.387 *	−0.113	−0.331
Simona	SY	0.004	0.020	0.093	−0.359 *	−0.448 *	0.484 **
	PH	−0.251	0.326	0.460 **	−0.280	−0.641 **	0.462 **
	NPP	−0.026	0.324	0.401 *	−0.466 **	−0.486 **	0.595 **
	NSP	−0.033	0.261	0.349	−0.473 **	−0.539 **	0.619 **
	WSP	0.223	−0.199	−0.134	−0.653 **	−0.553 **	0.804 **
	NPm	−0.054	0.562 **	0.605 **	−0.364 *	−0.284	0.432 *
	IFP	−0.507 **	0.300	0.487 **	−0.007	−0.753 **	0.232
	NRN	0.614 **	−0.776 **	−0.857 **	−0.472 **	0.079	0.444 **
	PC	−0.041	0.476 **	0.535 **	−0.422 *	−0.379 *	0.519 **
	TSW	0.839 **	−0.116	−0.366 *	−0.449 **	0.838 **	0.182
Respect	SY	0.373 *	0.171	0.149	−0.721 **	−0.190	0.755 **
	PH	−0.211	0.469 **	0.571 **	−0.281	−0.499 **	0.419 *
	NPP	−0.245	0.639 **	0.716 **	−0.176	−0.338	0.265
	NSP	−0.438 *	0.608 **	0.729 **	0.016	−0.445 *	0.112
	WSP	−0.404 *	0.370 *	0.516 **	−0.056	−0.597 **	0.231
	NPm	0.233	0.494 **	0.465 **	−0.566 **	−0.074	0.565 **
	IFP	−0.122	0.245	0.357 *	−0.401 *	−0.610 **	0.571 **
	NRN	0.747 **	−0.765 **	−0.884 **	−0.564 **	0.200	0.496 **
	PC	−0.592 **	0.717 **	0.791 **	0.487 **	−0.051	−0.466 **
	TSW	0.300	−0.815 **	−0.867 **	0.038	0.166	−0.077
Mean across varieties	SY	0.292 **	0.153	0.131	−0.550 **	−0.121	0.568 **
	PH	−0.150	0.360 **	0.458 **	−0.331 **	−0.512 **	0.472 **
	NPP	−0.026	0.410 **	0.458 **	−0.367 **	−0.314 **	0.446 **
	NSP	−0.096	0.330 **	0.408 **	−0.331 **	−0.431 **	0.448 **
	WSP	−0.037	−0.018	0.074	−0.349 **	−0.526 **	0.499 **
	NPm	0.141	0.467 **	0.462 **	−0.482 **	0.140	0.504 **
	IFP	−0.234 *	0.209 *	0.348 **	−0.280 **	−0.670 **	0.473 **
	NRN	0.574 **	−0.783 **	−0.855 **	−0.429 **	0.061	0.408 **
	PC	−0.076	0.401 **	0.443 **	−0.239 *	−0.241 *	0.300 **
	TSW	0.366 *	−0.494 **	−0.587 **	−0.056	0.322 **	−0.037

AGDD—accumulated growing degree days, HTC—hydrothermal coefficient; SY—seed yield; PH—plant height, NPP—number of pods per plant, NSP—number of seeds per plant, WSP—weight of seeds per plant, NPm—number of pods per m^2, IFP—insertion height of the first pod, NRN—number of root nodules, PC—protein content, TSW—thousand seed weight; * $p \leq 0.05$; ** $p \leq 0.01$.

Table 6. Correlation coefficient (R) of the multiple correlation between pea seed yield, morphological traits, quality indices, and meteorological indices.

Parameters (y)	Ieva DS		Simona		Respect		Mean across Varieties	
	R	$F_{fact.}$	R	$F_{fact.}$	R	$F_{fact.}$	R	$F_{fact.}$
SY	0.867	12.63 **	0.664	3.29 **	0.894	16.50 **	0.684	13.03 **
PH	0.982	115.25 **	0.991	216.48 **	0.983	119.22 **	0.931	97.14 **
NPP	0.928	25.65 **	0.971	67.87 **	0.965	56.53 **	0.841	35.88 **
NSP	0.864	12.32 **	0.973	72.88 **	0.943	33.37 **	0.839	35.28 **
WSP	0.648	30.20 *	0.906	18.97 **	0.886	125.27 **	0.711	15.18 **
NPm	0.857	11.54 **	0.958	46.80 **	0.935	28.72 **	0.873	47.36 **
IFP	0.959	47.60 **	0.974	76.07 **	0.974	77.88 **	0.935	103.14 **
NRN	0.869	12.88 **	0.885	15.00 **	0.954	42.07 **	0.873	47.30 **
PC	0.968	62.10 **	0.985	131.61 **	0.826	8.91 **	0.697	14.01 **
TSW	0.937	30.16 **	0.976	82.84 **	0.985	138.59 **	0.725	16.41 **

Multi regression equation $y = a + bx^1 + cx^2 + dx^3 + ex^4 + fx^5$, where y—SY, or morphological parameters, or PC, or TSW, x^1—precipitation, x^2—AGDD—accumulated growing degree days > 5 °C, x^3—AGDD—accumulated growing degree days > 10 °C, x^4—sunshine duration, x^5—relative air humidity. SY—seed yield, PH—plant height, NPP—number of pods per plant, NSP—number of seeds per plant, WSP—weight of seeds per plant, NPm—number of pods per m^2, IFP—insertion height of the first pod, NRN—number of root nodules, PC—protein content, TSW—thousand seed weight. *, **—relationship between indices is significant at $p \leq 0.05$ and $p \leq 0.01$, respectively.

3. Discussion

Fertilization is one of the basic ways to ameliorate the availability of soil nutrients to plants. The use of fertilizers is considered to be one of the most meaningful factors in enhancing crop yield. Fertilizing can positively change plant growth and productivity, but heavy uses of chemical fertilizers have created a variety of environmental and ecological problems. As a result, it is necessary to diversify crop rotation with legumes, such as peas, which are able to fix atmospheric nitrogen [6]. Their short growing period and ability to fix atmospheric nitrogen makes pea the best pre-crop for winter wheat. In addition, the plants not only ensure themselves on 2/3 of nitrogen, they also leave 60–100 kg of available nitrogen for subsequent culture [7]. There are different opinions about the fertilization of peas. Some researchers say that pea fertilizer rates depend on nutrients in the soil [20]. Huang et al. [17] found that when before sowing in soil and NO_3-N was low (10 kg·ha^{-1}), an application of starter N positively influenced pea yield (up to +19%); however, under higher initial soil NO_3-N (44 kg·ha^{-1}), the application of starter N reduced pea yield. Under no-tillage technology, $N_{25}P_{30}K_{40}$ fertilization was recommended for peas [26]. The results of the present investigation revealed that the biggest part of the variation in yield data was determined by the year's meteorological conditions. Weather conditions of the experimental year influenced the effectiveness of fertilizers. According to average data, the application of N_{15}–N_{45} rates increased seed yield by 11.1–18.3% under a background of $P_{40}K_{80}$, compared with unfertilized peas. After increasing the N rate to N_{60}, the seed yield did not increase. The highest $N_{60}P_{80}K_{160}$ rate increased the yield, but the difference was not significant, compared with N_{60} under a background of $P_{40}K_{80}$. The N_{15}–N_{45} rate was found to be economically and ecologically optimal for peas, grown under boreal conditions, which is close to the above-mentioned studies. There are conflicting data showing that doses of N from mineral fertilizers can vary from 73 to 97 kg·ha^{-1} N [27].

Abundant N fertilization has a negative effect on root nodule formation [17], which can affect subsequent nitrogen fixation [6]. Higher N rates reduce pea germination [17]. There are studies that show that germinating peas can only tolerate the 10 kg N ha^{-1} rates [16]. We found that fertilization had no significant effect on root nodules. This index was significantly influenced only by the weather conditions in the experimental year. The most favourable conditions for the formation of root nodules were at HTC 1.0–1.3, which means optimal irrigation. According to Khan et al. [23], the number of root nodules significantly differed between four tested varieties. In fact, in contrast with him, we found that the three tested pea varieties had no significant differences in the number of root nodules.

Pea varieties respond differently to fertilization [23,28] and environmental conditions [2]. The main traits that determine the level of adaptability in peas are a high harvest index, good ripening, disease resistance, resistance to shatter, high potential yield, seed weight, and seed number per pod [25,29]. In this study, the variety Ieva DS demonstrated the greatest adaptability and the ability to produce the highest seed yield of all studied varieties, regardless of the weather conditions. The highest values of yield components, such as NPP, WSP, and NPm, were also established for the variety Ieva DS. The highest TSW and protein content was for varieties Ieva DS and Simona.

The growth, development, and productivity of plants are strongly affected by weather conditions [30]. In a study with eight pea cultivars, it was found that seed yield had a strong relationship with weather factors and the reproductive phase was limited by stress [31]. The most important factor that determined seed yield and protein content was the sum of the rainfall over the vegetation period [32]. Kuznetsov et al. [30] established a moderate relationship ($r = 0.486$) between pea yield and the amount of precipitation. Grabowska and Banaszkiewicz [33] found that the influence of air temperature and atmospheric precipitations on the yield of different varieties differed and caused from 82% to 87% of seed yield variability. The results of the current experiment revealed that the most favourable conditions for the seed yield components were in the wet years. Correlation analysis among the yield components and meteorological indices revealed that HTC had the greatest positive influence on the seed yield for the variety Respect. HTC influenced seed yield by 57.1%, 43.4%, and 23.4% for varieties Respect, Ieva DS, and Simona, respectively. Our findings are in line with previous results [33], stating that water stress is the key cause of a decrease in pea yield in a temperate climate. Other studies have also found a positive correlation between pea yield and HTC [30].

We found that the relationship between sunshine duration and seed yield was negative and varied from weak to strong. Sunshine duration determined 12.9%, 51.1%, and 52.0% of seed yield data variation, respectively, for varieties Simona, Ieva DS, and Simona. Other researchers found that, on the contrary, the lowest average sunshine duration related to a lower seed yield for faba beans [34].

The results obtained in this study show that genotype had a significant influence on the levels of protein content in pea seeds. The variety Simona was distinguished from the others with the highest protein content, the protein content was very similar for the variety Ieva DS, and the variety Respect significantly accumulated the lowest content of protein. This confirms previous findings in the literature that differences in climate, soil, varieties, and agronomic practices may cause a different yield and protein content [22,32,35]. We found that fertilizers did not have a significant influence on protein content. This is in good agreement with previous findings in the literature [18].

4. Materials and Methods

4.1. Site and Soil Description

A field experiment was carried out during 2014–2017 at the Institute of Agriculture, Lithuanian Research Centre for Agriculture and Forestry in Central Lithuania (55°23′50″ N and 23°51′40″ E). The locality is situated in a boreal environmental zone, where the average annual air temperature is 6.4 °C and the long-term annual precipitation is 572 mm. The soil of the experimental site was Endocalcari-Epihypogleyic Cambisol. The mean soil characteristics (at 0–25 cm sampling depth), determined annually before sowing, were as follows: pH_{KCl} 5.4–7.2 (potentiometrically), available phosphorus 84–150 mg kg^{-1} (A-L method), available potassium 140–186 mg kg^{-1} (A-L method), humus 1.5–2.2% (Tyurin method). The content of mineral nitrogen was 43–59 $kg \cdot ha^{-1}$ in 0–40 cm soil layer (as sum of N-NO_3 and N-NH_4, N-NO_3—ionometrically, N-NH_4—spectrophotometrically.)

4.2. Experimental Details and Agronomic Management

The experiment involved three semi-leafless pea (*Pisum sativum* L.) varieties: Ieva DS, Simona, and Respect. Eight N fertilization treatments were used: (1) without fertilizers (N_0), (2) without N fertilizers (N_0), (3) 15 kg N·ha^{-1} (N_{15}), (4) 30 kg N·ha^{-1} (N_{30}), (5) 45 kg N·ha^{-1} (N_{45}), (6) 15 + 15 kg N·ha^{-1} (N_{15+15}), (7) 60 kg N·ha^{-1} (N_{60}), (8) 60 kg N·ha^{-1} (N_{60}). Nitrogen, as ammonium nitrate (NH_4NO_3) (34% N), was applied pre-sowing, with the exception of the 6th treatment, where additional 15 kg N·ha^{-1} was applied at stem elongation stage (BBCH 30–49). In conjunction with N fertilizer application, plots of 2nd–7th treatments received a base application of phosphate (P) fertilizer and potassium (K) fertilizer at 40 kg P·ha^{-1} and 80 kg K·ha^{-1}, and plots of 8 treatment received double rate of PK, i.e., 80 kg P·ha^{-1} and 160 kg K·ha^{-1}. P fertilizers were applied as granular superphosphate ($Ca(H_2PO_4)_2H_2O$) with P_2O_5 concentration of 20% and K as potassium chloride (KCl), with K_2O concentration of 60%. All fertilizers applied before sowing were incorporated into the soil.

The field trial was conducted using a split plot design with four replications, in which pea varieties were laid out in the main plot and the fertilization treatments in the subplot. The plot size was 15.0 m^2 (1.5 m × 10.0 m).

Spring barley (*Hordeum vulgare* L.) was a pre-crop in all experimental years. Peas were sown on 15 April in 2014 and 2015, 11 April in 2016, and 18 April in 2017 at a density of 1.2 million viable seeds per hectare with a 13 cm distance between rows and an 8 cm distance in the rows. After sowing, herbicide Fenix (a.i. aclonifen, 600 g L^{-1}) (3.0 L·ha^{-1}) was sprayed each year. Chemical insect control was performed using the following products at BBCH 14–16: Fastac 50 (a.i. alfa-cipermetrin 50 g L^{-1}) (0.20 L·ha^{-1}) in 2014 and 2015 and Decis Mega 50 EW (i.e., deltametrin 50 g L^{-1}) (0.15 L·ha^{-1}) in 2016 and 2017.

4.3. Plant Sampling and Measurements

The number of pods per plant (NPP), number of seeds per plant (NSP), and weight of seeds per plant (WSP) were established from randomly selected ten plants of each plot. The number of pulses per m^2 (NPm) was determined on plant samples collected from 100 cm long sections of two rows (0.50 m^2) from each plot. All indices were assessed at pea physiological maturity. Number of root nodules (NRN) per plant was determined from randomly selected five plants of each plot at inflorescence emergence (BBCH 51–59).

4.4. Seed Yield and Seed Quality Analyses

Each year, the plots were harvested within the first ten-day period of August at complete maturity (BBCH 89) using a plot harvester "Wintersteiger Delta" (Ried im Innkreis, Germany). Seed yield (SY) of pea was adjusted to a 15% moisture content. Protein content (PC) from each plot, from samples selected after harvesting, were measured using an Infratec 1241 grain analyzer (Foss, Hilleroed, Denmark). Thousand seed weight (TSW) was counted using a seed counter "Contador" (Pfeuffer GmbH, Kitzingen, Germany) from four samples of 250 seeds per each plot.

4.5. Meteorological Conditions

The locality is situated in boreal climatic zone, with an average annual air temperature of 6.4 °C and a long-term annual precipitation of 568 mm. Rainfall and mean air temperature at the experimental site over the two growing seasons are provided in Figure 3. The conditions of the plant growing season were described using the hydrothermal coefficient (HTC) as the agrometeorological indicator, which was calculated according to the formula [36]:

$$HTC = \Sigma\, p / 0.1\, \Sigma\, t, \quad (1)$$

where Σ p represents the sum of precipitation (mm) during the test period, when the average daily air temperature is above 10 °C, and Σ t denotes the sum of active temperatures (°C) during the same period. If HTC > 1.6, the irrigation is excessive; HTC = 1.0–1.5 optimal irrigation; HTC = 0.9–0.8 weak drought; HTC = 0.7–0.6 moderate drought (arid); HTC = 0.5–0.4 heavy drought; and HTC < 0.4 very heavy drought [36].

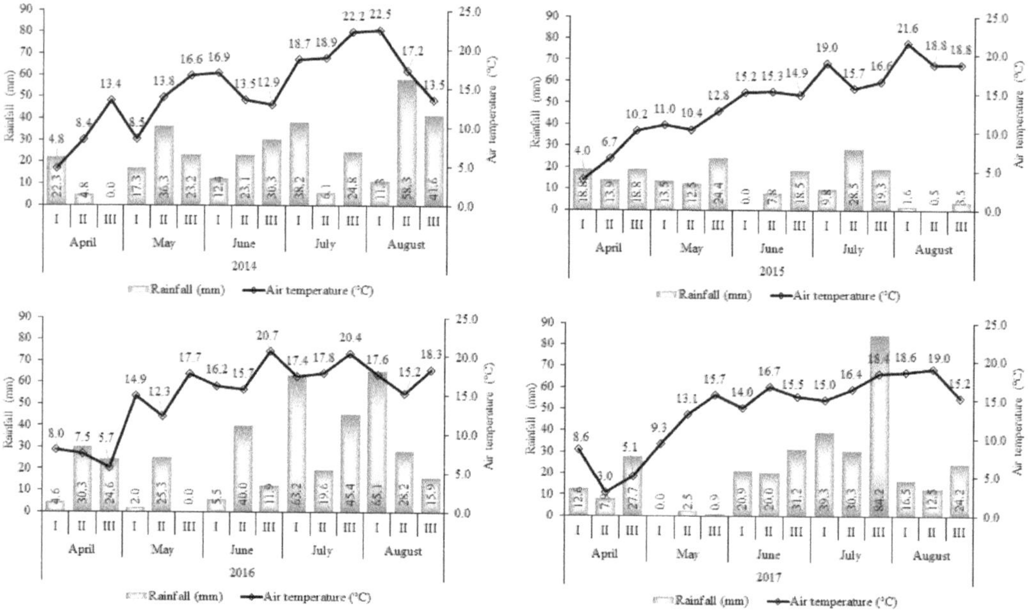

Figure 3. Rainfall and temperature distribution during the growing season.

Rainfall differed between the growing seasons, and the amount of rainfall totaled 212 mm (HTC = 1.6), 134 mm (HTC = 1.0), 215 mm (HTC = 1.3), 229 mm 212 mm (HTC = 1.4) in 2014, 2015, 2016 and 2017, respectively.

4.6. Statistical Analysis

A three-way ANOVA was used to determine the effects of year, variety, and fertilization treatment, on the morphological traits, SY, and seed quality parameters. The data were compared using Fisher's least significant difference (LSD) test at the probability levels $p \leq 0.05$ and $p \leq 0.01$. Standard statistical procedures were used for calculating simple correlation coefficients. The statistical analysis was performed using Stat Eng from the statistical data processing package Selekcija.

5. Conclusions

It was found that the experimental year determined 39.4%, variety—15.8%, and fertilization—11.2%, of the yield averaged data variations. The influence of fertilizers was more strongly expressed in wet conditions compared with dry conditions. The lack of moisture negatively influenced the pea seed yield. The new pea variety, Ieva DS, showed the highest seed yield of all varieties, compared with the trial mean, and is suitable for cultivation in the boreal environmental zone. According to seed yield, the varieties were arranged in the following descending order: Ieva DS > Respect > Simona. NPK fertilizers increased the seed yield by 0.40–0.78 t·ha^{-1} (or 10.6–12.9%) on average, compared with unfertilized peas. Under a background of $P_{40}K_{80}$, the application of N_{60} and splitting the N_{30} rate into N_{15+15} was not efficient. The optimal rate of $N_{15–45}P_{40}K_{80}$ fertilizer was determined when peas were grown in soil containing moderate amounts of available phosphorus and available potassium. The meteorological factors had a significant impact on yield components and pea seed quality, and significantly influenced the seed yield by 75.2%, 44.1%, and 79.9% for varieties Ieva DS, Simona, and Respect, respectively.

Funding: This research received no external funding.

Institutional Review Board Statement: Not applicable.

Informed Consent Statement: Not applicable.

Data Availability Statement: Not applicable.

Conflicts of Interest: The authors declare no conflict of interest.

References

1. Kumari, J.; Dikshit, H.K.; Singh, B.; Singh, D. Combining ability and character association of agronomic and biochemical traits in pea (*Pisum sativum* L.). *Sci. Hortic.* **2015**, *181*, 26–33. [CrossRef]
2. Maharjan, P.; Penny, J.; Partington, D.L.; Panozzo, J.F. Genotype and environment effects on the chemical composition and rheological properties of field peas. *J. Sci. Food Agric.* **2019**, *99*, 5409–5416. [CrossRef] [PubMed]
3. European Commission (EC). From Farm to Fork: Our Food, Our Health, Our Planet, Our Future. 2021. Available online: https://ec.europa.eu/food/horizontal-topics/farm-fork-strategy_en#Strategy (accessed on 21 March 2022).
4. Wysokinski, A.; Lozak, I. The dynamic of nitrogen uptake from different sources by pea (*Pisum sativum* L.). *Agriculture* **2021**, *11*, 81. [CrossRef]
5. Ntatsi, G.; Karkanis, A.; Yfantopoulos, D.; Pappa, V.; Konosonoka, I.H.; Travlos, I.; Bilalis, D.; Bebeli, P.; Savvas, D. Evaluation of the field performance, nitrogen fixation efficiency and competitive ability of pea landraces grown under organic and conventional farming systems. *Arch. Agron. Soil Sci.* **2019**, *65*, 294–307. [CrossRef]
6. Enrico, J.M.; Piccinetti, C.F.; Barraco, M.R.; Agosti, M.B.; Eclesia, R.P.; Salvagiotti, F. Biological nitrogen fixation in field pea and vetch: Response to inoculation and residual effect on maize in the Pampean region. *Eur. J. Agron.* **2020**, *115*, 126016. [CrossRef]
7. Faligowska, A.; Kalembasa, S.; Kalembasa, D.; Panasiewicz, K.; Szymanska, G.; Ratajczak, K.; Skrzypczak, G. The nitrogen fixation and yielding of pea in different soil tillage systems. *Agronomy* **2022**, *12*, 352. [CrossRef]
8. Peoples, M.B.; Brockwell, J.; Herridge, D.F.; Rochester, I.J.; Alves, B.J.R.; Urquiaga, S.; Boddey, R.M.; Dakora, F.D.; Bhattarai, S.; Maskey, S.L.; et al. The contributions of nitrogen-fixing crop legumes to the productivity of agricultural systems. *Symbiosis* **2009**, *48*, 1–17. [CrossRef]
9. Hajduk, E.; Wlaoeniewski, S.; Szpunar-Krok, E. Influence of legume crops on content of organic carbon in sandy soil. *Soil Sci. Annu.* **2015**, *66*, 52–56. [CrossRef]
10. Kumar, S.; Meena, R.S.; Lal, R.; Yadav, G.S.; Mitran, T.; Meena, B.L.; Dotaniya, M.L.; EL-Sabagh, A. Role of Legumes in Soil Carbon Sequestration. In *Legumes for Soil Health and Sustainable Management*; Meena, R., Das, A., Yadav, G., Lal, R., Eds.; Springer: Singapore, 2018; pp. 109–138.
11. Stagnari, F.; Maggio, A.; Galieni, A.; Pisante, M. Multiple benefits of legumes for agriculture sustainability: An overview. *Chem. Biol. Technol. Agric.* **2017**, *4*, 2. [CrossRef]
12. Zhao, Y.; Fan, Z.; Hu, F.; Yin, W.; Zhao, C.; Yu, A.; Chai, Q. Source-to-sink translocation of carbon and nitrogen is regulated by fertilization and plant population in maize-pea intercropping. *Front. Plant Sci.* **2019**, *10*, 891. [CrossRef]
13. Didinger, C.; Thompson, H.J. Defining nutritional and functional niches of legumes: A call for clarity to distinguish a future role for pulses in the dietary guidelines for Americans. *Nutrients* **2021**, *13*, 1100. [CrossRef] [PubMed]
14. McKenzie, R.H.; Middleton, A.B.; Solberg, E.D.; DeMulder, J.; Flore, N.; Clayton, G.W.; Bremer, E. Response of pea to rate and placement of triple superphosphate fertilizer in Alberta. *Can. J. Plant Sci.* **2001**, *81*, 645–649. [CrossRef]
15. Pampana, S.; Masoni, A.; Mariotti, M.; Ercoli, L.; Arduini, I. Nitrogen fixation of grain legumes differs in response to nitrogen fertilisation. *Exp. Agric.* **2018**, *54*, 66–82. [CrossRef]
16. Dona, W.H.G.; Schoenau, J.J.; King, T. Effect of starter fertilizer in seed-row on emergence, biomass and nutrient uptake by six pulse crops grown under controlled environment conditions. *J. Plant Nutr.* **2020**, *43*, 879–895. [CrossRef]
17. Huang, J.; Afshar, R.K.; Tao, A.; Chen, C. Efficacy of starter N fertilizer and rhizobia inoculant in dry pea (*Pisum sativum* Linn.) production in a semi-arid temperate environment. *Soil Sci. Plant Nutr.* **2017**, *63*, 248–253. [CrossRef]
18. Stevovic, V.; Djukic, D.; Djurovic, D.; Mandic, L.; Bokan, N. The effect of nitrogen fertilization on the yield and quality of field pea (*Pisum sativum*). In *Sustainable Grassland Productivity*; Lloveras, J., Ed.; EGF: Badajoz, Spain, 2006; pp. 721–723.
19. Sainju, U.M.; Lenssen, A.W.; Allen, B.L.; Jabro, J.D.; Stevens, W.B. Pea growth, yield, and quality in different crop rotations and cultural practices. *Agrosyst. Geosci. Environ.* **2019**, *2*, 180041. [CrossRef]
20. Gospodarek, J.; Klimek-Kopyra, A.; Rusin, M. Suitability of NDVI index to pea condition evaluation at diverse phosphorus fertilization. *Ital. J. Agron.* **2020**, *15*, 1418. [CrossRef]
21. Macák, M.; Candráková, E.; Đalović, I.; Prasad, P.V.; Farooq, M.; Korczyk-Szabó, J.; Šimanský, V. The influence of different fertilization strategies on the grain yield of field peas (*Pisum sativum* L.) under conventional and conservation tillage. *Agronomy* **2020**, *10*, 1728. [CrossRef]
22. Symanowicz, B.; Kalembasa, S.; Becher, M.; Toczko, M.; Skwarek, K. Effect of varied levels of fertilization with potassium on field pea yield and content and uptake of nitrogen. *Acta Sci. Pol. Agric.* **2017**, *16*, 163–173.
23. Khan, S.; Aman, F.; Ismaeel, M.; Ali, Z.; Alam, M.; Iqbal, S.; Khan, T. Growth and yield response of pea (*Pisum sativum* L.) cultivars to phosphorus fertilization. *Sarhad J. Agric.* **2021**, *37*, 369–376.

24. Mohammed, Y.A.; Chen, C.; Walia, M.K.; Torrion, J.A.; McVay, K.; Lamb, P.; Miller, P.; Eckhoff, J.; Miller, J.; Khan, Q. Dry pea (*Pisum sativum* L.) protein, starch, and ash concentrations as affected by cultivar and environment. *Can. J. Plant Sci.* **2018**, *98*, 1188–1198. [CrossRef]
25. Škarpa, P.; Školníková, M.; Antošovský, J.; Horký, P.; Smýkalová, I.; Horácek, J.; Dostálová, R.; Kozáková, Z. Response of normal and low-phytate genotypes of pea (*Pisum sativum* L.) on phosphorus foliar fertilization. *Plants* **2021**, *10*, 1608. [CrossRef] [PubMed]
26. Danilovič, M.; Šoltysova, B. Nutrient management analysis in field pea (*Pisum sativum* L.) cultivated in the no-tillage technology. *Agriculture (Pol'nohospodárstvo)* **2010**, *56*, 18–24.
27. Hlisnikovský, L.; Menšík, L.; Čermák, P.; Křížová, K.; Kunzová, E. Long-term effect of pig slurry and mineral fertilizer additions on soil nutrient content, field pea grain and straw yield under winter wheat–spring barley–field pea crop rotation on cambisol and luvisol. *Land* **2022**, *11*, 187. [CrossRef]
28. Husain, J.; Kashyap, P.; Prusty, A.K.; Dutta, D.; Sharma, S.S.; Panwar, A.S.; Kumar, S. Effect of phosphorus fertilization on growth, yield and quality of pea (*Pisum sativum*). *Indian J. Agric. Sci.* **2019**, *89*, 87–91.
29. Lykhochvor, V.; Andrushko, M.; Andrushko, O. Influence of variety, elements of the fertilization system, sowing rates of seeds on the pea yield (*Pisum sativum*). *Agric. Aliment. Pisc. Zootech.* **2020**, *355*, 23–30. [CrossRef]
30. Kuznetsov, I.; Davletov, F.; Anokhina, N.; Akhmadullina, I.; Safin, F. Influence of weather condtion on the field peas (*Pisum sativum* L. ssp. sativum) vegetation period and yield. *Agron. Res.* **2020**, *18*, 472–482.
31. Bueckert, R.A.; Wagenhoffer, S.; Hnatowich, G.; Warkentin, T.D. Effect of heat and precipitation on pea yield and reproductive performance in the field. *Can. J. Plant Sci.* **2015**, *95*, 629–639. [CrossRef]
32. Prusiński, J.; Borowska, M. Effect of planting density and row spacing on the yielding and morphological features of pea (*Pisum sativum* L.). *Agronomy* **2022**, *12*, 715. [CrossRef]
33. Grabowska, K.; Banaszkiewicz, B. Effect of air temperature and atmospheric precipitation on yielding of sowing pea in central Poland. *Acta Agroph.* **2009**, *13*, 113–120. (In Polish)
34. Skovbjerg, C.K.; Knudsen, J.N.; Füchtbauer, W.; Stougaard, J.; Stoddard, F.L.; Janss, L.; Andersen, S.U. Evaluation of yield, yield stability, and yield–protein relationship in 17 commercial faba bean cultivars. *Legume Sci.* **2020**, *2*, e39. [CrossRef]
35. Olle, M. The yield, height and content of protein of filed peas (*Pisum sativum* L.) in Estonian agroclimatic condition. *Agron. Res.* **2017**, *15*, 1725–1732.
36. Dirsė, A.; Taparauskienė, L. Humidity fluctuations in plant vegetation periods and a comparison of its assessment methods. *Žemės Ūkio Moksl.* **2010**, *17*, 9–17. (In Lithuanian)

plants

MDPI

Article

Influence of Variable Chloride/Sulfur Doses as Part of Potassium Fertilization on Nitrogen Use Efficiency by Coffee

Victor Hugo Ramirez-Builes [1,*], Jürgen Küsters [1], Ellen Thiele [1], Luis Alfredo Leal-Varon [2] and Jorge Arteta-Vizcaino [3]

1 Center for Plant Nutrition and Environmental Research Hanninghof, Yara International, 48249 Dülmen, Germany
2 Yara Colombia, Zona Industrial Mamonal, Cartagena 130009, Colombia
3 Department of Plant Science, National University, Palmira 763533, Colombia
* Correspondence: victor.ramirez@yara.com; Tel.: +49-0-2594-7980 (ext. 112)

Abstract: Chloride (Cl^-) is applied in coffee at rates as a "macronutrient" in the form of muriate of potash (MOP). Potassium (K^+) is one of the most demanded nutrients by the coffee plant, and MOP is one of the most used fertilizers in coffee production. No scientific evidence shows how Cl^- applied with MOP influences coffee growth, nutrient uptake, and nitrogen use efficiency (NUE). In order to address these questions, a greenhouse trial over two years and a field trial over four years were conducted. The trials were designed to test the influence of variable Cl^-/S ratios on biomass accumulation, nutrient uptake, and NUE. A significant effect of the Cl^- rates on growth was observed under greenhouse conditions but a non-significant effect on yield under field conditions. Cl^- and S significantly influenced the NUE in coffee. The results allow us to conclude that Cl^- rates need to be balanced with S rates, and that Cl^- applied at macronutrient rates can improve the NUE in coffee between 7 and 21% in greenhouse conditions and between 9% and 14% in field conditions, as long as the rates do not exceed 180 mg L^{-1} Cl^- and 80 mg$\cdot L^{-1}$ S in the greenhouse and 150 kg$\cdot ha^{-1}\cdot year^{-1}$ Cl^- and 50 kg $ha^{-1}\cdot year^{-1}$ S in the field. With the aim to improve the NUE in coffee, the Cl^- content in leaves in coffee should be lower than 0.33% of dry matter, and in soil lower than 30 mg$\cdot L^{-1}$. In practical terms, coffee farmers need to balance K-based fertilizers to avoid the excessive Cl^- applications that reduce the nutrient use efficiency, especially the NUE.

Keywords: coffee; chloride (Cl^-); sulfur (S); nutrients; NUE

1. Introduction

In crop production, chloride (Cl^-) is still being seen as an undesired anion rather than a plant nutrient, mainly because of its toxicity effects resulting from excessive Cl^- accumulation in sensitive plant organs under salt stress [1,2]. There is also a widespread belief that Cl^- and nitrate (NO_3^-) are antagonistic ions competing for plant uptake [3]. However, since 1954, chloride has been considered as an essential micronutrient, when Broyer et al. [4] demonstrated a direct influence of Cl^- on the growth of roots, leaves, stems, and petioles in tomato plants. Chloride is involved in important physiological processes, including stomatal regulation, water splitting or Hill reactions in photosystem II. Cl^- also has an osmoregulatory function in the vacuole, participating in phloem loading and unloading of sugars and stimulating membrane-bound proton-pumping ATPases and PPiases [4,5]. Other important functions are related to balancing the electrical charge of essential cations, such as K^+ and H^+, which both play a main role in stabilizing the cell membranes' electrical potential and which regulate pH gradients [6].

Being a non-assimilating highly mobile anion, Cl^- is the preferred molecule to balance the electrical charge of important cations such as K^+, Ca^{+2}, and protons (H^+), helping in the stabilization of the electrical potential of cell membranes and the regulation of the pH gradients and electrical excitability [2]. Cl^- is directly involved in cell division and, subsequently, leaf area formation [2,7]. Most recently, a positive influence of Cl^- on the vegetative yield per unit of nitrogen available to the crop (NUE) in different crop species has been documented [8].

Chloride occurs predominantly as Cl^- in the soil, does not form complexes readily, and tends to be repelled by the negatively charged sites on layer silicates. The concentration of Cl^- in bulk solutions is greater than in the diffuse layers surrounding the soil particles [1]. Cl^- redistribution in the soil profile is likely to be dominated by mass flow and convention processes associated with water fluxes [2]. The surface complexation of Cl^- by soil particles is relevant in variable-charge soils, where two types of soil complexes are distinguished: outer sphere and inner sphere complexes. In outer sphere complexes, the ion is separated from the particle surfaces by water molecules, and bonding depends solely on the presence of opposite charge on the particle surfaces. Cl^-, NO_3^- and many cations (K^+, Mg^{+2}, Na^+) form only outer-sphere complexes [9]. Permanent positive surface charge occurs, but outer-sphere anion complexation is important mainly in variable charged soil with a pH below the point zero charge [9].

N fertilizer application in coffee is relatively high, with rates of 200 to 500 $kg \cdot ha^{-1} \cdot year^{-1}$ [10–14]. The efficiency of mineral N fertilizer or NUE depends on the N rate applied and the crop N uptake at a given yield level. Bruno et al. [12] reported an N recovery for the whole coffee plant at harvest time of 66% when the N rate was 200 $kg\ N \cdot ha^{-1}$ and only 37% when a high N rate of 800 $kg\ N \cdot ha^{-1}$ was used. Cannavo et al. [15] reported that 60% of the N applied to the coffee crop remained in the top 60 cm of the soil, meaning that not more than 40% of the N applied was used by the coffee crop. In nursery coffee plants, Salamanca et al. [16] found that fertilizer N supplied approximately 20–29% of the total plant N after 4 months and an NUE of lower than 10%. The above reports demonstrate that the NUE in coffee needs to be improved and that integrated strategies must be sought from plant breeding to agronomic management. Within the agronomic management strategies, the appropriate use of sources and doses of fertilizers in conjunction with the times of application are part of the pool of alternatives that need to be considered with the aim to improve the NUE in coffee.

The most often used potassium source by coffee farmers is the muriate of potassium (MOP), also known as potassium chloride (60% K_2O, 46% Cl), because of its high K concentration, high solubility, and relatively low costs compared to other K sources such as potassium sulfate (SOP) and potassium nitrate (PN). The potassium demand of the coffee crop is similar to the nitrogen demand, and in some phenological stages, it is even higher. The mean application of K by the coffee production systems changes from region to region, with K fertilization rates between 100 and 400 kg $K_2O\ ha^{-1} \cdot year^{-1}$ depending on the expected yield and K content in the soil [14,17–19]. In Colombia, for example, the economic optimum coffee yield is achieved with a mean rate of 273 kg $K_2O\ ha^{-1} \cdot year^{-1}$ in productive coffee plantations [13]. High K_2O rates applied with MOP result in high inputs of Cl^-, reaching up to 300 kg $Cl^- \cdot ha^{-1} \cdot year^{-1}$.

In weathered tropical soils, the natural content of Cl^- tends to be low. Cl^- as an anion is not adsorbed by the soil particles and hence is very mobile in the soil. It can be easily lost by leaching under freely drained conditions [2,20].

Laboratories in coffee regions usually have well-standardized analytical methods to quantify Cl^- content in the soil and tissues; however, despite its role as a plant nutrient, this element is not included as a key nutrient in any of the nutritional programs in the coffee regions. Only in fertigation is special emphasis made that Cl^- should not exceed 350 $mg \cdot L^{-1}$ to avoid a reduction in the production capacity of the trees [14].

In rainfed conditions, Cl^- is the third most applied nutrient in coffee when MOP is used to supply the K^+ needs of the coffee crop. Excess supply and uptake of Cl^- in plants is a serious problem, mainly in crops that grow on salt-affected soils [1,20]. Recently, Santos et al. [21] demonstrated the negative effect of chloride applied via fertilizers in coffee with respect to productivity and coffee cup quality in the Minas Gerais state of Brazil, but no references are available regarding the effects of high Cl^- supplies on nitrogen use efficiency (NUE) in coffee growing under rainfed conditions. Hence, this research aimed to evaluate the influence of Cl^- applied at rates as a macronutrient for coffee regarding growth, productivity, nutrient uptake, and NUE in coffee.

2. Results

2.1. Influence of Cl and S Rates on Biomass Accumulation, Nutrient Uptake, and NUE under Greenhouse Conditions

The dry biomass was significantly affected by the Cl/S rates (p value < 0.001). The treatments without S and with a Cl^- rate of 300 $mg \cdot L^{-1}$ showed 2,2 times less biomass (37.9 $g \cdot plant^{-1}$) with respect to the treatments without Cl and 200 $mg \cdot L^{-1}$ of S that accumulated at 82.1 $g \cdot plant^{-1}$ (Table 1). This reduction can be basically attributed to the strong S deficit observed in this treatment, where this low dry biomass accumulation reduces the total N uptake, NU_TE, and NUE, clearly indicating the importance of S on coffee growth and NUE. Otherwise, the treatment without Cl^- and the highest S rate (200 $mg \cdot L^{-1}$) showed the highest biomass accumulation and NU_TE but not the highest NUE. The intermediate treatment with 180 $mg \cdot L^{-1}$ of Cl^- and 80 $mg \cdot L^{-1}$ of S showed the highest total N uptake (1.2811 $mg \cdot plant^{-1}$) and NUE (0.89 g shoot $DW \cdot g^{-1}$ of N applied) but significantly lower NU_TE (0.038 g $DW \cdot mg^{-1}$ N).

Table 1. Influence of different chloride (Cl^-) and sulfur (S) rates on coffee plant biomass accumulation, nutrient uptake, and nitrogen use efficiency under greenhouse conditions.

Cl/S Rates	Total DW	Root DW	N	Cl	N Uptake	S Uptake	Total Shoot Cl Uptake	NU_TE ¶	NUE ¥
$mg\ L^{-1}$	$g\ Plant^{-1}$		Content in Leaves %		$mg\ Plant^{-1}$			$g^2\ DW \cdot mg^{-1}\ N$	$g\ DW \cdot g^{-1}$ of N
0/200	82.1 d	27.0 c	2.58 a	0.03 a	1234.7 b	174.2 c	19.3 a	0.047 c	0.83 c
60/160	74.1 bc	20.3 bc	2.68 a	0.16 b	1230.0 b	137.2 c	89.1 b	0.044 c	0.86 cb
120/120	79.4 cd	23.3 bc	2.59 a	0.33 c	1236.5 b	106.0 c	173.4 c	0.046 c	0.86 cb
180/80	74.8 bc	18.5 b	2.69 a	0.55 d	1281.1 c	90.4 b	276.3 d	0.038 b	0.89 b
240/40	72.1 b	16.2 ab	2.67 a	0.69 e	1306.3 c	84.9 b	361.7 e	0.043 c	0.89 b
300/0	37.9 a	10.2 a	4.43 b	2.98 f	1095.8 a	19.4 a	587.2 f	0.029 a	0.73 a
p value	***	**	***	***	***	***	***	***	***

Different letters denote statistically significant differences according to Tukey's test alfalfa = 0.05 ** p value < 0.05; *** p value < 0.01. ¶ Nitrogen utilization efficiency (NU_TE total DW/total N content); ¥ Nitrogen use efficiency (NUE total N uptake by the shoot biomass/total N applied N).

2.2. Cl/S Ratios Influence Nutrient Concentration and Uptake by Coffee Beans under Field Conditions

In the field conditions where the present study was carried out, it was possible to observe the occurrence of two dry periods during the year, represented by a soil moisture index (SMI) lower than 1.0, during the months of July to September and from December to March. The greatest flowering occurred in the months of February to March associated with a greater water deficit, with SMI values between 0.7 and 0.5, indicating that between 70% and 50% of the pore space of the soil is occupied by water (Figure 1).

In field conditions on productive coffee plants, the nutrient concentration of the coffee cherries after flowering was reduced in the way that the fresh cherries develop and gain weight. Nitrogen (N) concentrations move from a mean value of 3.6% 30 days after flowering to 1.8% 240 days after flowering or harvest time. At harvest time, the N concentration was significantly different between treatments (Table 2), where the treatment with the higher Cl/S ratio (201 kg Cl/33 kg S) showed the lowest N concentration with a mean value of 1.69% compared with the treatment with 150 kg Cl/50 kg S, which showed a mean N concentration of 1.89% (Figure 2A).

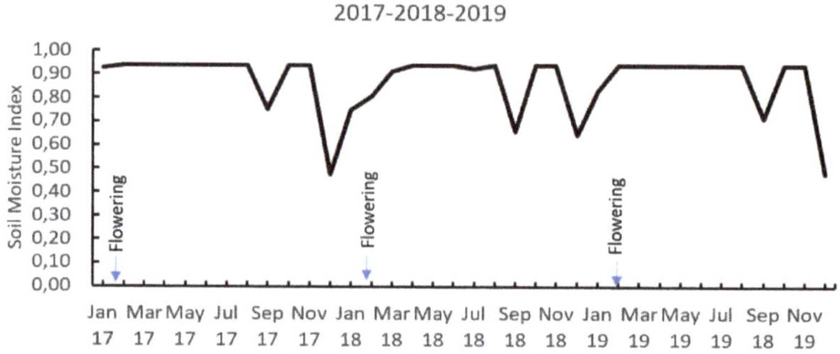

Figure 1. Soil moisture index distribution for 3 years (2017, 2018, and 2019). Soil moisture index means the fraction of the porous media of the soil that is occupied by water, SMI = 1.0 means that 100% of the soil porous is occupied by water, and SMI = 0 means that the soil is completely dry. (SMI was calculated seeing the methodology described by Ramirez et al. [22]).

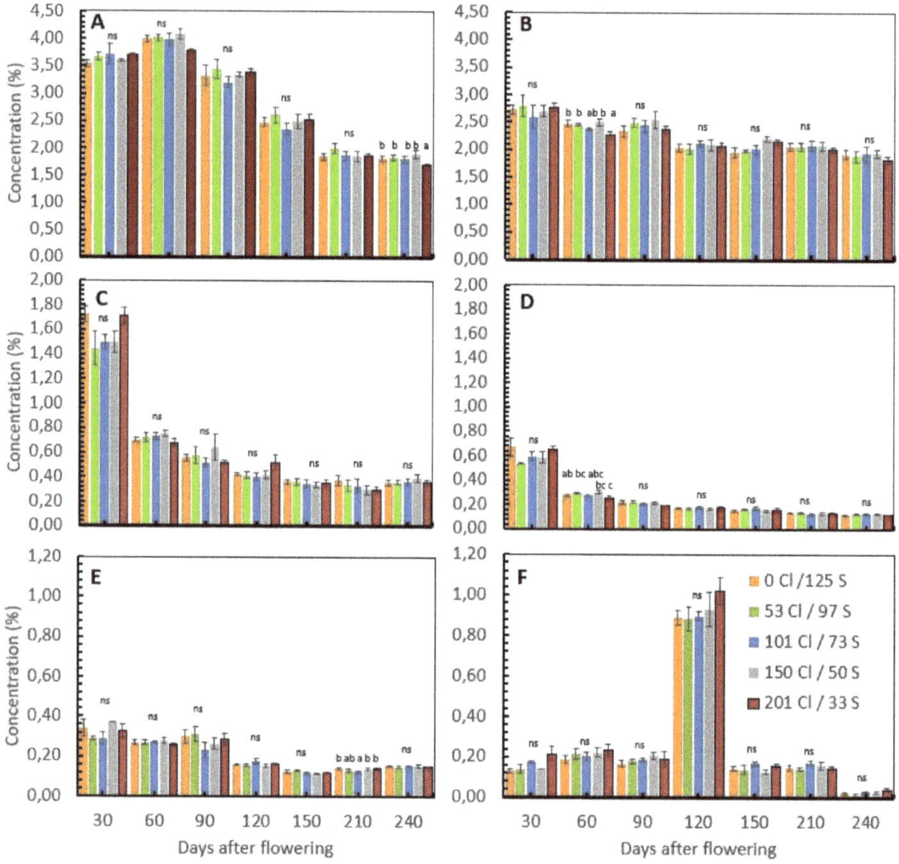

Figure 2. Influence of chloride/sulfur (Cl/S) rates on nutrient concentration in coffee cherries after flowering. Nitrogen (**A**), potassium (**B**), calcium (**C**), magnesium (**D**), sulfur (**E**), and chlorite (**F**). Values are the means of four replications ± standard error. Different letters denote statistically significant differences according to Fisher's LSD test with $p < 0.05$, and ns denotes non-significant differences.

Table 2. Pr > F values from the statistical output (ANOVA) for the nutrient concentration after flowering and nutrient uptake at harvest time in the field trial.

Nutrient	Nutrient Concentration (%)							Nutrient Uptake at Harvest Time (t·ha^{-1})
	Days After Flowering							
	30	60	90	120	150	210	240	240
	p Value							
N	0.67	0.13	0.48	0.37	0.61	-	0.099	0.028 **
K	0.90	0.02 **	0.64	0.93	0.03 **	0.94	0.83	0.239
Ca	0.13	0.50	0.46	0.25	0.91	0.42	0.82	0.220
Mg	0.38	0.04 **	0.52	0.85	0.49	0.83	0.19	0.398
S	0.38	0.68	0.18	0.36	0.60	0.081 *	0.54	0.459
Cl	0.14	0.69	0.78	0.53	0.11	0.21	0.27	0.006 **

* p value < 0.05; ** p value < 0.01.

The potassium (K^+) concentration of the coffee cherries moves from average values of 2.7% 30 days after flowering to 1.9% 240 days after flowering. Significant differences between treatments were observed in potassium (K^+) at 60 and 150 days after flowering (Table 2). Sixty days after flowering, the treatment with 201 kg Cl/33S showed a significantly lower K^+ concentration of 2.27% than the treatment with 150 kg Cl/50, which shows 2.5% of K^+ on the cherries. Furthermore, 150 days after flowering, the treatment without Cl^- and high S (0 kg Cl/125 kg S) showed the lowest and most significant different K^+ concentration for the cherries at 1.95% compared with the treatment with a higher Cl/S ratio of 201 kg Cl/33 kg S with 2.14% of K^+ concentration for the cherries (Figure 2B). At harvest time, no significant differences were observed in K concentration for the coffee cherries, but the treatment with a higher Cl/S ratio (201 kg Cl/33 kg S) showed the lowest K^+ concentration for the cherries with 1.82% concerning the other treatments, where the K^+ concentration tended to be higher.

The calcium (Ca^{+2}) concentration of the coffee cherries moves from a mean value of 1.6% 30 days after flowering to 0.36% 240 days after flowering, without any significant differences between treatments during the coffee cherry's development period (Figure 2C).

The magnesium (Mg^{+2}) concentration in the coffee cherries moves from a mean value of 0.61% 30 days after flowering to 0.12% 240 days after flowering (Figure 2D), with significant differences between treatments 60 days after flowering (Table 2), where the treatment with a higher Cl/S ratio showed the lowest Mg^{+2} concentration of 0.25% with respect to the treatment with 150 kg Cl/50 kg S, which registered a concentration of Mg^{+2} of 0.30% at the development stage. At harvest time, no significant differences were observed in the Mg^{+2} concentration in the coffee cherries.

The sulfur (S) concentration of the coffee cherries moves from a mean value of 0.32% 30 days after flowering to 0.15% 240 days after flowering (Figure 2E), with significant differences between treatments observed 210 days after flowering (Table 2). The treatment with a medium Cl/S ratio (101 kg Cl and 73 kg S) showed a lower S concentration of 0.12%. No significant differences in the S concentration for the cherries were observed at harvest time (240 days after flowering).

The Cl^- was the only nutrient that did not reduce the concentration during the cherries' development, as was described with the other nutrients previously. At 30 days after flowering, the Cl^- concentration for the coffee cherries showed a mean value of 0.16%, while 120 days after flowering, the Cl^- concentration was six times higher, reaching a mean value of 0.92%, with a subsequent reduction 240 days after flowering, reaching a mean value of 0.028% (Figure 2F).

The Cl/S ratios significantly influence the total N and Cl^- uptake per ton of green coffee beans. For N, the treatment with 150 kg Cl/50 kg S showed a mean N uptake of 39.13 kg t^{-1} of green coffee beans, 14% more N uptake per ton of green coffee than the treatment with 201 kg Cl/33 kg S, which showed a mean N uptake of 34.2 kg N·t^{-1}, followed by the treatments with 101 kg Cl/73S with 37.5 kg N·t^{-1} and the treatment with 53 kg Cl/97 kg S with 37.9 kg N·t^{-1}, representing 9% and 11% more N uptake per ton of green coffee bean, respectively (Figure 3A).

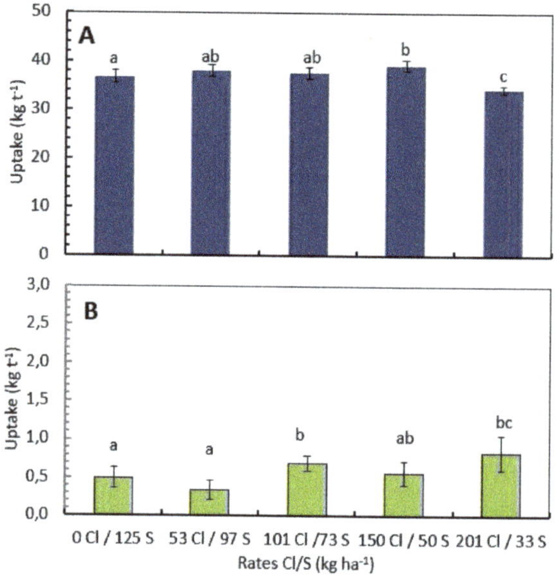

Figure 3. Influence of the chloride/sulfur rates on the nutrient demand per ton of green coffee beans. Nitrogen uptake (**A**), and chloride uptake (**B**). Values are the mean of four replications ± standard error; different letters denote statistically significant differences according to Fisher's LSD test with $p < 0.05$.

The Cl$^-$ uptake per ton of green coffee bean increased proportionately with the increase in the Cl/S ratio, treatments without and with low Cl$^-$ presented lower and significantly different Cl$^-$ uptake per ton of green coffee, with 0.497 kg Cl·t^{-1} of green coffee bean for the treatment with 0 kg Cl/123 kg S, and 0.336 kg Cl·t^{-1} of green coffee beans for the treatment with 53 kg Cl/97 kg S. This represents 65% and 144% less Cl$^-$ uptake with respect to the treatment with the higher Cl$^-$ rate (201 kg Cl/53S) where the mean uptake was 0.82 kg Cl·t^{-1} of green coffee beans (Figure 3B).

2.3. *Cl/S Ratios Influence Nitrogen Use Efficiency for Coffee Beans at the Field Level*

The NUE in the field trial, described as the ratio of N uptake by the green coffee beans with respect to the mineral N applied, was significantly influenced by the age of the plantation and by yield (Figure 4). For instance, the NUE for the first harvest after stem pruning was lower than 0.25 kg N uptake per kg of N applied, while in the second harvest, it increased until a mean value of 0.45 kg N uptake per kg of N applied reached the highest level during the third harvest year, with a mean value of 0.75 kg N uptakekg N applied^{-1}.

The NUE for the productive plants shows a polynomial correlation with respect to Cl$^-$ rates (Figure 4) during the second and third harvest years. A significant polynomial correlation between the Cl$^-$ rates and the NUE was observed (R^2 = 0.54 for the second harvest and R^2 = 0.97 for the third harvest). The polynomial function reaches an optimal NUE with Cl$^-$ rates of 73, 113, and 110 kg·ha^{-1} during the first, second, and third harvests, respectively (Figure 4).

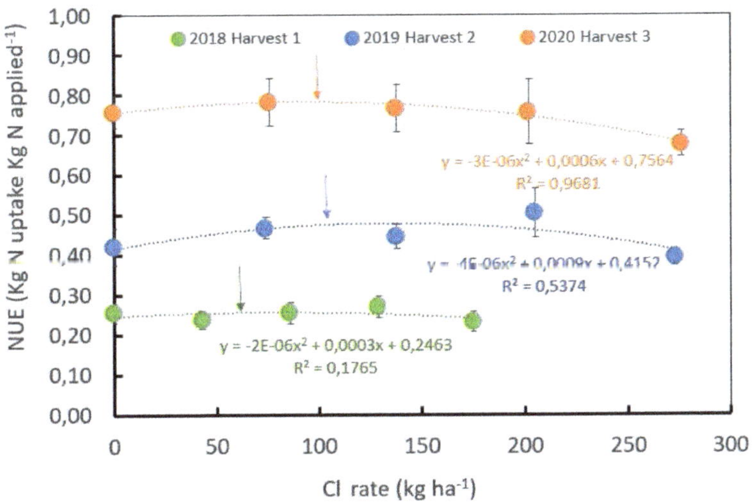

Figure 4. Influence of the chloride rates during three years of harvests on nitrogen use efficiency. Optimum Cl^- rates estimated from the first derivate of the function were 73, 113 and 110 kg·ha^{-1} for the harvest years of 2018, 2019 and 2020, respectively, ± standard error.

The increase in NUE over the years is directly related to the increase in yield. During the first and second year after pruning (first harvest), the Cl^- rates were low due to the low yield potential of the crop and the low K^+ demand. After the second year after pruning, yield potential increases, and crop nutrient demands, including K^+, also increase.

The Cl/S rates do not have a significant effect on coffee yield (Table 3). The treatment with a higher Cl/S ratio (201 kg Cl/33 kg S) for 4 years and three harvests showed only a 3% lower yield compared with the treatments with lower Cl/S ratios such as 145 kg Cl/50 kg S or 53 kg Cl/97 kg S, but those treatments showed an improvement in NUE with a mean value of 14% and 16%, respectively.

Table 3. Influence of tCl/S rates on coffee yield, nitrogen uptake, and use efficiency in coffee.

Cl/S Rate $^{\pm}$ kg·ha^{-1}	Yield $^{+}$	N-Uptake per Ton	Total N^- Uptake per ha	Mean N. Applied per ha $^{++}$	NUE $^{+++}$
	t·ha^{-1}	kg·t^{-1}	kg·ha^{-1} year^{-1}		kg·kg^{-1}
0/125	4.02 ±2.05)	36.7 (±2.04)	147.4 (±74.5)	308 (±20.75)	0.48 (±0.23)
53/97	4.01 (±1.85)	39.1 (±2.43)	157.3 (±75.3)	308 (±20.75)	0.51 (±0.24)
101/73	4.03 (±1.96)	37.5 (±2.32)	151.4 (±75.9)	308 (±20.75)	0.49 (±0.23)
150/50	4.03 (±2.11)	37.9 (±2.08)	152.8 (±78.5)	308 (±20.75)	0.50 (±0.24)
201/33	3.92 (±1.87)	34.3 (±1.31)	134.3 (±64.21)	308 (±20.75)	0.44 (±0.20)
p value	ns	*	*		ns

$^{\pm}$ The chlorate rate is the mean Cl^- applied during the whole season (2017–2020); $^{+}$ Mean yield for three years of harvest: 2018, 2019, 2020; $^{++}$ Mean nitrogen applied for four years (2017–2020); $^{+++}$ NUE is the ratio between N uptake by the coffee cherries/N applied during the season. NUE is total N uptake by the coffee cherries per ha/total N applied per ha; ns is not significant, and * is significant at p value < 0.10.

On average, the NUE ranges between 0.44 and 0.51 kg of green coffee per kg of N applied in the treatments with a higher Cl/S ratio (201 kg Cl/33 kg S) and a lower Cl/S ratio (53 kg Cl/97 kg S), respectively (Table 3).

2.4. Cl/S Ratios Influence Cl and S Distribution in the Soil Profile

Significant differences in the Cl^- concentration in the soil were observed between treatments at soil depths of 10 and 20 cm. The treatment with a higher Cl/S ratio (201 kg Cl/33 kg S) showed a higher concentration with a mean value of 95.8 mg·L^{-1} of Cl^- at 10 cm depth and 66.4 mg.L^{-1} at 20 cm depth, which was significantly different from the other treatments were the mean Cl^- concentrations at those depths ranged between 27 mg·L^{-1} at 10 cm depth and 18 mg·L^{-1} at 20 cm depth in the treatment with 101 kg Cl/73 kg S, reaching the lowest concentration in the treatment with 0 kg Cl/125 kg S with a mean Cl^- concentration of 9.4 mg·L^{-1} at 10 cm depth and 10.18 mg·L^{-1} at 20 cm depth (Figure 5).

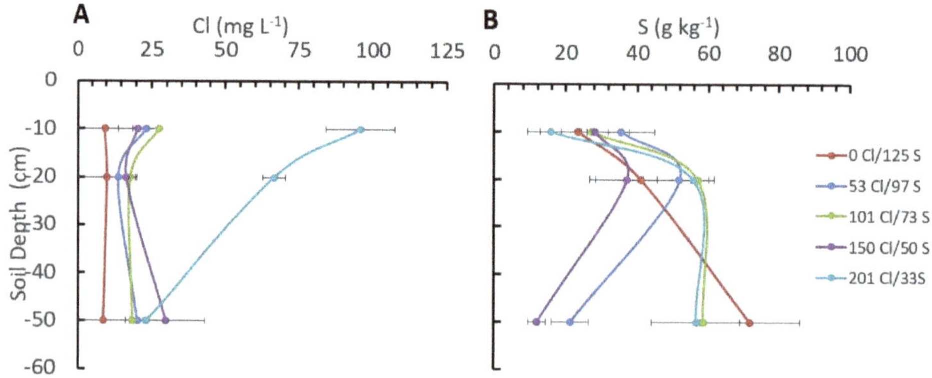

Figure 5. Influence of the chloride and sulfur rates on chloride (**A**) and sulfur (**B**) distribution in the soil profile. The bars in the figure represent the standard error.

These results clearly show a reduction in Cl^- concentration in the soil for the treatments with less Cl^- applied within a specific soil depth. At 50 cm depth, no significant differences in the Cl^- concentration were observed between treatments (Figure 5).

In the case of S, no significant differences were observed between treatments, but there were significant differences between soil depths. In the first 20 cm, the treatment with the higher Cl/S ratio (201 kg Cl/33 kg S) had a mean S content of 15.6 mg·kg^{-1}, while the treatment without Cl^- and the higher S rate (0 kg Cl/125 kg S), showing a mean value of 23.4 mg·kg^{-1}, but at a 50 cm depth, the treatment without Cl^- and the higher S rates (0 kg Cl/125 kg S) increase the S concentration significantly until 71.8 mg·kg^{-1}, while the treatments with lower S rates and medium Cl^- rates (150 kg Cl/50 kg S) showed a lower S concentration of 11.92 mg·kg^{-1} at 50 cm depth.

3. Discussion

3.1. Cl^- Concentration on Tissues and Influence on Growth and Productivity

Cl^- does not appear as a typical micronutrient since the actual Cl^- concentration in plants is in the range of 0.2% to 2% of dry matter [2,6]. In most plant species, the Cl^- requirements for optimal plant growth, however, are in the range of 0.02% to 0.04% of dry matter [3,5], which corresponds to the content of a micronutrient, and the Cl^- content available in nature is sufficient to fulfill these requirements [3]. The critical tissue Cl^- concentration for toxicity is about 0.4% to 0.7% for Cl^--sensitive and 1.5% to 5.0% for Cl^--tolerant plant species [1]. In this research, the Cl^- concentration on the leaves changes according to the Cl^- rate from 0.03% in the treatment without Cl^- to 2.98% of dry matter for the treatment with 300 mg L^{-1} Cl^- without S. These increases in Cl^- concentration in the greenhouse trial had a significant effect on dry biomass accumulation, with a significant reduction on increasing Cl^- rates (Table 1). The significant reduction in biomass accumulation in the greenhouse trial with leaf Cl^- content above 0.33% places coffee in the group of Cl^--sensitive glycophytic plants.

According to Chen et al. [2], when Cl^- levels are high enough to be toxic, cation absorption, such as for K^+, decreases because of the disordered cell metabolism. In the greenhouse trial, we see this tendency with the significant reduction in the dry biomass accumulation when the Cl^- rates increase from 0 to 60 mg L^{-1} (Table 1), and in the field conditions, a significant reduction in the K^+ concentration on the coffee cherries was observed 60 days after flowering but with no significant effect on K^+ uptake at harvest time. (Table 2, Figure 2).

The Cl^- content in grains, fruits, and seeds is very low and is hardly affected by the Cl^- concentration of the soil solution [23]. In the case of the field trial, we observe strong changes in the Cl^- concentration in the coffee cherries during the development process that was directly linked with the fertilizer application time and the soil moisture changes during the year (Figures 1 and 2). In the work of Silva et al. [24], who compared the influence of K sources on coffee productivity and quality, it is possible to observe 30% less Cl^- concentration in the coffee beans on the treatments with SOP compared with the MOP, with significant impact on the coffee quality parameters such as total sugar content and polyphenol oxidase activity.

Cl^- application stimulates plant growth when it is supplied at macronutrient levels [8]. Root Cl^- uptake and long-distance transport require considerable use of metabolic energy, clearly indicating that shoot Cl^- accumulation on macronutrient levels responds to specific biological adaptation [3]. The increase in biomass production induced by the higher rates of Cl^- application as a macronutrient is associated with the stimulation of higher turgor, cell size, and shoot expansion [3,6]. Studies in soils with high to very high extractable K^+ levels on wheat and alfalfa in Argentina showed a positive influence of Cl^- fertilization using MOP and ammonium chloride as a fertilizer source [25]. In both cases, the authors report 50% yield increases independently of the fertilizer sources with Cl^- rates between 23 and 56 kg ha^{-1}. In anion crops, Cl^- supplying higher levels up to 500 mg kg^{-1} on the nutrient solution shows that Cl^- on average is the fourth most utilized essential element, superseded only by N, K, and P [26]. In durum wheat under field conditions, it has been proven that a number of physiological disorders impairing growth and yield are especially due to Cl^- deficiency [27]. Plants such as kiwi fruit and palm trees have higher Cl^- requirements, which cannot be alleviated through NO_3^- addition, and the reasons for such high Cl^- demand is still unknown [3,28,29].

The concept of beneficial elements became popular around the early 1980s to include elements that stimulate plant growth or health but have not been shown thus far to meet the strict essentiality criteria [30], or not essential in certain plant species, or under specific conditions [31]. In this investigation, in both coffee trials, no stimulating effect of the application of Cl^- on growth and yield was observed (Tables 1 and 3), nor an increase in biomass accumulation in the greenhouse with Cl^- rates higher than 60 mg$\cdot L^{-1}$. At field conditions in Brazil, Santos et al. [21] reported a yield reduction with an application of 100% of the K^+ as a MOP, which represented an average Cl^- dose of 230 kg Cl$\cdot ha^{-1}\cdot year^{-1}$.

A negative effect could be observed in some crops when the rates of Cl^- increase to 200–400 mg$\cdot kg^{-1}$. For most crops, the negative effect was obvious when the applied amount increased to 400–500 mg$\cdot kg^{-1}$, and the yield of most crops decreased rapidly when the applied Cl^- exceeded 800 mg$\cdot kg^{-1}$ [2].

3.2. Cl^- *Influence on Nutrients Uptake and NUE*

Cl^- plays a quantitatively important role in ion balance when Cl^- is abundant, but other anions (nitrate, malate) can fulfill this role when Cl^- supply is reduced. Competitive effects in uptake between Cl^- and N-NO_3^- and Cl^- and SO_4^{2-} were documented by De

Wit et al. [32], who recently reported that Cl^- somewhat affects the uptakes and utilities of N, P, K, Ca, Mn, Si, S, Zn, Mg, Fe, and Cu in potatoes, with the most extreme competitive effects of N-NO_3^- in crops such as rice, corn, soybean, cabbage, tomato, strawberry, melon and lettuce, peanut, barley, citrus, and spring wheat [2,23].

Cl^- in excess can strongly reduce the NUE specifically interfering with its uptake, transport, and loading into the root xylem, since it uses the same anion channels used by NO_3^- [33,34]. NO_3^- and Cl^- are the most abundant inorganic anions in plants and share similar physical properties and transport mechanisms, which is the origin of the strong dynamic interactions between these two monovalent anions and which frequently explains why the higher accumulation of Cl^- leads to lower NO_3^- content in plants [3]. This antagonistic interaction between Cl^- and NO_3^- has been reported by several authors [20,23,35] and is one of the reasons why Cl^- is considered a detrimental nutrient in agriculture.

Colmenero-Flores et al. [3] have shown that prolongated exposure to a nutrient solution containing Cl^- at a concentration of 4–5 mM (140–180 ppm) may cause a gradual non-toxic accumulation of Cl^- at values ranging between 2.5% and 5.0% DW (macronutrient levels), without any interferences on plant growth and stress symptoms. According to Carillo and Rouphael [34], when Cl^- is in excess, it is passively transported into the cortical cell and the xylem by anion channels, such as the NO_3^- transporter NPF7.3 and S-type anion heteromeric channel SLAH1/SLAH3, where high Cl^- concentration at the leaf level is less controlled and more dangerous than that of sodium due to the lower capacity of the leaf blade to exclude Cl^- and its limited basipetal phloem transport toward the roots.

When Cl^- is accumulated in high concentration in the leaf tissues, it initially decreases the apoplast osmotic potential interfering with the cellular water relations [35]. Thereafter, it diffuses into the symplast by using anion (e.g., nitrate and phosphate) uptake symporters competing with these beneficial nutrients for uptake within the cell [35].

In coffee growing under greenhouse conditions, the highest doses of Cl^- without S strongly reduce N uptake, the utilization efficiency—NU_TE, and NUE. Treatments without Cl^- showed the highest NU_TE but not the highest N uptake and NUE. Treatments with low to medium Cl^- rates (60 to 180 mg L^{-1}) significantly increased NUE without significant changes in NU_TE (Table 3).

At the field level, the results showed significant differences in N uptake by the coffee cherries (Figure 3) according to the Cl^- and S rates, the treatment with high Cl^- (201 kg Cl/33 kg S) showed significantly lower N uptake than the treatment without Cl^- (0 kg Cl/125 kg S), but the higher and significantly different nitrogen uptake was achieved in the treatment with medium Cl^- rate (between 100–150 kg Cl/73–50 kg S), in which a similar tendency was observed for NUE (Figure 4).

Nitrate uptake and allocation are key factors regulating NUE [36], given the close interaction between Cl^- and NO_3^-, and it is expected that Cl^- can significantly influence the NUE, as can be seen in both coffee trials, where the NUE was reduced in the higher Cl^- rates, but the question is, why in medium or low Cl^- rates was the NUE improved? Can we say that Cl^- at a certain point can improve the NUE? The potential reason could be related to the net NO_3^- uptake dynamic, which results from the differences between NO_3^- influx mediated by active transport and its passive efflux through anion channels [3]. Root anion efflux to the rhizosphere might be important in regulating H^+-ATPase activity, maintaining the H^+ charge balance [37], or regulating plant cell growth [3]. The release of Cl^- from root cells through anion channels, replacing NO_3^- efflux, could be an important mechanism for preventing N loss [38], which is expected to improve NUE [3].

Rosales et al. [8] suggest that Cl^- nutrition reduces NO_3^- sequestration in plant leaf tissues (e.g., vacuolar compartmentalization), making this valuable N source available for assimilation and biosynthesis of organic N. Cl^- can improve the NUE, despite significantly reducing foliar NO_3^- storage, which represents a radical change in the perception of the Cl^- and NO_3^- antagonist. The most likely scenario is that when NO_3^- is available, the active transport mechanism that is frequently more selective for NO_3^- than for Cl^- prioritizes NO_3^- influx by inhibiting Cl^- uptake. When little NO_3^- is available, Cl^- influx is less inhibited, thus increasing root uptake and intracellular Cl^- concentration, which is expected to replace NO_3^- in serving an osmotic function, allowing for more efficient use of the available N [2].

The stimulatory effect of Cl^- on the asparagine synthetase activity has also been suggested as another mechanism of interaction between Cl^- and NUE [3]. Chloride increases the affinity of asparagine synthetase for glutamate, its substrate [39]. Asparagine is a major compound in the long-distance transport of soluble N in many plant species [3,5], which explains the higher accumulation of N in coffee cherries with medium Cl^- rates and the higher NUE compared with the treatments without Cl^-.

Rosales et al. [8] reported that Cl^- significantly increases the NUE in different crops such as tobacco, olive, mandarin, lettuce, spinach, and chard when accumulated at the macronutrient level. Finely modulating the Cl^- dose for decreasing the NO_3^- accumulation in leaves or improving its uptake and assimilation without decreasing the growth and productivity of the plants is necessary [34].

In the case of coffee at the field level, this fine modulation of Cl^- and S rates and ratios shows potential benefits to improving the NUE, as is indicated by these results, creating a fine balance for the Cl/S rates with mean values on a productive coffee plantation between 100 and 150 kg Cl^- ha^{-1}. $year^{-1}$ and 73–50 kg S·ha^{-1}·$year^{-1}$. We can hypothesize in this study that Cl^- improved NUE when applied at medium doses at the "macronutrient" level, through a reduction in N compartmentalization and improved transport through stimulation of asparagine synthetase, represented in higher N uptake by coffee cherries without a significant increase in yield in field conditions and biomass in greenhouse conditions (Figure 4, Table 1). However, such macronutrient rates of Cl^- did not improve NO_3^- assimilation represented by the NU_TE (Table 1), similarly reported in Cl^- excluding species such as olive and citrus rootstock Cleopatra mandarin plants [8].

In the field trial over four years, a significant yield reduction was not observed, likely related to the high soil moisture during the study time, with long periods of water excess between March and July and between October and December over the years (Figure 1), allowing for Cl^- movement into the soil profile. Although the Cl^- contents in the soil in the first 20 cm of depth were significantly higher in the treatment with 201 kg Cl and 33 kg S, the electrical conductivity (EC) was lower than 1.2 dS·m^{-1} in the soil's first 10 cm and less than 0.5 dS·m^{-1} at 30 and 50 cm depths.

The discussion on the effect of Cl^- on the quality of coffee is still open. Silva et al. [24] demonstrated more than ten years ago a reduction in quality parameters with the application of MOP in coffee, and recently, Santos et al. [21] demonstrated a significant reduction in the quality assessment of coffee in treatments that had low and high proportions of Cl^- in mineral K fertilization. However, an interesting note in the study by Santos et al. [21] is that the treatment without Cl^- was not the one that presented the highest yields in coffee, while the treatment with 25 to 50% of K^+ as MOP and 75% to 50% as SOP (57 to 115 kg Cl·ha^{-1}·$year^{-1}$) presented the highest yield, probably related to the effect of Cl^- on the NUE.

3.3. *Cl^- on the Soil*

The Cl^- content of the soil is not an intrinsic property of the soil but rather a result of soil management [23]. As Cl^- can move freely with soil water, soil Cl^- levels can be highly variable and can increase or decrease from year to year, depending on the water table and the location in the landscape [1,40]. Soils considered low in Cl^- are below 2 mg·kg^{-1}. Regarding wheat in several soils of the United States, Fixen et al. [41] reported that higher Cl^- levels in the soil of 43.5 kg·ha^{-1} (0–0.6 m) were adequate for near-maximum wheat yield; in Argentina, Diaz-Zorita et al. [25] reported that the Cl^- levels in the soil higher than 13.2 mg·Kg^{-1} (0.0 to 0.2 m) were adequate for maximum grain yield.

In the present research on coffee, the Cl^- concentration on the treatments with 201 kg Cl/33 S showed a mean value for Cl^- of 98.5 mg·L^{-1} at 0–0.10 m depth and 66.4 mg·L^{-1} at 0.10–0.20 m depth, without significant differences on crop yield during the

4 years but with significant influence on N uptake and efficiency. These results indicate that for coffee, the ideal Cl^- concentration on the soil should be below 30 mg L^{-1}. Regarding coffee, Brazil Santos et al. [21] found a significant difference in the Cl^- content in the soil, where the treatments with 100% K^+ applied as MOP had an average Cl^- content in the soil in the first 20 cm of the depth of 130 kg·ha^{-1} (approx. 65 mg·L^{-1} of Cl^-), while the treatment without Cl^- with 100% K^+ applied as SOP had a significantly lower Cl^- content in the soil at the same depth equivalent to 50 kg·ha^{-1} (approx. 25 mg·L^{-1} of Cl^-).

When Cl^- is applied to coffee at average doses of 100 to 150 kg·ha^{-1} in balance with S without exceeding foliar concentrations of 0.33% of DW and soil contents of 30 mg L^{-1}, it has positive effects on the efficient use of nitrogen, increasing it from 9% to 14%. We can affirm that the effect of Cl^- on NUE is related to a reduction in compartmentalization and an increase in $NO_3{}^-$ translocation, but with no effect on $NO_3{}^-$ assimilation, due to the negative effect of high doses of Cl^- on nitrogen utilization efficiency (NU_TE). High doses of Cl^- both in the greenhouse (>180 mg·L^{-1}) and in the field (>150 kg·ha^{-1}) reduce NUE due to a Cl^- saturation effect in the vacuole at the physiological level, but also due to Cl^- saturation at the soil level, generating competition with anion uptake, especially with $NO_3{}^-$. The positive effects of moderate doses of Cl^- on NUE may not be observable or reproducible under conditions where coffee plants grow with prolonged periods of water deficit, generating soil and tissue Cl^- concentrations above normal levels, as previously mention.

4. Materials and Methods

During the period of four years, two trials were carried out under greenhouse and field conditions, with the aim to test rates and ratios of Cl^- and S in coffee and their influences on growth, nutrient uptake, productivity, and NUE.

4.1. Greenhouse Trial

The greenhouse trial was located in Dülmen, Germany, at the Hanninghof Research Center of Yara International. The mean air temperature was 23.1 °C (±2.2 °C) with a maximal air temperature of 31.4 °C and minimal air temperature of 15.3 °C, mean relative humidity of 64% (±10%), and mean light intensity of 20.0 Klux during summertime. Supplemental light (300 mmol m^{-2} s^{-1} photosynthetic photon flux density) over a period of 12–14 h was given when natural light became insufficient.

Coffee seeds from the *Coffea arabica* var. Cenicafé 1 were pre-germinated in dark conditions with a mean temperature of 28 °C for 6 weeks using disinfected sphagnum-moos as a germination medium. Before the radicle emerged (BBCH scale 03-Arcila et al., 2002 [42], the pre-germinated seeds were moved to small containers with perlite as a growing medium. The seeds were allowed to germinate for 6 months. During this germination process, the plants received a nutrient solution once per week containing: N (7.6 mM), P (0.3 mM), K (1.7 mM), Mg (0.2 mM), Ca (0.9 mM), Fe (5.0 mM), Mn (2.9 mM), Zn (1.5 mM), Cu (0.6 mM), B (9.2 mM) and Mo (0.2 mM).

Once the plants reached three pairs of leaves that were completely open (BBCH scale 13; Arcila et al. [42]), they were transplanted in pots of 4.5 L. The aim of the greenhouse trials was to evaluate the influence of different Cl^- and S rates and proportions on growth and nutrient uptake and NUE. In this trial, several Cl^- and S rates and ratios were tested as follows: 0/200; 60/160; 120/120; 180/80; 240/40 and 300/0 Cl/S in mg.L^{-1}. In this trial, the coffee plants were grown in coarse sand as a growing medium. All other nutrients were applied as a nutrient solution to the soil surfaces, without any foliar application. Pots were watered with a complete nutrient solution containing N (28.6 mM), P (1.5 mM), Mg (6 mM), Ca (4.5 mM), Fe (14.7 mM), Mn (8.4 mM), Zn (5.3 mM), B (5.2 mM), Cu (14.2 mM) and Mo (1.5 mM). The nutrient solution was applied once per week with application volumes between 60 and 120 mL, according to the water demand of the plants. The soil moisture was monitored daily with the aim of avoiding water deficit or excess, keeping it between 60% and 70% of the water holding capacity.

After 9 months of transplanting, the coffee plants were trimmed, and the tissues (leaves, stems, and branches) were dried in an oven at 65 °C until a constant weight was attained. The dried material was then finely ground for nutrient analysis in the lab.

Finely milled plant materials were used for elemental analysis after wet digestion in a microwave digester (MLS mega; MLS GmbH, Leutkirch, Germany). All the micro- and macronutrients (excluding nitrogen) were analyzed using inductively coupled plasma optical emission spectrometry (Perkin-Elmer Optima 3000 ICP-OES; Perking-Elmer Corp, Shelton, CT, USA). The nitrogen was determined by the micro-Kjeldahl method.

Two nitrogen use efficiency (NUE) indicators were calculated: (i) N use efficiency (NUE), calculated as the total nitrogen uptake by the shoot plant divided by total N applied during the growing period (g shoot DW g^{-1} N), and (ii) N utilization efficiency (NU_TE), calculated as the total shoot dry biomass divided by the total nitrogen content in the shoot (g DW mg^{-1} N). The NU_TE indicator allows for an understanding of how efficiently the transported N is used by the plant [8].

4.2. Field Trial

For four years, from January 2017 to December 2020, a field trial was carried out in the southeast region of Colombia, in Garzón-Huila on a farm located at 2°11′ N–75°34′ W at 1.437 m elevation. The soil predominantly comprised consolidated materials, was mainly granitic in nature, and had moderately low soil fertility, with characteristics as follows: pH 4,71; organic matter at 3.08%; and P, K, Ca, and Mg contents at 11, 125, 530, and 52 mg·kg^{-1}, respectively; and soil particle distribution of 48% sand, 18% silt, and 34% clay. The pH was determined in water (1:1), organic matter by Walkley–Black, P by Bray-II, and the exchangeable fraction of K, Mg, and Ca with 1 N ammonium acetate extraction (1 N $NH_4C_2H_3O_2$, pH 7.0). The cations in the extracts were detected using an ICP (Perkin Elmer, Optima 8000, Shelton, CT, USA), and soil texture analysis was performed using the hygrometer Bouyoucos method. The mean climatic conditions observed in the region from 2017 to 2019 are shown in Table 4.

Table 4. Climate conditions obtained from the weather station *.

Year	T·min (°C)	T·max (°C)	T. med (°C)	R.H (%)	Rainfall (mm)	Sunshine (h)	Rainfall in the Trial Area (mm)
2017	16.9	24.6	20.2	77.1	1.575	1.1467	2.483
2018	16.6 ±	25.1 ±	20.3 ±	72.6 ±	1.234	-	2.053
2019	16.1	23.8	19.4	-	1.426	1.314,0	2.319
2020	-	-	-	-	1.206	-	2.100
Mean	16.5	24.5	20.0	74.8	1.360	1.2303	2.238
1955–2010	16.2	24.5	19.7	-	1.330	1.2500	-

* Jorge Villamil weather station, 2°20′ N–75°31′ W, provided by the National Coffee Research Center—Meteorological Network. T. min, average minimum air temperature; T. max, average maximum air temperature; T. med, average mean air temperature; R.H., average relative humidity. ± Mean value for three months.

The trial was run using the *Coffea arabica* L. variety Castillo® with resistance to the coffee leave rust (CLR) disease generated by the fungi *Hemileia vastatrix* Berkeley and Brome [42]. The plantation was established in 2012 in a full sunshine condition (FS) planted at a density of 6.666 plants·ha^{-1} at 1.0 m distance between plants and 1.5 m distance between rows. Before the treatment application, the plantation was stem trimmed at 30 cm height in January 2017, aiming to rejuvenate the plantation and to initiate a new productive cycle from 2018 to 2020. Five treatments differing in their Cl/S ratio were installed in the trial. Rates of Cl^- and S increased over the 4 years according to increments in growth and yield formation (Table 5).

Table 5. Treatment description during the four-year trial.

Treatment	2017	2018	2019	2020	Average
			Cl/S Rates (kg·ha^{-1})		
1	0/41	0/96	0/174	0/188	0/125
2	21/30	43/62	74/140	74/156	53/97
3	40/20	86/41	138/109	138/123	101/73
4	61/10	129/22	205/78	202/91	150/50
5	80/0	175/30	276/44	276/59	201/33
			K_2O Rates (kg·ha^{-1})		
1 to 5	110	230	360	360	265
			N Rates (kg·ha^{-1})		
1 to 5	160	163	280	320	231
			P_2O_5 Rates (kg·ha^{-1})		
1 to 5	90	46	92	92	80
			CaO Rates (kg·ha^{-1})		
1 to 5	130	86	104	121	110
			MgO Rates (kg·ha^{-1})		
1 to 5	29	50	54	72	51

The sources of Cl^- used in the trial were potassium muriate (MOP with 60% K_2O and 46% Cl), and the S source was potassium sulfate (SOP with 50% K_2O and 18% S). The Cl^- and S rates were adjusted based on the potassium requirement of the crop, considering crop K_2O demand and K_2O content in the soil [43–46]. The mean nutrient rates over 4 years were: 231 kg N, 265 kg K_2O, 80 kg P_2O_5, 110 kg CaO, and 51 kg MgO kg·ha^{-1} (Table 5). The total nutrient rate was split into 3 applications during the year: at pre-flowering, and at 30 and 100 days after flowering. Fertilizer rates and application time are currently and commonly used by coffee farmers.

The experiment was set up in a randomized complete block design with four replications; each plot in the block had 45.0 m^2 with 12 effective plants. Five plants of each compelling lot were selected for yield assessments, and the other 7 plants of the plot were chosen to make destructive sampling of the coffee cherries every 30 days after flowering until harvest with the aim to evaluate the nutrient concentration and nutrient uptake of the coffee cherries and beans over time. In each of the 7 plants, branch number 10, counted from the apex to the base, was labeled at the pre-flowering stage (BBCH scale 54–57), and the cherries of the labeled branch were harvested every 30 days from 30 to 240 days after flowering. Each effective plant served as a replicate.

The other 5 effective plants were used for the assessments at harvest. On average, the coffee was harvested over 8 months in a year with the main harvest (70% of the year) collected from September to November. The coffee cherries were harvested when they reached maturity at BBCH scale 88 at 240 days after flowering.

All cherry samples were dried at 60 °C for a few days until a constant weight was attained and finely ground for nutrient analysis. The nutrients K, Ca, Mg, S, and Cl were analyzed using inductively coupled plasma optical emission spectrometry (Perkin-Elmer 400; Perking-Elmer Corp., Norwalk, CT, USA). N was analyzed by Micro-Kjeldahl method.

The nitrogen use efficiency by the coffee beans (NUE) was estimated using an index between the total nitrogen uptake by the coffee beans at harvest (BBCH scale 88) divided by the total amount of nitrogen applied during the crop season. This NUE indicator is known as a fertilizer-based indicator [47] and can be considered similar to the partial N balance (PNB = plant N content per unit of fertilizer N applied) described by Doberman [48].

With the aim to evaluate the Cl^- concentration and distribution in the soil profile after three years of treatment application, soil samples were collected per treatment and replicated at a 25 cm lateral distance to the stem, where nearly 86% of the coffee roots are located [49]. Samples were taken at three soil depths (10, 30, and 50 cm).

A soil moisture index (SMI) was estimated using daily weather data. The SMI is defined as the ratio between actual volumetric soil moisture and volumetric saturation humidity [50], following the water balance approach described by Ramirez and Küsters [22].

All data were submitted to statistical analysis (ANOVA) according to the experimental design using the Statgraphics Centurion XV software package (Statgraphics Technologies, Inc.). The Shapiro–Wilk modified test was applied for normality testing and the residual vs. prediction test to evaluate the heterogeneity of variances. Fischer's test was used to detect the treatments that significantly affected the ANOVA.

5. Conclusions

This work presents novel results regarding the influence of Cl^- application on coffee and its influence on growth, productivity, and NUE. After four years of research in the greenhouse and on the field, we can conclude that:

- A balance between Cl^- and S is necessary for coffee with the aim to improve N uptake and NUE. For greenhouse conditions, there should be between 60 and 180 $mg \cdot L^{-1}$ of Cl^- and between 160 and 80 $mg \cdot L^{-1}$ of S. Plants without Cl^- or without S significantly reduce the NUE. At the field level, the balances should be between 100 and 150 kg Cl^- $ha^{-1} \cdot year^{-1}$ and between 73 and 50 kg $S \cdot ha^{-1} \cdot year^{-1}$.
- Cl^- is applied to coffee at rates as a "macronutrient", and the coffee plants can take higher amounts of Cl^-, reaching concentrations on the leaves of higher than 0.69% without toxicity symptoms, but with a significant reduction in dry biomass accumulation. With the aim to keep a balance between biomass accumulation and NUE, the Cl^- content in the leaves should be lower than 0.33%, placing the coffee into the group of glycophyte Cl^--sensitive plants.
- Cl^- rates at the field level that are higher than 150 kg Cl^- $ha^{-1} \cdot year^{-1}$ significantly increase the Cl^- content in the soil and Cl^- uptake by the coffee cherries and reduces N uptake by the coffee cherries, reducing the NUE on average by 9% to 14%.
- In terms of the NUE, the Cl^- content in the soil in coffee in the first 20 cm depth should not exceed the concentration of 30 $mg \cdot L^{-1}$.
- The NUE in coffee could be improved by using agronomical management practices such as the selection of mineral potassium fertilizer sources, with the main aim to reduce the Cl^- application rates and to limit them to no more than 100 to 150 kg Cl^- ha^{-1} with a fine balance with S rates, as mentioned before.

Author Contributions: Trial conceptualization, data analysis, writing, V.H.R.-B. and J.K.; field trial implementation, data acquisition J.A.-V. and L.A.L.-V.; greenhouse trial implementation, data acquisition, E.T. All authors have read and agreed to the published version of the manuscript.

Funding: This work was supported by Yara International.

Data Availability Statement: No applicable.

Acknowledgments: The authors thank the Plant Science Department from the National University of Colombia for support the graduate students, especially Maria Sara Mejía de Tafur, and Jaime Figueroa-Cantillo, the coffee grower, for supporting the field trial, and finally the reviewers, for their comments and suggestions that helped to improve the quality of the manuscript.

Conflicts of Interest: The authors declare no conflict of interest.

References

1. White, J.P.; Broadley, R.M. Chloride in Soils and its Uptake and Movement within the Plant. A Review. *Ann. Bot.* **2001**, *88*, 967–988. [CrossRef]
2. Chen, E.; He, L.Z.; Yang, E.X.; Mishra, S.; Stoffella, J.P. Chlorine nutrition of higher plants: Progress and perspectives. *J. Plant Nutr.* **2010**, *33*, 943–952. [CrossRef]

3. Colmenero-Flores, J.M.; Franco-Navarro, J.D.; Cubero-Front, P.; Peinado-Torrubia, P.; Rosales, M.A. Chloride as a beneficial Macronutrient in Higher Plants: New Roles and Regulation. *Int. J. Mol. Sci.* **2019**, *20*, 4686. [CrossRef]
4. Broyer, T.C.; Carlton, A.B.; Johnson, C.M.; Stout, P.R. Chlorine—A micronutrient element for higher plants. *Plant Physiol.* **1954**, *29*, 526–532. [CrossRef]
5. Marschner, P. *Marschner's Mineral Nutrition of Higher Plants*, 3rd ed.; Marschner, P., Ed.; Elsevier: Amsterdam, The Netherlands; Academic Press: San Diego, CA, USA, 2012; 668p.
6. Franco-Navarro, J.D.; Brúmos, J.; Rosales, M.A.; Cubero-Font, P.; Talón, M.; Colmero-Flores, J.M. Chloride regulates leaf cell size and water relations in tobacco plants. *J. Exp. Bot.* **2015**, *67*, 873–891. [CrossRef] [PubMed]
7. Terry, N. Photosynthesis, Growth, and the Role of Chloride. *Plant Physiol.* **1977**, *60*, 69–75. [CrossRef]
8. Rosales, M.A.; Franco-Navarro, J.D.; Peinado-Torrubia, P.; Diaz-Rueda, P.; Alvarez, R.; Colmero-Flores, J.M. Chloride Improves Nitrates Utilization and NUE in Plants. *Front. Plant Sci.* **2020**, *11*, 442. [CrossRef]
9. Sollins, P.; Robertson, P.G.; Uehara, G. Nutrient mobility in variable- and permanent-charge soils. *Biogeochemistry* **1988**, *6*, 181–199. [CrossRef]
10. Neto, A.P.; Favarin, J.L.; de Almeida, R.E.M.; dos Santos Dias, C.T.; Tezotto, T.; Ferreira, M.T. Changes of Nutritional Status during a Phenological Cycle of Coffee under High Nitrogen Supply by Fertigation. *Commun. Soil Sci. Plant Anal.* **2011**, *42*, 2414–2425. [CrossRef]
11. Neto, A.P.; Favarin, J.L.; do Reis, A.R.; Tezotto, T.; de Almeida, R.E.M.; Levres, J.; Gallo, L.A. Nitrogen metabolism in coffee plants in response to nitrogen supply by fertigation. *Theor. Exp. Plant Physiol.* **2015**, *27*, 41–50. [CrossRef]
12. Bruno, P.I.; Unkovich, J.M.; Bortolotto, P.R.; Bacchi, O.S.; Dourado, N.D.; Reichardt, K. Fertilizer nitrogen in fertigated coffee crop: Absorption changes in plant compartments over time. *Field Crops Res.* **2011**, *124*, 369–377. [CrossRef]
13. Sadeghian, K.S.; Duque, O.H. Optimal doses of nutrients for coffee plantations in production: Economical considerations. *Adv. Técnicos Cenicafé* **2021**, *533*, 1–8. (In Spanish) [CrossRef]
14. Favarin, J.L.; de Souza, L.T.; Mazzafera, P.; Dimenstein, L. Soil correction and fertilization of irrigated coffee plants in production. In *Cafeicultura do Cerrado/Editores Técnicos Gladyston Rodrigues/Carvalho et al-Belo Horizonte*; EPAMIG: Minas Gerais, Brazil, 2021; 564p. (In Portuguese)
15. Cannavo, P.; Harmand, J.M.; Zeller, B.; Vaast, F.; Ramirez, J.E.; Dambrine, E. Low nitrogen use efficiency and high nitrate leaching in a highly fertilized *Coffea arabica–Inga densiflora* agroforestry system: A ^{15}N labeled fertilizer study. *Nutr. Cycl. Agroecosyst.* **2013**, *92*, 377–394. [CrossRef]
16. Salamanca, J.A.; Doane, T.; Horwath, R.W. Nitrogen Use Efficiency of Coffee at the Vegetative Stage as Influenced by Fertilizer Application Method. *Front. Plant Sci.* **2017**, *8*, 223. [CrossRef] [PubMed]
17. Silva, E.B.; Nogueira, F.D.; Guimarães, P.T.G.; Furtini Neto, A.E. Responds of the coffee to the potassium fertilization. *Pesq. Agropec. Bras.* **2001**, *36*, 133–1337. (In Portuguese) [CrossRef]
18. Sadeghian, K.S. Soil fertility and coffee nutrition in Colombia. *Boletín Técnico Cenicafé.* **2008**, 43. (In Spanish)
19. Cuzato-Mancuso, M.A.; Peres, S.R.; Costa, C.C.A.; Amaral, C.G.S. Effect of the potassium sources and rates on arabica coffee yield, nutrition, and macronutrient export. *Rev. Bras. Cienc. Solo.* **2014**, *38*, 1448–1456. [CrossRef]
20. Mengel, K.; Kirkby, E.A. *Principles of Plant Nutrition*; CH-3048; International Potash Institute: Worblaufen-Bern, Switzerland, 1978; 593p.
21. Santos, C.; Malta, M.R.; Gonzalves, M.G.M.; Borém, F.M.; Pozza, A.A.A.; Martinez, H.E.P.; de Souza, T.I.; Chagas, W.F.T.; de Melo, M.E.A.; Oliveira, D.P.; et al. Chloride Applied via Fertilizer Affects Plant Nutrition and Coffee Quality. *Plants* **2023**, *12*, 885. [CrossRef]
22. Ramirez, B.V.H.; Küsters, J. Calcium and Potassium nutrition increase the water use efficiency in coffee: A promising strategy to adapt to climate change. *Hydrology* **2021**, *8*, 75. [CrossRef]
23. Xu, G.; Magen, H.; Tarchitzky, J.; Kafkafi, U. Advances in chloride nutrition of plants. In *Advances in Agronomy*; Sparks, D.L., Ed.; Academic Press: San Diego, CA, USA, 2000; Volume 68, pp. 97–150.
24. Silva, E.B.; Nogueira, F.D.; Guimarães, P.T.G.; Chagas, S.J.R.; Costa, L. Sources and rates of potassium influence on productivity and quality of the coffee. *Pesq. Agropec. Bras.* **1999**, *34*, 133–1337. (In Portuguese)
25. Diaz-Zorita, M.; Duarte, A.G.; Barraco, M. Effects of Chloride Fertilization on Wheat (*Triticum aestivum* L.) Productivity in the Sandy Pampas Region, Argentina. *Agron. J.* **2004**, *96*, 839–844. [CrossRef]
26. Randle, M.W. Chloride Requirements in Onion: Clarifying a Widespread Misunderstanding. *Better Crops* **2004**, *88*, 101.
27. Schwenke, G.D.; Simpfendorfer, S.R.; Collard, B.C.Y. Confirmation of chloride deficiency as the cause of leaf spotting in durum wheat grown in the Australian northern grains regions. *Crop Pasture Sci.* **2015**, *66*, 122–134. [CrossRef]
28. Smith, G.S.; Clark, C.J.; Holland, P.T. Chlorine requirements of kiwifruit (*Actinidia deliciosa*). *New Phytol.* **1987**, *106*, 71–80. [CrossRef]
29. Braconnier, S.; Dauzac, J. Chloride and stomatal conductance in coconut. *Plant Physiol. Biochem.* **1990**, *28*, 105–111.
30. Brown, H.P.; Zhao, J.F.; Dobermann, A. What is a plant nutrient? Changing definitions to advance science and innovation in plant nutrition. *Plant Soil* **2022**, *476*, 11–23. [CrossRef]
31. Broadley, M.; Brown, H.P.; Cakmak, I.; Ma, J.F.; Rengel, Z.; Zhao, J.F. Beneficial Elements. In *Marschner's Mineral Nutrition of Higher Plants*, 3rd ed.; Marchner, P., Ed.; Academic Press: San Diego, CA, USA, 2012; pp. 249–269.

32. De Wit, C.T.; Dijkshoorn, W.; Noggle, J.C. *Ionic Balance and Growth of Plants*; Verlagen van Landbouwkundige Onderzoekingen: Wageningen, The Netherlands, 1963; 68p.
33. Diatlofft, E.; Roberts, M.; Sanders, D.; Roberts, K.S. Characterization of Anion Channels in the Plasma Membrane of Arabidopsis Epidermal Roots Cells and the Identifications of a Citrate-Permeable Channel Induced by Phosphate Starvation. *Plant. Phys.* **2004**, *136*, 4136–4149. [CrossRef]
34. Carillo, P.; Rouphael, Y. Nitrate Uptake and Use Efficiency: Pros and Cons of Chloride interference in the Vegetable Crops. *Front. Plant. Sci.* **2022**, *13*, 899552. [CrossRef]
35. Kafkafi, U.; Valoras, N.; Letey, J. Chloride interaction with nitrate and phosphate nutrition in tomato (*Lycopersicon esculentum* L.). *J. Plant. Nutr.* **2008**, *5*, 1369–1385. [CrossRef]
36. Wang, Y.Y.; Hsu, P.K.; Tsay, Y.F. Uptake, allocation and signaling of nitrate. *Trends Plant Sci.* **2012**, *17*, 458–467. [CrossRef]
37. Hawkesford, M.; Horst, W.; Kichey, T.; Lambers, H.; Schjoerring, J.; Moller, I.S.; White, P. Functions of macronutrients. In *Marschner's Mineral Nutrition of Higher Plants*, 3rd ed.; Marchner, P., Ed.; Academic Press: San Diego, CA, USA, 2012; pp. 135–189.
38. Wege, S.; Gilliham, M.; Henderson, S.W. Chloride: Not simply a "Cheap osmoticum" but a beneficial plant macronutrient. *J. Exp. Bot.* **2017**, *68*, 3057–3069. [CrossRef]
39. Rognes, S.E. Anion regulation of lupin asparagine synthetase-Chloride activation of the glutamine-utilizing reactions. *Phytochemistry* **1980**, *19*, 2287–2293. [CrossRef]
40. Imas, P.; Bansal, S.K. Use of MOP as fertilizer—Some Myths and Reality. *Fertil. News* **2004**, *49*, 51–55.
41. Fixen, P.E.; Buchenau, W.G.; Gelderman, H.R.; Schumacher, E.T.; Gerwing, E.T.; Cholick, A.F.; Farber, G.B. Influences of Soil and Applied Chloride on Several Wheat Parameters. *Agron. J.* **1989**, *78*, 736–740. [CrossRef]
42. Arcila, P.J.; Buhr, L.; Bleiholder, H.; Hack, H.; Meier, U.; Wicke, H. Application of the extended BBCH scale for the description of the growing stages of coffee (*Coffea* spp.). *Ann. Appl. Biol.* **2002**, *141*, 19–27. [CrossRef]
43. Alvarado, G.; Posada, H.E.; Cortina, H.A. CASTILLO a New Coffee Variety with Rust Resistance. *Adv. Técnicos Cenicafé* **2005**, *337*, 1–8. (In Spanish)
44. Riaño, H.N.M.; Arcila, P.J.; Jaramillo, R.A.; Chaves, C.B. Dry matter accumulation and nutrient extraction by *Coffea arabica* L. cv. Colombia in three localities of the central coffee zone. *Cenicafé* **2004**, *55*, 265–276. (In Spanish)
45. Sadeghian, K.S.; Mejia, M.B.; Arcila, P.J. Elemental composition of coffee fruits and extraction of nutrients by the harvest in the coffee zone of Colombia. *Cenicafé* **2006**, *57*, 251–261. (In Spanish)
46. Sadeghian, K.S.; Mejia, M.B.; González, O.H. Accumulation of nitrogen, phosphorus, and potassium in coffee fruits. *Adv. Técnicos Cenicafé* **2013**, *429*, 1–8. (In Spanish)
47. Congreves, K.A.; Otchere, O.; Ferland, D.; Farzadfar, S.; Williams, S.; Arcand, M.M. Nitrogen Use Efficiency Definitions of Today and Tomorrow. *Front. Plant Sci.* **2021**, *12*, 637108. [CrossRef]
48. Dobermann, A. *Nutrient Uses Efficient Measurement and Management*; Agronomy & Horticulture; University of Nebraska: Lincoln, NE, USA, 2007; p. 1442. Available online: https://digitalcommons.unl.edu/agronomyfacpub/1442 (accessed on 10 January 2023).
49. Rendón, S.J.R.; Giraldo, H. Distribution of roots in coffee variety Castillo® under two spatial arrangements. *Cenicafé* **2019**, *70*, 7–17. (In Spanish) [CrossRef]
50. Ramirez, B.V.H.; Jaramillo, R.A.; Arcila, P.J. Indices to evaluate water status in coffee plantations. *Cenicafé* **2010**, *61*, 55–66. (In Spanish)

plants

Article

Influence of Organic and Inorganic Fertilizers on Tea Growth and Quality and Soil Properties of Tea Orchards' Top Rhizosphere Soil

Manzoor [1,2,3], Lifeng Ma [2], Kang Ni [2] and Jianyun Ruan [2,*]

[1] Graduate School, Chinese Academy of Agricultural Sciences, Beijing 100081, China; manzoorali3755@gmail.com
[2] Key Laboratory of Tea Biology and Resource Utilization of Tea, Tea Research Institute, Chinese Academy of Agriculture Sciences, The Ministry of Agriculture, Hangzhou 310008, China
[3] Key Laboratory of Tropical Forest Ecology, Xishuangbanna Tropical Botanical Garden, Chinese Academy of Sciences, Xishuangbanna 666303, China
* Correspondence: jruan@mail.tricaas.com; Tel.: +86-571-86653938

Abstract: Organic-based fertilizers have been ratified to be effective in ameliorating tea growth and the fertility of soil. However, the effect of integrated fertilization on tea growth and quality and the chemical properties of the soil in tea gardens are unclear. To address this, from 2020 to 2021, five different treatments were carried out in the greenhouse of the Tea Research Institute, Hangzhou, CAAS, including CK (control), NPK (chemical fertilizers), RC (rapeseed cake), NPK+B (chemical fertilizer + biochar), and NPK+RC, to investigate the effects of different fertilizations on soil chemistry and tea growth and quality. The results indicated that NPK+B and NPK+RC significantly improved the different amino acid and catechin concentrations in the young shoots, stems, and roots of the tea compared to the CK. The plant growth parameters, e.g., the plant height, no. of leaves, mid-stem girth, and fresh weights of stems and leaves, were significantly increased with integrated fertilization (NPK+B and NPK+RC) compared to the CK and solo organic and inorganic fertilizers. The chlorophyll contents (Chl a, Chl b, and Chl a+b) were generally higher with NPK+RC than with the CK (37%, 35%, and 36%), RC (14%, 26%, and 18%), and NPK (9%, 13%, and 11%) treatments. Integrated fertilization buffered the acidic soil of the tea garden and decreased the soil C:N ratio. NPK+RC also significantly increased the soil's total C (31% and 16%), N (43% and 31%), P (65% and 40%), available P (31% and 58%), K (70% and 25%), nitrate (504% and 188%), and ammonium (267% and 146%) concentrations compared to the CK and RC. The soil macro- (Mg and Ca) and micronutrients (Mn, Fe, Zn, and Cu) were significantly improved by the RC (100% and 72%) (49%, 161%, 112%, and 40%) and NPK+RC (88% and 48%) (47%, 75%, 45%, and 14%) compared to the CK. The chlorophyll contents and soil macro- and micronutrients were all significantly positively correlated with tea quality (amino acids and catechin contents) and growth. These results indicated that integrated fertilization improved the soil nutrient status, which is associated with the improvement of tea growth and quality. Thus, integrated nutrient management is a feasible tool for improving tea growth, quality, and low nutrient levels in the soil.

Keywords: tea plant growth; chlorophyll; integrated fertilization; amino acids; catechins; macro and micronutrients; soil properties

1. Introduction

After water, tea is the second-most popular non-alcoholic beverage in the world. Tea is made from the fresh leaves of a tea tree [1,2]. Tea is a very important cash crop that is largely cultivated in tropical and sub-tropical regions, such as Japan, China, Vietnam, and India [3]. Over the past few decades, the popularity of tea has grown steadily due to its abundance of beneficial amino acids, vitamins, and antioxidants [4]. Tea contains many secondary metabolites that are closely related to its quality, such as tea alkaloids, tea polyphenols, and free amino acids, which are rich in aroma, flavor, and health benefits [5]. In general, the umami taste of tea is caused by free amino acids, especially theanine, and as the amino acid concentration increases, the intensity of the flavor is augmented. Theanine is the most abundant amino acid in tea, typically accounting for 60% to 70% of the total free amino acid content in tea and about 0.2% to 2% of the dry weight of tea [6,7]. Tea polyphenols, such as catechins, are one of the principal contributors to tea's astringent properties [1,8]. The catechins in tea are not only responsible for the ultimate organoleptic properties, but they are also closely linked to many health benefits, such as anti-cancer [9], anti-inflammatory [10], and antibacterial effects [11]. Numerous research studies have shown that catechins are abundant in green tea, accounting for about 30% of the dry weight of green tea leaves [12]. Catechins are flavan-3-alcohols that occur naturally in tea. Epicatechin and its gallate derivatives are the main flavan-3-ols in green tea [13]; EGCG, the main catechin, accounts for more than 10% of green tea leaves' dry weight [14,15]. Furthermore, some research studies have shown that gallate-type catechins are more bitter and astringent than free catechins. The concentrations of compounds affecting tea quality vary significantly due to cultivation practices and growing environments (soil, climate, and altitude) [16,17]. As the demand for tea production continues to grow, farmers are increasingly using fertilizer to maximize tea production [18]. All the same, many research studies have shown that the long-term overuse of chemical fertilizers in tea cultivation can lead to soil degradation problems, such as a reduction in beneficial bacteria, nutrient loss, structural disruption, and acidification [19,20]. Tea plants prefer acidic soils, and the optimum pH for tea growth ranges from 4.5 to 5.5 [21]. However, in a nationwide soil survey of major tea-growing regions in China, about 46% of the tea soil samples had a pH of less than 4.5 [22]. Due to these issues, the quality and yield of green tea were reduced, leading to proven limitations in sustainable tea production [23]. Therefore, there is an acute need to adjust and optimize fertilization strategies to ameliorate soil fertility and tea garden traits, thereby improving tea quality and yield.

It is established that applying organic amendments is the best option for mitigating the negative effect of mineral fertilizer by improving organic carbon and soil porosity [24]. Organic fertilizer has the advantages of a negative surface charge, high porosity, higher alkalinity, and higher carbon content [25]. Over the past decade, many researchers have delineated the benefits of organic fertilizers on soil physicochemical characteristics, such as increasing the soil pH, reducing the bulk density, and improving the nutrient availability [26,27]. Furthermore, organic fertilizers have been shown to improve soil enzyme activities and alter the community composition [28,29]. However, due to the limited nutrient contents of organic fertilizers, their impact on crop yields is not particularly pronounced [30]. Therefore, the combined use of organic and inorganic fertilizers has received more advertence recently [31,32]. The combined application of chemical and organic fertilization was more effective than organic fertilizers alone in improving soil nutrients and yield. For example, N fertilizer plus biochar improves the soil nutrient availability and rain-fed rice yields while decreasing soil NO_3^- leaching [33]. Faloye et al., 2019 [34] indicated that the application of chemical and organic fertilization increased the nutrient uptake in maize and soil nutrients compared to organic or inorganic fertilizers alone. Although integrated fertilization applications have been reported in several crops, vegetables, and medicinal plants [35–37], few research studies were carried out to investigate the effects of organic and mineral fertilizers on tea growth and quality and improved soil properties in tea gardens.

Despite the growing understanding of the importance of organic fertilizers, previous research has focused on improving crop yields and soil nutrients [33,38]. Little is known about the effect of integrated fertilization on soil characteristics and its role in improving plant growth and quality. Particularly in highly acidic soils, such as tea garden soils, organic fertilizers play an important role in soil OM turnover. [24,39]. The influence of integrated fertilization on soil properties and its role in improving tea growth and quality are unclear.

A comprehensive understanding of the impact of organic fertilizers on tea growth, quality, and soil characteristics may provide new insights into improving tea production, quality, and soil management practices in tea gardens.

In this greenhouse experiment, five treatments (CK, NPK, NPK+B, NPK+RC, and RC) were used for greenhouse experiments to examine (1) the influence of different chemical and organic fertilization's on tea growth and quality, (2) the changes in soil chemical properties due to different fertilizer management methods, and (3) the effect of soil chemical properties on tea quality and growth. We also assume that using organic or organic-based fertilizers changes soil chemistry. We also hypothesize that changes in tea quality and growth correlate with changes in soil chemical properties.

2. Materials and Methods

2.1. Experimental Site

The greenhouse experiment was carried out in the 2020–2021 period at the Tea Research Institute, CAAS (30°10′ N, 120°05′ E, 22 m MSL), situated in the main green tea growing area in China. The research site has a humid sub-tropical climate with an average RH of about 70%. The average maximum temperature is 22.8 °C (73 °F), and the average minimum temperature is 13.5 °C (56.3 °F) (Figure 1). According to the GSCC, research soils are classified as red-yellow soils, equivalent to Alfisol in the World Reference Base for Soil Resources [40,41], taken from the garden of Royal Tea Village Co., Ltd., Shaoxing, China, 20 cm deep (top rhizosphere soil). According to Bao 2005 [42], properties of the top rhizosphere soil of the tea garden were determined before the experiment began (Table 1).

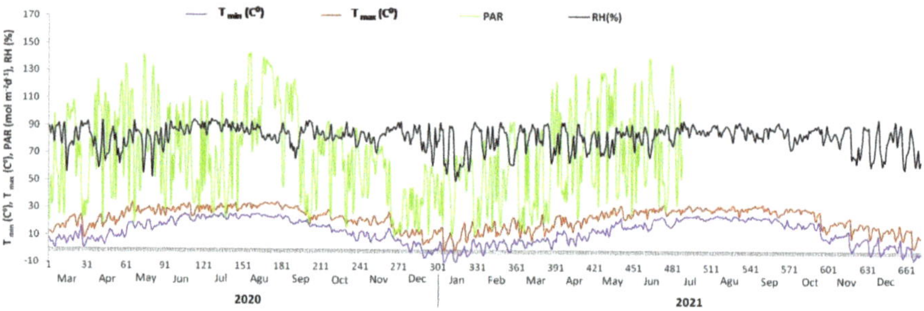

Figure 1. Weather data during entire experimental period from 2020 to 2021. PAR = Photosynthetic Active Radiation (mol m^{-2} d^{-1}); T_{max} (C°) = Maximum Temperature; T_{min} (C°) = Minimum Temperature; RH (%) = Relative humidity.

Table 1. Basic characteristics of tea orchard top rhizosphere soil (0–20 cm soil depth).

Soil Properties	Means
Soil texture	Clay
Clay (%)	63.0 ± 2.2
Sand (%)	9.0 ± 0.2
Silt (%)	28.0 ± 1.0
EC (μS cm^{-1})	108.0 ± 4.6
pH	4.4 ± 0.01
Available K (mg kg^{-1})	114.0 ± 2.7
Available P (mg kg^{-1})	30.0 ± 0.06
NO_3^- (mg kg^{-1})	16.0 ± 0.4
NH_4^+(mg kg^{-1})	33.0 ± 2.2
P (g kg^{-1})	0.8 ± 0.01
N (g kg^{-1})	2.7 ± 0.01
K (g kg^{-1})	12.0 ± 0.1
C (g kg^{-1})	18.0 ± 0.02

Mean ± standard deviation.

2.2. Seedling Establishment and Treatment Design

Tea seedlings germinate from the seed of the tea variety "Longjing-43" [43]. The soil for the greenhouse experiment was taken from a tea garden and air dried, sifted through a 2 mm sieve, and then thoroughly mixed, and 4.6 kg of soil was loaded into pots (14 × 24 cm diameter and depth) with 3 cm of perlite at the bottom. In March 2020, 3 tea seedlings were planted in small pots. After one month, healthy seedlings were left, and the remaining seedlings were discarded. In October 2020, the tea plant was transplanted into large pots.

An amount of 8 kg of soil was placed into a pot (25 cm inner diameter × 25 cm depth). Experimental treatments included control (CK) (no fertilizers), 100% rapeseed cake (RC), 100% NPK, 50% NPK + 50% rapeseed cake (NPK+RC), and 100% NPK + biochar (NPK+B).

NPK was used at a ratio of 75:10:25 mg per kg of soil. The application doses of rapeseed cake and biochar were 1.26 and 1.25 g per kg soil, respectively. KH_2PO_4 (39.09% K_2O and 30.97% P_2O_5) was used as a compound fertilizer for P and K, Urea (46% N) was used as N fertilizer, and potassium chloride (KCl) was used as K fertilizer (60% K_2O). Rice straw was heated for 2 h at 500 °C with limited O_2 to make biochar. The nitrogen, phosphorus, and potassium contents of RC were about 5.98%, 3.42%, and 1.96%, respectively. Biochar and RC were amalgamated with soil and loaded into pots. Urea, KH_2PO_4, and KCl solutions were prepared and applied in three split doses at a 2-week interval before transplanting in the July–August 2020 period and 62 days after transplanting from December 2020 to January 2021. A 30% (*w*/*w*) soil moisture was maintained. A completely random design with four replicates (CRD) was exploited for the treatment arrangements. Plant, root, and soil samples were collected in October 2021. The research duration was 480 days.

2.3. Determination of Different Amino Acid and Catechin Contents by HPLC

We weighed the milled sample to obtain 250 mg and added 1.5 mL of a 75% methanol extract solution to the Eppendorf tube. We shook the tube with a vortex shaker and placed it in an ultrasonic shaker water bath for 15 min, shook it up and down twice at 7 min intervals, and then centrifuged at 11,000 rpm at 25 °C for 10 min. We filtered the extract through a 0.22 μm PTFE filter and divided the sample solution into two parts for catechin and amino acid analysis in a DRAM glass tube.

The e2695-connected and 2998-Photodiode Array Detector System (Waters) was used for high performance liquid chromatography analysis, which injected 25 μL and 10 μL of free amino acid and catechin sample solutions, respectively. For catechins, mobile phase A consisted of ddH_2O containing 2% formic acid, while mobile phase B consisted of acetonitrile (ACN), a high-performance liquid chromatography solvent (Sigma-Aldrich Co., St. Louis, MO, USA). Sample was eluted at a flow rate of 1 mL/min with a column temperature of 40 ± 1 °C and monitored at 278 nm. For amino acids, mobile phase A had AccQ. Tag eluent of water, while mobile phase B was acetonitrile, and the column temperature was set to 37 ± 2 °C. The rest of the process was carried out according to the AccQ. Label Chemical Packaging Instruction Manual. We compared the retention time of the sample with the standard, and identified the peak of catechins and amino acids, as described in the manual.

2.4. Determination of Chlorophyll

With minor modifications, chlorophyll (Chl) contents were measured according to the protocol of Ni et al., 2009 [44]. We placed the fresh leaves in a mortar and ground them into a fine powder. We added 400 mg of sample and 5 mL of 80% acetone to a 15 mL falcon tube, mixed it for 5 min, and then kept it at 4 °C for 15 min in the dark (chlorophyll is hydrolyzed under light). Then, we centrifuged the mixture at 3000 rpm for 15 min, transferred the supernatant to a new centrifuge tube, and protected it from light. This process was repeated twice. We combined the supernatant and obtained a total volume of 25 mL with 80% acetone. We shook the solution thoroughly and determined the absorbance of Chl using a spectrophotometer (varian carries winning UV) with 80% acetone as a blank (control). The Chl contents were measured using the following equation:

$$\text{Chlorophyll a}_{(mg/g)} = \frac{[12.7\times A_{663}-2.69\times A_{645}]\times V}{1000\times W}$$
$$\text{Chlorophyll b}_{(mg/g)} = \frac{[22.9\times A_{645}-4.68\times A_{663}]\times V}{1000\times W}$$
$$\text{Chlorophyll a}+\text{b}_{(mg/g)} = \text{Chlorophyll a}+\text{Chlorophyll b}$$
$$\text{Chlorophyll a/b ratio} = \frac{\text{Chlorophyll a}}{\text{Chlorophyll b}}$$

where W = sample weights (g); V = volume of supernatant (ml); A_{663} = 663 nm wavelength absorbance; and A_{645} = 645 nm wavelength absorbance.

2.5. Plant Growth Parameters

We randomly selected 4 potted plants from each treatment unit, cut off the shoots at the level of the crowns (the point of separation between the root system and the stem), plucked off all leaves (counted), and separated the plant into leaves, stems, and roots. We measured the plant height with a ruler and stem diameter with a digital caliper. We recorded the fresh weights of leaves and stems (cut stem into 7–8 cm parts). The young shoots (1 bud and 2 leaves), stems, and roots were dried, grinded, and passed through a mesh sieve with a 0.7–1.0 mm size. Before analysis, we stored the powdered young shoots, stems, and roots in a zipper-lock bag.

2.6. Soil Macro$^-$ and Micronutrients

In late October 2021, soil samples were collected. We mixed the soil from the same treatment pots into one sample and removed any stones, plant residues, and roots by hand. The collected fresh soil sample was passed through a 2 mm sieve and divided into two parts. Portions of the mineral nitrogen (NO_3^- and NH_4^+) were immediately analyzed, and the other part was air-dried before other soil properties were determined.

Orion 3 STAR pH meter (Thermoer Inc., Boston, MA, USA) was used to measure soil pH at a ratio of 1:1 (*w*/*v*) in double-distilled water (ddH_2O). A C/N elemental analyzer (Vario Max, Elementar, Frankfurt, Germany) was used to measure soil's total N and C contents. Available soil nutrients (K, P, Ca, Fe, Mg, Mn, Zn, and Cu) were extracted from the soil using the Mehlich-3 method (Mehlich, 2008), and their contents were determined using ICP-AES (Thermo Jarrell Ash Ltd., Franklin, MS, USA). TP and TK were determined by digestion. After digestion, the sample was diluted in a 50 mL volumetric flask with 50 mL of ddH_2O, and the contents of TP and TK were measured using ICP-AES (Thermo Jarrell, Ash. Ltd., Franklin, MA, USA). NO_3^-–N and NH_4^+–N were extracted at a soil/water ratio of 1:10 with 2 M KCl solution at 25 °C, and NO_3^- and NH_4^+ concentrations were determined using a smart Continuous Flow Analyzer (Smart Chem 140, Shenzhen, China).

2.7. Statistical Analysis

Analysis of plant and soil data was carried out using ANOVA. To determine the significance among the treatments, F-test was used, and to find the significance of difference between the means of 2 treatments, the least significant difference (LSD) test was used. LSD, F-test, and ANOVA were analyzed using "Statistix 8.1.1, 2008" and MS Excel, 2019. IBM SPSS Statistics 20 was used for Pearson's correlation analysis. The heatmaps were created on the R-platform using the "ggplot2" package (version 3.5.1.).

3. Results

3.1. Plant Growth Parameters

Compared with the control, NPK+RC, NPK+B, NPK, and RC treatments significantly ($p \leq 0.05$) improved the no. of leaves per plant^{-1} and the plant height (Table 2). Compared with the NPK+B, NPK, and RC treatments, the no. of leaves per plant^{-1} and plant height were also significantly ($p \leq 0.05$) increased by the NPK+RC treatment. The highest no. of leaves per plant^{-1} and plant height were observed in NPK+RC, followed by the NPK+B and NPK treatments. There was no statistically significant difference between NPK and NPK+B. In addition, the application of chemical fertilizers with rapeseed cake significantly improved the stem diameter of tea plants compared with the NPK+B, NPK, RC, and control treatments. The NPK+RC treatment resulted in the thickest stem diameter, followed by the NPK+B and NPK fertilizers, and the CK treatment resulted in the thinnest stem diameter. No significant difference was recorded between the NPK and RC treatments. The plant height, number of leaves per plant^{-1}, and mid stem diameter under the NPK+RC treatment increased by 47.1%, 58.6%, and 28.1%, respectively, compared with the CK (Table 2).

Table 2. Effect of integrated fertilization on plant height, no. of leaves per plant^{-1}, leaf and stem fresh weights, and stem diameter.

Treatments	Plant Height (cm)	No. of Leaves per Plant^{-1}	Leaf Fresh Weight (g Plant^{-1})	Stem Fresh Weight (g Plant^{-1})	Stem Mid Girth (mm)
CK	22.63 ± 1.80 c	46.0 ± 8.2 c	6.22 ± 1.23 d	2.73 ± 0.95 c	2.18 ± 0.09 c
NPK	32.75 ± 3.30 b	71.0 ± 22.8 b	12.04 ± 0.80 bc	4.65 ± 1.50 bc	2.76 ± 0.17 b
NPK+B	34.00 ± 4.08 b	80.0 ± 19.9 b	13.62 ± 1.52 b	5.87 ± 1.48 ab	2.83 ± 0.19 ab
NPK+RC	42.75 ± 4.57 a	111.0 ± 10.5 a	18.47 ± 3.98 a	7.16 ± 1.32 a	3.03 ± 0.10 a
RC	30.50 ± 6.66 b	58.0 ± 9.1 bc	10.38 ± 1.28 c	4.54 ± 1.23 bc	2.58 ± 0.14 b

Control (CK); Chemical Fertilizers (NPK); Chemical Fertilizer + Biochar (NPK+B); Chemical Fertilizer + Rapeseed Cake (NPK+RC); Rapeseed Cake (RC). Data are shown as means ± standard deviation, and different lowercase letters indicate significant differences among different treatments at $p \leq 0.05$ levels in the same line (LSD test).

In this study, it was found that the organic and inorganic fertilization treatments significantly improved the leaf fresh weight per plant^{-1} (66.3% and 54.3%) and the stem fresh weight per plant^{-1} (61.9% and 53.5%), respectively, compared to the control (CK). Compared to the RC treatment, the NPK+RC treatment significantly ($p \leq 0.05$) improved the leaf and stem fresh weights per plant by 43.8% and 36.6%, respectively (Table 2). The highest leaf and stem fresh weights per plant^{-1} were noted in NPK with RC fertilization, followed by the NPK+B treatment, while the control treatment had the lowest leaf and stem fresh weights per plant^{-1}.

3.2. Catechin Contents in Tea Young Shoots, Stems, and Roots

The catechin compositions determined by HPLC in the young shoots (two leaves and one bud), stems, and roots are presented in Figure 2a–c. Various catechin contents were detected in the roots, stems, and young shoots, namely CG, GC, GCG, GA, EGC, EGCG, EC, ECG, C, and EC (matery) gallate; caffeine (CF); and total polyphenol (TPL). It was found that the various catechin contents in different parts of the tea plant were in the order of young shoots> stems >roots. The catechin contents in the roots were significantly lower than those in the other parts of the tea plant. Integrated fertilization had a significant effect on the catechin contents (CG, GC, EGC, EGCG, EC, ECG, EC (matery) gallate, and CF) in the young shoots of tea except GCG, GA, and TPL. The highest catechin contents (CG, GC, EGC, EGCG, EC, ECG, C, EC (matery) gallate, and CF) were recorded in the NPK+RC treatment (0.08, 0.14, 11.7, 34.3, 6.7, 13.2, 4.4, 3.4, and 11.0 mg g^{-1}) and the NPK+B treatment (0.08, 0.10, 11.3, 31.9, 6.3, 13.0, 3.3, 3.2, and 10.3 mg g^{-1}), while the lowest catechin contents were recorded in the control (0.01, 0.03, 8.3, 19.6, 2.09, 5.7, 1.9, 2.2, and 7.6 mg g^{-1}). There was a non-significant difference between the NPK+RC and NPK+B treatments in the majority of catechin contents in young shoots. In the tea stems, the catechin contents (CG, GC, EGC, EGCG, EC, ECG, and C) were significantly affected by organic and inorganic fertilizers, while GCG, GA, TPL, EC (matery) gallate, and CF were not significantly affected by integrated fertilization. The maximum catechin concentrations were established in NPK+RC, while the minimum catechin concentrations were found in the control treatment. The catechin contents between the NPK+RC and NPK+B treatments were not significant (specifically, EGC, EGCG, EC, ECG, and C). The catechin contents of the roots were significantly affected by integrated fertilization, while organic and inorganic fertilization had non-significant effects on the roots' CG, GC, GCG, GA, EC (matery) gallate, TPL, and CF contents. The concentrations of catechin (EGC, EGCG, EC, ECG, and C) in the NPK+RC treatment were 17%, 115%, 147%, 555%, and 227% higher, and in the NPK+B treatment, they were 13%, 87%, 125%, 473%, and 186% higher than the control treatment, respectively, in the tea roots.

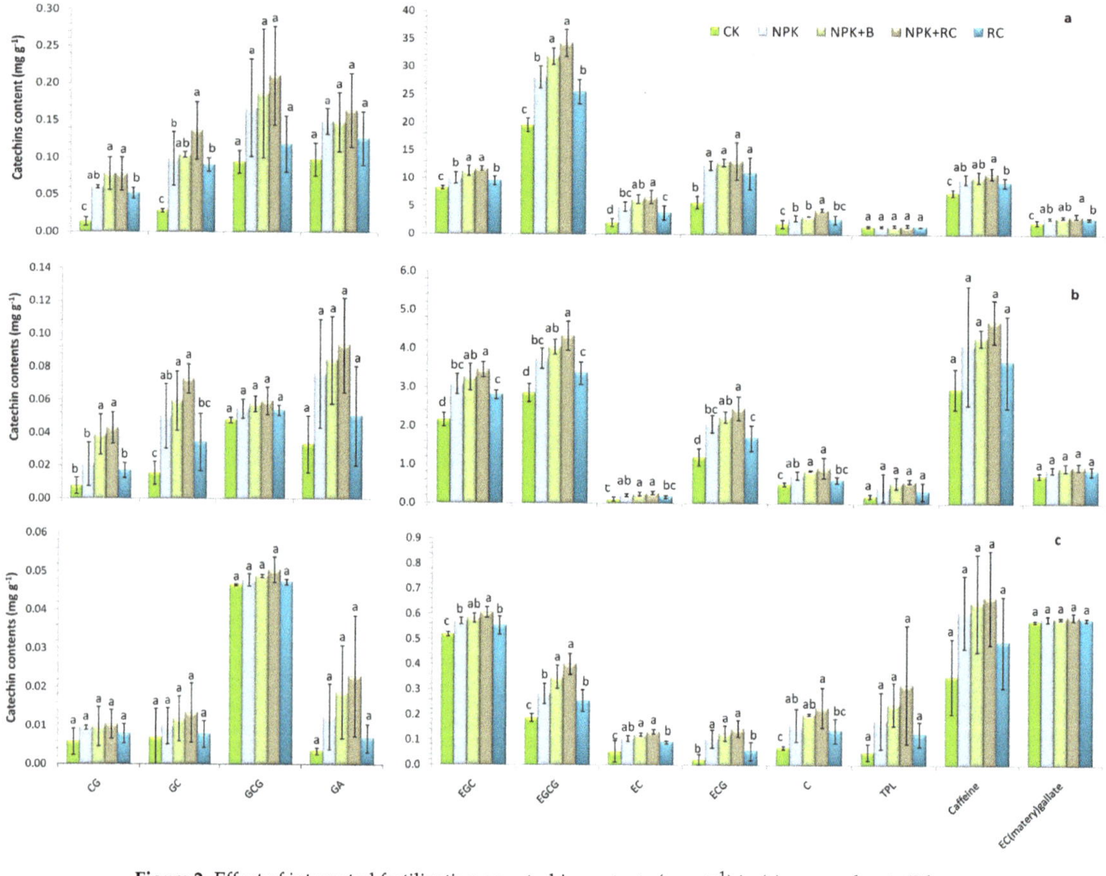

Figure 2. Effect of integrated fertilization on catechin contents (mg g^{-1}) in (**a**) young shoots (2 leaves and 1 bud), (**b**) stems, and (**c**) roots of tea (Longjing 43), as determined by HPLC. Gallocatechin (GC), Epicatechin (EC), Gallocatechin Gallate (GCG), Gallic Acid (GA), Catechin Gallate (CG), Epigallocatechin (EGC), Epigallocatechin Gallate (EGCG), Catechins (C), Epicatechin Gallate (ECG). Different lowercase letters indicate significant differences among different treatments at $p \leq 0.05$, and vertical bars represent the standard deviation of the mean (n = 4).

3.3. Targeted Metabolites' Amino Acid Concentrations in Tea Young Shoots, Stems, and Roots

The concentrations of different free amino acids in the young shoots, stems, and roots of tea (Longjing 43), due to different organic and inorganic fertilizer management methods, are given in Figure 3a–c. It was established that Thea (49–65%), Glu (3–10%), Asp (4–7%), and Pro (2–6%) were the major amino acids in the tea samples. Among the 18 amino acids in the tea young shoots and roots, 16 amino acids (Pro, Ala, Thr, Asp, His, Ser, Gly, Glu, Arg, The, Tyr, Met, Val, Leu, Ile, and Phe) were significantly affected by integrated nutrient management, while 2 amino acids (Cys and Lys) were not significantly affected. The concentrations of amino acids (Asp, Ser, Glu, Gly, His, Arg, Thr, Ala, Pro, The, Tyr,

Val, Met, Ile, Leu, and Phe) in the NPK+RC treatment were 41%, 75%, 72%, 114%, 77%, 103%, 85%, 42%, 63%, 42%, 86%, 154%, 109%, 138%, 19%, and 112% higher than the control, and in the NPK+B treatment, they were 33%, 60%, 52%, 90%, 38%, 58%, 68%, 36%, 49%, 38%, 68%, 122%, 72%, 106%, 111%, and 81% higher compared to the control treatment, respectively, in the young shoots. In the roots, the maximum amino acid concentrations were observed in NPK+RC, followed by NPK+B and NPK fertilization, and the minimum amino acid concentrations were observed in the control treatment. In the stems, integrated nutrient management had a significant effect on the amino acid concentrations (Ser, Glu, Thr, Gly, Asp, Pro, His, The, Met, and Val) except for those of the Arg, Ala, Cys, Tyr, Lys, Ile, Leu, and Phe amino acids. The highest amino acid concentrations were observed in the NPK+RC treatment, followed by the NPK+B treatment, while the lowest amino acid concentrations were recorded in the control. There were no significant differences recorded for the amino acid concentrations among the NPK and RC, and among the NPK+B and NPK+RC fertilizer treatments in the roots, stems, and young shoots of green tea. The amino acid concentrations in different parts of the green tea plant were recorded as follows: roots > young shoots > stems.

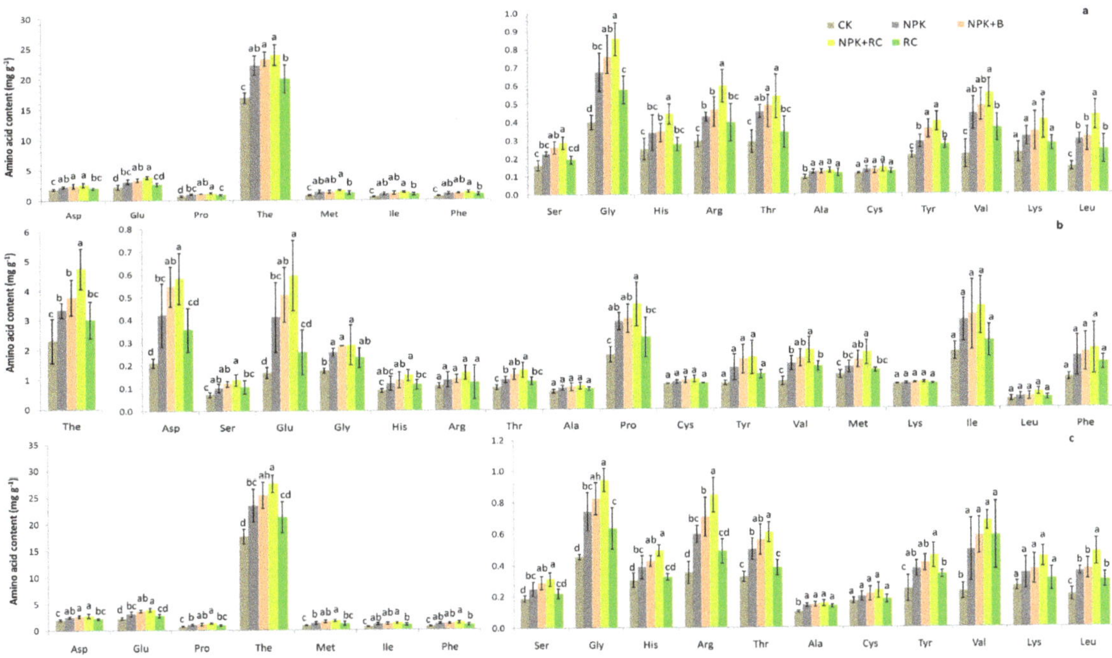

Figure 3. Effect of integrated nutrient management on amino acids (mg g^{-1}) in (**a**) young shoots (2 leaves and 1 bud), (**b**) stems, and (**c**) roots of tea (Longjing 43), as assessed by HPLC. Aspartate (Asp), Glutamate (Glu), Proline (Pro), Theanine (The), Isoleucine (Ile), Methionine (Met), Phenylalanine (Phe), Glycine (Gly), Serine (Ser), Histidine (His), Threonine (Thr), Arginine (Arg), Alanine (Ala), Tyrosine (Tyr), Cysteine (Cys), Valine (Val), Leucine (Leu), Lysine (Lys). Different lowercase letters indicate significant differences among different treatments at $p \leq 0.05$, and vertical bars represent the standard deviation of the mean (n = 4).

3.4. Chlorophyll Contents of Leaves

The effects of integrated fertilization on the chlorophyll contents (Chl) of tea are shown in Figure 4. The research outcome indicated that the concentrations of Chl a, b, and a+b in the tea leaves were significantly affected by organic and inorganic fertilizer management, while for the Chl ratio (Chl a/b), there were no significant differences among all treatments. The chlorophyll a content was higher in the plants treated with NPK+RC, followed by the plants treated with NPK+B integrated fertilization. Furthermore, the lowest chlorophyll a content was found in the control pots and plants treated with RC. The maximum chlorophyll b content was found in the pots treated with integrated fertilization (NPK+RC), followed by those that underwent NPK+B treatment, while the minimum chlorophyll b content was measured in the control treatment pots. There was no significant difference between the NPK and NPK+B fertilizer treatments. The maximum Chl a+b concentration was recorded with the use of mineral fertilizer with RC and mineral fertilizers with biochar. The lowest Chl a+b content was noted in the control treatment. The concentrations of Chl a, b, and a+b in the NPK+RC treatment were 35.5%, 34.1%, and 37.3% higher compared to the control, and 14.3%, 25.8%, and 17.7% higher compared to rapeseed cake, respectively, while NPK+B had 28.9%, 25%, and 27.6% higher Chl a, Chl b, and Chl a+b concentrations compared to the control treatment, respectively.

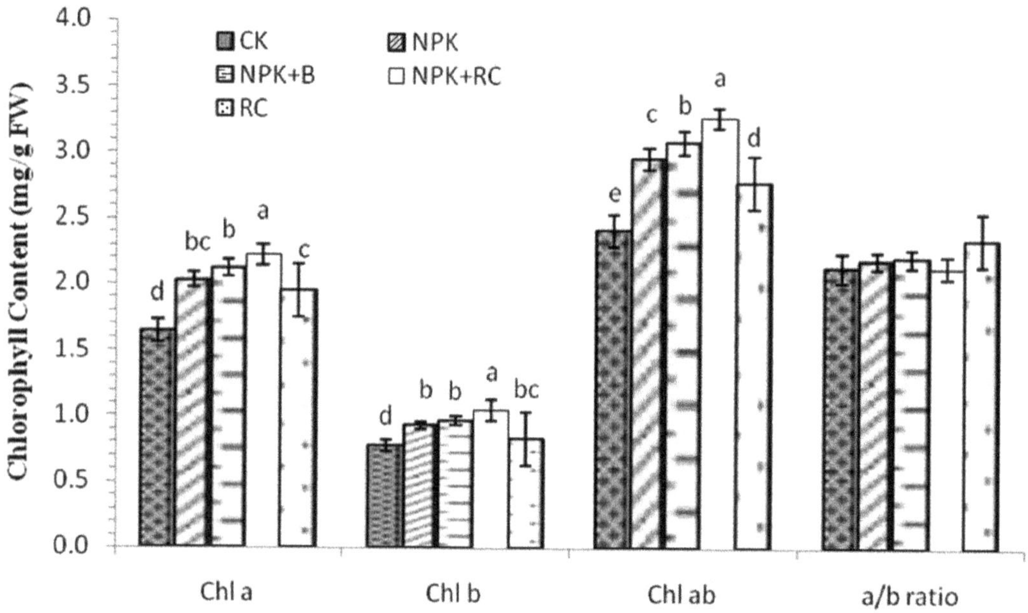

Figure 4. Chlorophyll contents of tea leaves as influenced by integrated nutrient management. Chlorophyll a (Chl a), chlorophyll b (Chl b), chlorophyll ab (Chl ab), chlorophyll a/b ratio (a/b ratio). Different lowercase letters indicate significant differences among different treatments at $p \leq 0.05$, and vertical bars represent the standard deviation of the mean (n = 4).

3.5. Soil pH and NH_4^+ and NO_3^- Concentrations

The application of integrated nutrient management had a significant effect on the soil pH. Rapeseed cake, alone or in combination, and biochar significantly increased the pH of tea orchard soil (Table 3). The soil pH in the control treatment was 4.41, which is 2.6% higher than that of the NPK treatment. The lowest soil pH (4.30) was recorded in inorganic fertilizers (NPK), while the highest soil pH levels (4.70, 4.64, and 4.55) were recorded in rapeseed cake (RC), NPK+RC, and NPK+B fertilization. Hence, the addition of organic fertilizers, irrespective of their nature, increased the soil pH in acidic soil. In this study, the rapeseed treatments (RC and NPK+RC) significantly changed the pH of the soil from 4.35 to 4.70 and 4.64, respectively.

Table 3. Effect of organic and inorganic fertilization on soil pH and macro and micronutrients.

Treatments	pH	N (%)	C (%)	CN Ratio	Ca (mg kg^{-1})	Mg (mg kg^{-1})	Fe (mg kg^{-1})	Mn (mg kg^{-1})	Cu (mg kg^{-1})	Zn (mg kg^{-1})
CK	4.41 ± 0.09 bc	0.19 ± 0.005 c	1.56 ± 0.03 c	9.57 ± 0.33 a	154.66 ± 54 c	24.73 ± 3.4 b	267.63 ± 31 b	10.48 ± 2.5 c	1.01 ± 0.06 b	1.74 ± 0.13 b
NPK	4.35 ± 0.03 c	0.21 ± 0.03 bc	1.61 ± 0.06 c	8.78 ± 0.93 ab	202.57 ± 41 bc	34.00 ± 5.8 ab	358.39 ± 32 a	16.48 ± 2.3 bc	1.18 ± 0.11 b	1.68 ± 0.22 b
NPK+B	4.55 ± 0.17 ab	0.23 ± 0.02 ab	2.01 ± 0.07 a	7.61 ± 0.96 b	275.61 ± 34 ab	30.70 ± 3.4 ab	336.12 ± 24 ab	17.99 ± 8.6 b	1.27 ± 0.21 b	1.70 ± 0.08 b
NPK+RC	4.64 ± 0.13 a	0.26 ± 0.03 a	2.05 ± 0.04 a	7.74 ± 0.19 b	290.44 ± 43 ab	36.68 ± 11.5 a	393.99 ± 40 a	18.39 ± 4.3 b	1.46 ± 0.43 b	1.98 ± 0.24 b
RC	4.70 ± 0.09 a	0.20 ± 0.01 c	1.77 ± 0.06 b	7.69 ± 1.08 b	324.30 ± 116 a	42.45 ± 10.9 a	399.85 ± 100 a	27.39 ± 3.8 a	2.14 ± 0.59 a	2.43 ± 0.34 a

Control (CK); Chemical Fertilizers (NPK); Chemical Fertilizer + Biochar (NPK+B); Chemical Fertilizer + Rapeseed Cake (NPK+RC); Rapeseed Cake (RC). N: Nitrogen; C: Carbone; Mg: Magnesium; Ca: Calcium; Fe: Iron; Cu: Copper; Mn: Manganese; and Zn: Zinc. Data are shown as means ± standard deviation, and different lowercase letters indicate significant differences among different treatments at the $p \leq 0.05$ level in the same line (LSD test).

Significant differences were recorded between the soil's NH_4^+ and NO_3^- concentrations after the application of organic and inorganic fertilizer treatments (Figure 5e,f). The solo application of chemical and organic fertilizers or integrated fertilization application significantly improved the overall soil NH_4^+ and NO_3^- concentrations. The maximum concentrations of NH_4^+ and NO_3^- were found in NPK+RC, followed by NPK+B and NPK, and the minimum contents of NH_4^+ and NO_3^- were found in the control and RC treatments. NPK+RC increased the NO_3^- and NH_4^+ concentrations by 504% and 267% compared to the control treatment, and by 188% and 146% compared to the RC treatment, respectively. Similarly, the NPK+B and NPK treatments also improved the NO_3^--N concentration by 221% and 33% compared to the control treatment, while the NH_4^+-N concentration showed increases of 129% and 93% in the NPK+B and NPK fertilizer treatments relative to the control.

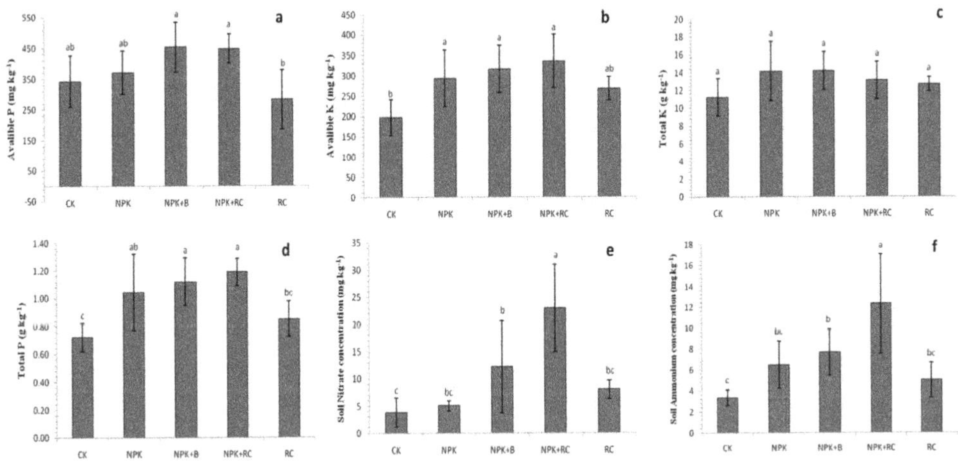

Figure 5. Effect of integrated nutrient management on soil. (**a**) Available P, (**b**) Available K, (**c**) Total P (**d**), Total K (**e**), Nitrate, and (**f**) Ammonium. Different lowercase letters indicate significant differences among different treatments at $p \leq 0.05$, and vertical bars represent the standard deviation of the mean ($n = 4$).

3.6. Soil Micro and Macronutrients

Organic and inorganic fertilizers significantly affected the soil micro- and macronutrients. The application of mineral fertilizers (NPK), along with RC and B amendments, effectively improved the total N, P, and C concentrations of the soil. The potassium content of the soil was not significantly improved by integrated fertilization (Table 3; Figure 5c,d). The highest total N, P, and C contents of the soil were observed in the NPK+RC treatment, while the lowest were recorded in the CK and RC treatments. The NPK+B treatment also improved the total N, P, and C concentrations of the soil. The NPK+RC treatment increased the macronutrient (N, P, and C) concentrations by 43%, 65%, and 31% compared to the control treatment, respectively, followed by the NPK+B treatment. The carbon content in the rapeseed cake was significantly higher than that of the mineral fertilizer (NPK) treatment. The N content was significantly higher in the chemical fertilizers (NPK) compared to the solo application of rapeseed cake fertilizer. Integrated fertilization significantly ($p \leq 0.05$) affected the C:N ratio of the soil (Table 3). Compared with organic fertilizer alone, integrated fertilization showed a lower C:N ratio. Among the treatments, the C:N ratio trend was as follows: CK > RC > NPK > NPK+B > NPK+RC. The C/N ratios of the NPK+RC and NPK+B treatments were low, indicating that the N content was higher than that of NPK and RC. The rapeseed cake and NPK+RC treatments had higher Mg and Ca contents, followed by the NPK+B treatment. RC improved the Mg and Ca contents by 72% and 100% compared to CK, and by 25% and 60% compared to chemical fertilizer. NPK+RC also improved the Mg and Ca contents by 48% and 88% compared to the control, respectively (Table 3).

An analysis of the soil test data revealed that the available P and K concentrations of the soil responded significantly ($p \leq 0.05$) to different organic and inorganic fertilizer applications (Figure 5a,b). After the application of treatments, the available P and K concentrations of the soil ranged from 281.9 to 452.8 mg kg^{-1} and from 196.6 to 334.5 mg kg^{-1}. The soil's available P and K concentrations of rapeseed cake with chemical fertilizers (NPK+RC) were significantly higher than those of the soil with the solo application of RC and inorganic fertilizer (NPK). NPK+B and NPK+RC fertilization improved the available P concentrations of the soil by 32% and 31%, while the available K concentrations of the soil improved by 70% and 60% compared to the control treatment, respectively.

This study also demonstrated that organic fertilizers alone or in combination with mineral fertilizer treatments significantly ($p \leq 0.05$) improved the contents of the soil micronutrients (Mn, Fe, Zn, and Cu) (Table 3). The organic fertilizers had significant high amounts of micronutrients compared to the inorganic fertilizers. Therefore, the highest soil micronutrients (Mn, Fe, Zn, and Cu) were observed in rapeseed cake, followed by the NPK+RC and NPK+B fertilizer treatments, while the lowest soil micronutrients were recorded in the control and chemical fertilizer (NPK) treatments. Rapeseed cake (RC) significantly increased the micronutrients (Fe, Mn, Cu, and Zn) of the soil by 49%, 161%, 112%, and 40% compared to the control, and by 12%, 66%, 82%, and 45% compared to chemical fertilizer treatment, respectively. The soil micronutrients (Fe, Mn, Cu, and Zn) were also improved by NPK+RC (47%, 75%, 45%, and 14%) and NPK+B (26%, 72%, 25%, and 2%) compared to the CK treatment, respectively. There was no significant difference in the soil micronutrients between the NPK+RC and RC treatments. So, the organic fertilization or the incorporation of integration fertilization improved the soil micronutrient statutes in the tea plantations, which promoted tea growth and quality.

3.7. Pearson's Correlation Analysis of Soil Chemical Properties and Plant Growth

A Pearson's correlation matrix for the relation of the soil chemical characteristics (pH and soil macro- and micronutrients) and plant growth factors (plant height, no. of leaves per plant, avg. stem diameter, and leaf and stem fresh weights per plant) of tea are presented in Figure 6. Most of the correlations between soil macro- and micronutrients were significantly positively correlated with each other, except for the soil's C:N ratio, which had significant negative correlations with the plant growth parameters and soil macro- and micronutrients. The soil macro- and micronutrients also had significant positive correlations with the plant growth parameters, except the total K did not have a significant positive correlation with the plant growth parameters; Fe and Cu did not have significant correlations with the no. of leaves per plant and the leaf fresh weight per plant; and Mn and Zn did not have significant correlations with the no. of leaves per plant and the stem fresh weight per plant. The soil pH was significantly positively correlated with the soil macro- and micronutrients (Mg, Ca, Mn, Fe, Zn, and Cu), and among the soil macro- and micronutrients, Ca, Mg, Fe, and Zn had very strong and significant positive correlations with the soil pH. The soil's total K, P, N, and C did not have a significant correlation with the soil micronutrients, except N, which had a significant positive correlation with Ca,

Fe, and Zn. The available P and K had significant correlations with the plant growth parameters; the soil's total N, C, and P; the soil's NO_3^- and NH_4^+; and the soil's Ca, Fe, and Mn concentrations. Nitrate was significantly positively correlated with NH_4^+, Mg, Cu, and the plant growth parameters. Ammonium was significantly positively correlated with only Ca and Mg and the plant growth parameters. All of the soil micronutrients were strongly and significantly positively correlated with each other's, except Ca, which was not significantly correlated with the soil Cu. The tea growth variables (plant height, no. of leaves per plant, avg. stem diameter, and leaf and stem fresh weights per plant) were also strongly and significantly positively correlated with each other.

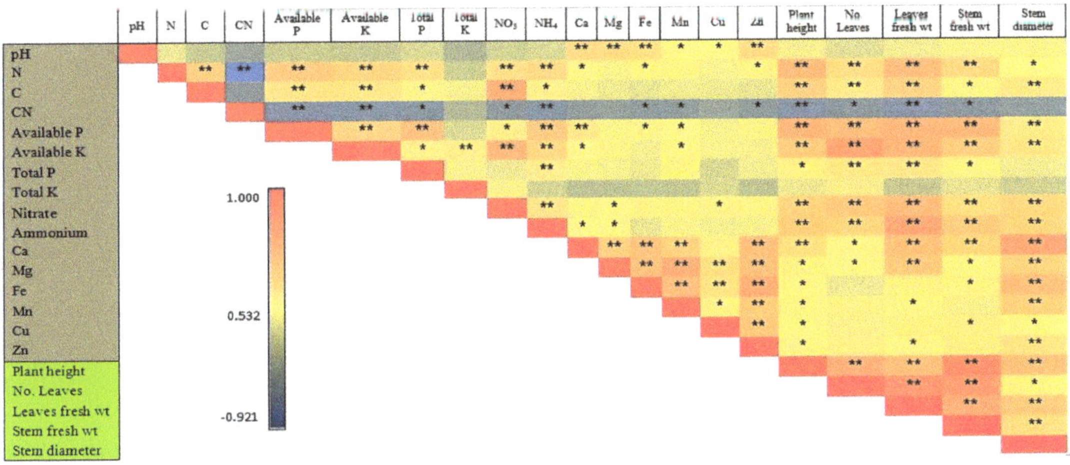

Figure 6. The Pearson's correlation analysis of soil chemical properties and plant growth. An asterisk indicates the significance of the correlations at p levels of ≤ 0.01 (**) and ≤ 0.05 (*).

3.8. Pearson's Correlation Analysis of Chlorophyll Content and Soil Chemical Properties with Different Catechin Contents in Tea Young Shoots

The leaves' Chl contents (Chl a, Chl b, and Chl a+b) and the soil chemical properties (total N; available P and K; and NO_3^-, NH_4^+, Ca, Fe, Mg, Cu, Mn, and Zn) were significantly positively correlated with different catechin contents of young shoots (Figure 7). The Chl contents (Chl a, Chl b, and Chl (a+b)) were significantly positively correlated with all catechin types of young shoots, except GCG and EC (matery) gallate, which were not significantly correlated with Chl b and a, respectively. The TPL is also not significantly correlated with the chlorophyll contents and soil chemical properties. The pH of the soil did not have significant correlations with the different catechin types. The soil macronutrients, including the total N, nitrate, ammonium, and available P and K, were significantly positively correlated with different catechin types, except GCG and GA. The soil macro- and micronutrients (Mg, Ca, Mn, Fe, Zn, and Cu) also had significant positive correlations with some catechin types, and some catechin types had non-significant positive correlations, for example, EGC, GCG, GA, TPL, EC (matery) gallate, and caffeine. Generally, significant positive correlations exist between soil macro- and micronutrients and different catechin types of tea young shoots.

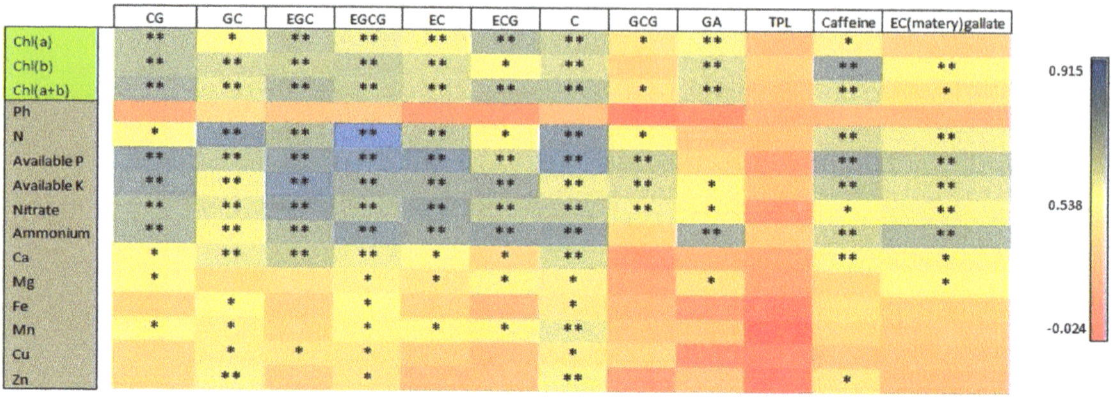

Figure 7. The Pearson's correlation analysis of chlorophyll content and soil chemical properties with young shoot catechin contents. An asterisk indicates the significance of the correlations at p levels of $\leq$0.01 (**) and $\leq$0.05 (*).

3.9. Pearson's Correlation Analysis of Chlorophyll Contents and Soil Chemical Properties with Different Targeted Amino Acids of Tea Young Shoots

The correlation matrix of the leaf chlorophyll contents and soil macro- and micronutrients with different targeted amino acid metabolites of tea young shoots is shown in Figure 8. The chlorophyll contents and soil macro- and micronutrients were significantly positively correlated with the different amino acids of the tea young shoots. Ala was correlated with Chl a, Cys was correlated with Chl a, and Chl b and Chl a+b were not significantly correlated, respectively. The soil macronutrients were significantly positively correlated with most of the targeted amino acids of the young shoots compared to the soil micronutrients. The soil pH only had a significant positive correlation with Arg and Tyr. The nitrate and ammonium of the soil did not have a significant correlation with the Cys and Lys amino acids of the tea young shoots, respectively. In the case of micronutrients, a significant positive correlation existed between the soil micronutrients and 17–39% of different targeted amino acids, except Ca and Zn, which had significant positive correlations with 83% and 50% of different amino acids in the tea young shoots, respectively.

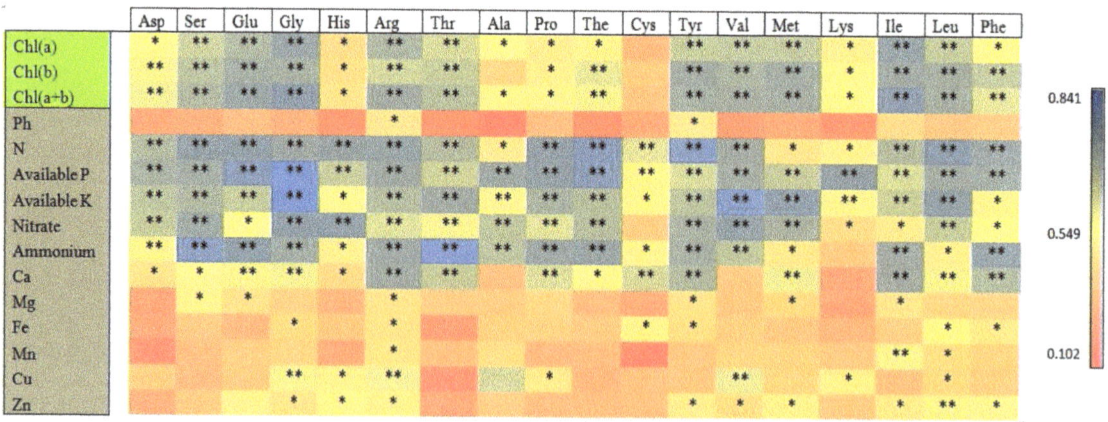

Figure 8. The Pearson's correlation analysis of chlorophyll content and soil chemical properties with young shoot's targeted amino acid metabolites. An asterisk indicates the significance of the correlations at p levels of $\leq$0.01 (**) and $\leq$0.05 (*).

4. Discussion

4.1. Effects of Different Fertilization Amendments on Plant Growth

Compared with the CK, the NPK+RC and NPK+B fertilizations significantly improved the plant height, stem circumference, leaf number, and fresh biomasses of the leaves and stems. The application of integrated fertilization has been reported to promote plant growth. For example, Rafael et al., 2019 [35] reported that the use of NPK+B resulted in a 4–5-fold increase in the cowpea biomass yield compared to the control, and it resulted in a 70% improvement compared to chemical fertilizer (NPK) treatments. Saha et al., 2019 [37] exemplified the synergetic effects of inorganic fertilizers and biochar on pasture yield. Soil characteristics, such as the microbial activity, pH, and CEC, were significantly improved by biochar [33,35]. After the application of biochar and RC to inorganic fertilizers, the enhancement of tea growth factors may be connected to the enhancement of the soil chemistry (pH and nutrient concentration) and microbial properties (soil fungal and bacterial diversity index). This finding is in accordance with the study by Bass et al., 2016 [45], which may indicate that organic fertilizers do not meet the nutritional needs of tea plants during growth. Integrated fertilization utilizes chemical fertilizers as a nutrient source and biochar as a soil amendment [33]. In addition, many studies have reported that B improves the nutrient use efficiency of mineral fertilizers [37,46]. We think this may be related to the adsorption of macronutrients (P, K, and N) by RC and B, which leads to a sustained released of nutrients and provides more macro- and micronutrients for tea plant growth. Particularly in sub-tropical tea-growing regions with higher rainfall and lower pH levels, organic fertilizers play a vital role in sustaining soil macro- and micronutrients and decreasing nutrient loss and the leaching of chemical fertilizers [47]. The increase in Casuarina height after biochar application can be attributed to augmented plant nutrient availability, such as potassium and phosphorus, and improved soil physical properties, such as reduced BD and improved soil water storage [48,49]. Nevertheless, research on the effects of biochar on tree growth remains prominent [50]. Research has further shown that soil improved with 3% biochar resulted in the highest increases in the leaf count and plant height.

The favorable effect of optimal nutrient levels obtained through inorganic and organic nutrient sources on the increasing leaf number per plant^{-1} could be one of the reasons for the higher leaf fresh weight per plant^{-1}. Under chemical fertilization, an insufficient nutrient supply throughout the growing phase may limit the growth rates of the leaves, thereby limiting the number of leaves per plant^{-1} due to low inadequate cell expansion or photosynthesis rates or both [51]. Ramana et al., 2002 [52] also reported that inadequate and unbalanced nutrient supply are often the main reasons why plants are unable to achieve their potential growth. The use of integrated fertilization increases the intrinsic nutrient supply capacity of the soil to macro- and micronutrients [53] and improves the soil physical properties, thereby promoting better rooting, improving crop nutrient uptake, and promoting plant growth.

4.2. Effect of Integrated Fertilization on Catechin Contents in Roots, Stems, and Young Shoots of Tea

Catechin is the main secondary metabolite in Chinese tea, which has a significant impact on the tea taste and quality. In this research, the concentrations of various catechin components, namely CG, GC, GCG, GA, EGC, EGCG, EC, ECG, C, and EC (matery) gallate; caffeine (CF); and total polyphenol (TPL) were determined in the young shoots, stems, and roots of tea plants under different integrated nutrient management strategies. The tea young shoots (two leaves and one bud), followed by the stems, had the highest catechin contents, while the tea roots had the lowest catechin contents, which is in accordance with prior findings [54]. Due to the higher availability of micro- and macronutrients, the application of mineral fertilizers (NPK) and organic fertilizers (RC and B) increases the content of catechins in tea. Organic fertilizer improved the biological, chemical, and physical soil properties, thereby augmenting the yield and quality of tea [4]. Bagchi et al.,

2022 [55] also found maximum contents of catechins in organic-based fertilizers compared to inorganic fertilizers and the control. Both the EC and ECG catechins belong to the class of dehydroxylated catechins, signifying that dehydroxylated catechin biosynthesis plays a key role in the root system, whereas EGC and EGCG were the major catechin constituents in the leaves (63%) and stems (64% in total).

Catechins are synthesized via the flavonoid and phenylpropanoid biosynthetic pathways. However, the individual catechins' exact biosynthetic pathways remain unknown. Stafford, 1990 [56] stated that C is an isomeric transition for the conversion of (+)-dihydroquercetin to EC. However, recent studies have found that the synthesis of EC and C is catalyzed by anthocyanin synthase (ANS) and leucocyanin reductase (LAR) from leucocyanin, respectively [54]. In addition, Wellmann et al., 2006 [57] reported that ANS can convert C into two major products and one minor product, suggesting that C may be involved in multiple biosynthetic pathways as an intermediate. The properties of catechins vary. For example, compared to other catechins, the (+)-catechin (C) significantly reduces the bitterness and astringency of tea.

4.3. Effect of Integrated Fertilization on Amino Acid Concentration in Tea Young Shoots, Stems, and Roots

In this research, compound fertilization (NPK+RC and NPK+B) significantly improved the amino acid contents of the young shoots (two leaves and one bud), roots, and stems. The amino acid content was higher in the roots compared to the stems. Former studies have reported that many amino acids were synthesized primarily in tea roots and then transported from the roots to the stems [58,59]. In this research, the application of organic-based fertilization enhanced nutrient uptake in tea a process that may affect the production of amino acid contents [60]. Consequently, integrated fertilization may improve amino acid synthesis in tea. In addition, rapeseed cake and biochar can increase the soil water holding capacity, thereby improving the leaf water fluidity and increasing the production of amino acids [37]. More importantly, B with NPK fertilizers can interact with the roots and influence the production of root secondary metabolites, finally influencing the quality of tea [22,36]. Organic fertilizers provide micronutrients that help in the improvement of tea quality and different amino acid concentrations with inorganic fertilizer application. In addition, integrated fertilization affected the soil microbiota, thereby increasing the amino acid contents of the tea. Studies have shown that microorganisms have positive correlations with tea yield and quality. Several genera, such as Rhodosporidium, Sphingomonas, Rhodanobacter, Chloridiums, Devosa, and Inocybes, have antecedently been reported to be participating in plant growth promotion and nutrient cycling [61,62]. Among the various metabolites, amino acids contribute a lot to the quality of tea. Studies have shown that nitrogen form and nitrogen levels significantly affect amino acid metabolism, thereby regulating the amino acid contents in the tea roots and aerial parts. There is growing evidence that nitrogen forms alter the amino acid content in tea tree leaves and roots [21,63,64]. Huang et al., 2018 [65] showed that providing different forms of nitrogen to tea plants significantly increased the Glu, Thea, and Pro contents in tea shoots compared to the control group. Through theanine synthetase, theanine is synthesized from glutamic and ethylamine acids. The decarboxylase of alanine in tea produced ethylamine. The N in these organic compounds comes from NO_3^- or NH_4^+. Tea plants prefer ammonium ions as inorganic N sources for growth [21].

Present studies have shown that amino acids provide different characteristics for the taste quality of tea infusion [66,67]. Thea, the most abundant amino acid, imparts complex flavors, including umami, bitterness, and sweetness, and the intensity level of the taste increase with an increasing Thea concentration [6,66]. Glu and Asp also contribute to the umami flavor. Gly, Ala, Thr, Pro, and Ser produce a sweet taste, while His, Ile, Arg, Met, Phe, Leu, Val, and Trp produce a bitter taste. The umami taste of tea infusion is primarily attributed to amino acid, especially Thea [68].

4.4. Chlorophyll Content of Leaves as Affected by Organic and Inorganic Fertilizer Management

The maximum concentrations of Chl a, b, and Chl a+b were found in the NPK+B and NPK+RC treatments, while the control and RC treatments had a minimum chlorophyll concentration. Similar studies were conducted at the seedling stage of peanuts, where the chlorophyll content increased with the application of biochar-based fertilizers [69]. Another similar result was found in soybean [70], sorghum, and maize plants [71], where inorganic fertilizers added to organic fertilizers increased the total chlorophyll content. The application of NPK-free organic fertilizer does not adequately provide the nutrients required for Chl biosynthesis. However, the addition of biochar and rapeseed cake in NPK applications significantly improved nutrient uptake, thereby optimizing chlorophyll biosynthesis. Large amounts of Chl are needed to maintain photosynthetic pigments and synthesized enzymes involved in tea growth, biomass, and quality. Higher photosynthetic rate mechanisms increase the chlorophyll concentration of tea, which induces higher plant growth.

The chemical structures of chlorophyll a ($C_{55}H_{72}O_5N_4Mg$) and chlorophyll b ($C_{55}H_{70}O_6$ N_4Mg) have different functional groups. In addition, the color of Chl b is light green, and Chl a has a dark green color. The light energy is converted into chemical energy by the principal pigment, Chl a. Chl b acts as a co-pigment and plays an indirect role in photosynthesis by transporting the light it absorbs to Chl a [72]. Green tea products and the green color of fresh tea leaves are created by the Chl compound. Water-soluble green pigments, chlorophyllides, and chlorophyll derivatives are derived from chlorophyll through the catalysis of chlorophyllase, which is dissolved during the tea brewing process, resulting in green tea-colored infusions [73]. The chlorophyll content gives green tea products a grassy flavor [74].

4.5. Effect of Integrated Fertilization on Soil pH and NH_4^+ and NO_3^- Concentrations

In this research, the soil pH was significantly influenced by combined fertilization, where the soil pH of inorganic fertilizer (NPK) decreased; rapeseed cake alone or in combination with inorganic fertilizer (NPK+RC) increased the soil pH; and the incorporation of biochar and chemical fertilizer (NPK+B) also improved the soil pH due to its liming properties. These results show that organic fertilizers have greater potential to increase the soil pH than nitrogen, phosphorus, and potassium fertilizers. This means that organic fertilizers can be used as good materials for improving acidic soils. The results are consistent with those of many similar studies, such as Yang et al., 2021 and Xie et al., 2021 [75,76], who noted that the pH of acidic soils increases with the incorporation of organic fertilizers.

The nationwide average pH of tea garden soils is 4.68, and the soil pH in different provinces in China varies from 3.96 to 5.48 [22]. Tea garden soils are acidic, with a pH of about 4.0 to 5.5, unlike other crops [77,78]. When the pH of the soil is less than 4, it is not conducive to the growth and quality of tea plants [79]. Therefore, increasing the soil pH is essential to improving tea soil fertility. In general, the addition of biochar reduces stress on plants due to its lime properties [80]. Similarly, Zhang et al., 2019 and Saha et al., 2019 [37,81] also indicated that the use of biochar (B) or biochar-based fertilizers increases the soil pH. Therefore, applying RC and biochar plus mineral fertilizer is a suitable way to alleviate acidification in tea orchard soil.

Combined fertilization increased the soil NO_3^- and NH_4^+ concentrations. The application of RC and B with NPK significantly increased the soil NO_3^- and NH_4^+ contents, while the NO_3^- and NH_4^+ contents of the RC and control treatments were lower. The content of ammonium in soil was significantly higher than that of nitrate, and the content of ammonium in RC and biochar with chemical fertilizer treatments was significantly higher than that in the mineral fertilizer treatments (Figure 5e,f). Therefore, RC is more beneficial to the growth of tea plants than mineral fertilizers that prefer NH_4^+. At the same time, the nitrate content in the overflow water was higher than the ammonium content. Ammonium in soil is generally adsorbed by the soil organic matter and soil particles, and ammonium is rapidly converted to nitrate under aerobic soil conditions [76]. The utilization of cow manure biochar (CMB) and straw biochar+ fertilizer (RSBF) increased the NH_4^+ content by 45.1% and 36.5%, respectively, compared to chemical fertilizers [82].

4.6. Effect of Integrated Fertilization on Soil Macro and Micronutrients

The application of B and RC with chemical fertilizers (NPK+B and NPK+RC) significantly improved soil macro- and micronutrient statuses. Regarding the soil nutrient status, chemical fertilizer and organic fertilizer jointly improved plant growth and significantly increased the N, P, and K contents in the soil. Organic amendments using chemical fertilizers may result in higher microbial activity and nutrient availability compared to no fertilization (controls) and mineral fertilization [61,62,83]. When organic fertilizer is applied together with chemical fertilizers, the application of organic amendments increases the soil's N, P, and K concentrations [83]. Organic fertilizers have a greater impact on the soil quality than chemical fertilizers, thereby improving nutrient release and their availability to tea plants [84]. Rapeseed cake significantly improved the exogenous input of the soil OM, which is more likely to retain nutrients in a form that is more easily absorbed by plants for a longer period of time. The nutrients in chemical fertilizers are already mineralized, which provides a ready source of nutrients for the soil. It is implied that the nutrient release time of NPK fertilizers was short because the leaching of nutrients from soils treated with NPK fertilizers may be higher than that of soils treated with organic fertilizers. Several studies have reported a longer residual effect of organic fertilization on soils [85].

Our results also indicated that the soil's available contents of K, P, N, Ca, etc., were significantly improved by the NPK+B treatment, which is also consistent with previous studies [29,33,75]. The available soil C content has been shown to play a critical role in conserving the fertility of soil. Biochar adsorbs OM and promotes the establishment of humus soil, which increases the amount of available soil carbon and nitrogen [86]. As for other nutrients, biochar itself carries certain nutrients, such as phosphorus, potassium, and calcium, which can directly provide nutrients. In addition, biochar is able to interact with soil and indirectly affect the availability of nutrients in soil through adsorption, precipitation, and dissolution [87]. To store nutrients in soil, biochar plays an important role in reducing nutrient leaching and nutrient loss from inorganic fertilizers [47,75]. These outputs may be due to the higher macronutrient content in the NPK+B treatment. Similar results have previously been reported; biochar-based fertilizers produce a higher soil nutrient value than biochar alone [29,81].

Synthetic chemical fertilizers are primarily used as nitrogen sources for plant growth, while RC fertilizer not only supplies nitrogen but also helps to maintain the carbon content in the soils. Therefore, chemical fertilizer and RC organic fertilizer have different effects on the soil's carbon and nitrogen cycles in tea gardens. Compared with rapeseed cake and biochar alone, the combined application of chemical and organic fertilization has a lower C:N ratio. However, the CK and NPK treatments had higher C:N ratios. The change in the proportion of carbon and nitrogen sources in the soil after fertilization may be due to the fact that the chemical fertilizer is a synthetic fast-acting N fertilizer, which is primarily used to supply the nitrogen needed by the plant, so little carbon is input. Rapeseed is a kind of biomass fertilizer that also contains a large quantity of OM, which provides a huge quantity of carbon to the soil. As a slow-release fertilizer, RC has a relatively small and slow nitrogen input. Similar research has also shown that organic N in organic fertilizer is more effective than chemical nitrogen in synthetic fertilizer, particularly in low-yield agricultural production systems [76,88]. The lower C:N ratio of the soil treated with integrated fertilization may be attributed to higher nitrogen availability due to the accumulation of chemical and organic fertilizers and their retention in the soil, which is associated with lower nitrogen loss rates than carbon loss rates during organic matter degradation [89]. Soil organic carbon was significantly improved by organic amendments, which has a significant influence on soil micro- and macronutrient availability, microbial activity, and nutrient uptake, potentially altering the C:N ratio. However, the incorporation of external organic matter with higher C:N ratios may also speed up the mineralization of existing OM, releasing the N trapped in existing OM—a process known as the priming effect [90]. These outcomes are confirmed by former studies [83].

The application of integrated fertilization improved NUE and the recovery of micro- and macronutrients. The improvement of OM may be due to the application of organic fertilizers, which improve soil properties and plant growth. Adediran et al., 2005 [91] reported that plots treated with organic fertilizers increased organic carbon by 11–20%, calcium by 26–96%, magnesium by 6–19%, and zinc by 4–8% compared with chemical fertilizers. The application of integrated nutrient management increases the content of soil micronutrients compared with the solo application of chemical fertilizers. Similarly, organic fertilizer sources also contributed to the supply of soil macro- and micronutrients through mineralization processes and improved the soil physicochemical and microbial properties. Furthermore, these fertilizers are able to gradually release nutrients (macro and micronutrients) and make them available to plants throughout the growing period [91]. Thus, the application of integrated fertilization was considered to be a good choice for promoting nutrient recovery and plant growth [83].

5. Conclusions

This study investigated the effects of integrated fertilization on soil chemistry, tea growth and quality, and the relationship between soil chemistry and tea growth and quality. Studies have shown that NPK+RC and NPK+B amendments increase the soil pH, macronutrients, and micronutrients, especially soil-available nutrients. Additionally, changes in the soil chemistry caused by integrated fertilization were key factors in improving the growth parameters of the tea plant. Furthermore, integrated fertilization significantly improves the quality of tea compared to the solo application of mineral fertilizers or organic fertilizers. Combined fertilization also significantly improved the Chl concentrations (Chl a, Chl b, and Chl a+b). The growth parameters and quality of tea were significantly positively correlated with soil macronutrients (N, P, K, Ca, and Mg), nitrate, ammonium, and soil micronutrients (Cu, Mn, Fe, and Zn). The chlorophyll content was also significantly positively correlated with tea quality and growth. The present study suggested that, compared with organic or inorganic fertilizer alone, integrated fertilization (NPK+RC and NPK+B) shows better performance in promoting soil properties and tea growth and quality, and it has important potential to improve the chemical properties of degraded tea garden soil and further improve the growth and quality of green tea.

Author Contributions: Conceptualization and design of the study, M. and J.R.; methodology, K.N., L.M. and M.; investigation, M.; resources, K.N.; data collection and analyses, M.; writing of manuscript, M.; editing and review, J.R. and K.N.; supervision, J.R.; project administration, J.R.; funding, J.R. All authors have read and agreed to the published version of the manuscript.

Funding: This research was supported by the Earmarked Fund for China Agriculture Research System (CARS 19), the National Key R&D Project (2021YFD1601100), the Department of Agriculture of Zhejiang Province (through the contract no. 2021SNLF032), and the Agricultural Science and Technology Innovation Program of the Chinese Academy of Agricultural Sciences (CAAS-ASTIP-TRICAAS).

Data Availability Statement: Data will be available upon request.

Acknowledgments: We would like to thank Ruan Jianyun, Kang Ni, and Lifeng Ma for their valuable advice and technical expertise. Additionally, special thanks to Ayesha Haq for improving the English language in the manuscript.

Conflicts of Interest: The authors declare no conflict of interest.

References

1. Ji, H.-G.; Lee, Y.-R.; Lee, M.-S.; Hwang, K.H.; Kim, E.-H.; Park, J.S.; Hong, Y.-S. Metabolic phenotyping of various tea (*Camellia sinensis* L.) cultivars and understanding of their intrinsic metabolism. *Food Chem.* **2017**, *233*, 321–330. [CrossRef]
2. Fraser, K.; Harrison, S.; Lane, G.; Otter, D.; Quek, S.Y.; Rasmussen, S. Analysis of Low Molecular Weight Metabolites in Tea Using Mass Spectrometry-Based Analytical Methods. *Crit. Rev. Food Sci. Nutr.* **2014**, *54*, 924–937. [CrossRef]

3. Kavitha, S.; Prapagar, K.; Gunarathne, G.P. Nutrient Availability of Tea Growing Soil Influenced by Different Rates of Dolomite. *J. Tea Sci.* **2016**, *6*, 1–13. [CrossRef]
4. Lin, W.; Manhong, L.; Zhou, H.; Wu, H.; Li, Z.; Lin, W. The effects of chemical and organic fertilizer usage on rhizosphere soil in tea orchards. *PLoS ONE* **2019**, *14*, e0217018. [CrossRef] [PubMed]
5. Too, J.; Thomas, K.; Kanyiri, W.; Wachira, F. Effect of Sunlight Exposure and Different Withering Durations on Theanine Levels in Tea (*Camellia sinensis*). *Food Nutr. Sci.* **2015**, *6*, 1014–1021.
6. Horanni, R.; Engelhardt, U.H. Determination of amino acids in white, green, black, oolong, pu-erh teas and tea products. *J. Food Compos. Anal.* **2013**, *31*, 94–100. [CrossRef]
7. Fang, R.; Redfern, S.P.; Kirkup, D.; Porter, E.A.; Kite, G.C.; Terry, L.A.; Berry, M.J.; Simmonds, M.S.J. Variation of theanine, phenolic, and methylxanthine compounds in 21 cultivars of *Camellia sinensis* harvested in different seasons. *Food Chem.* **2017**, *220*, 517–526. [CrossRef] [PubMed]
8. Chen, Q.; Zhao, J.; Guo, Z.; Wang, X. Determination of caffeine content and main catechins contents in green tea (*Camellia sinensis* L.) using taste sensor technique and multivariate calibration. *J. Food Compos. Anal.* **2010**, *23*, 353–358. [CrossRef]
9. Thomasset, S.C.; Berry, D.P.; Garcea, G.; Marczylo, T.; Steward, W.P.; Gescher, A.J. Dietary polyphenolic phytochemicals—Promising cancer chemopreventive agents in humans? A review of their clinical properties. *Int. J. Cancer* **2007**, *120*, 451–458. [CrossRef]
10. Cavet, M.E.; Harrington, K.L.; Vollmer, T.R.; Ward, K.W.; Zhang, J.-Z. Anti-inflammatory and anti-oxidative effects of the green tea polyphenol epigallocatechin gallate in human corneal epithelial cells. *Mol. Vis.* **2011**, *17*, 533–542. [PubMed]
11. Cho, Y.-S.; Oh, J.J.; Oh, K.-H. Synergistic anti-bacterial and proteomic effects of epigallocatechin gallate on clinical isolates of imipenem-resistant Klebsiella pneumoniae. *Phytomedicine* **2011**, *18*, 941–946. [CrossRef] [PubMed]
12. Crozier, A.; Jaganath, I.B.; Clifford, M.N. Dietary phenolics: Chemistry, bioavailability and effects on health. *Nat. Prod. Rep.* **2009**, *26*, 1001–1043. [CrossRef] [PubMed]
13. Del Rio, D.; Stewart, A.; Mullen, W.; Burns, J.; Lean, M.; Brighenti, F.; Crozier, A. HPLC-MSn Analysis of Phenolic Compounds and Purine Alkaloids in Green and Black Tea. *J. Agric. Food Chem.* **2004**, *52*, 2807–2815. [CrossRef]
14. Zhao, Y.; Chen, P.; Lin, L.; Harnly, J.M.; Yu, L.; Li, Z. Tentative identification, quantitation, and principal component analysis of green pu-erh, green, and white teas using UPLC/DAD/MS. *Food Chem.* **2011**, *126*, 1269–1277. [CrossRef] [PubMed]
15. Friedman, M.; Mackey, B.; Kim, H.-J.; Lee, I.-S.; Lee, K.-R.; Lee, S.-U.; Kozukue, E.; Kozukue, N. Structure–Activity Relationships of Tea Compounds against Human Cancer Cells. *J. Agric. Food Chem.* **2007**, *55*, 243–253. [CrossRef]
16. Zheng, X.-Q.; Nie, Y.; Gao, Y.; Huang, B.; Ye, J.-H.; Lu, J.-L.; Liang, Y.-R. Screening the cultivar and processing factors based on the flavonoid profiles of dry teas using principal component analysis. *J. Food Compos. Anal.* **2018**, *67*, 29–37. [CrossRef]
17. Daglia, M.; Antiochia, R.; Sobolev, A.P.; Mannina, L. Untargeted and targeted methodologies in the study of tea (*Camellia sinensis* L.). *Food Res. Int.* **2014**, *63*, 275–289. [CrossRef]
18. Ji, L.; Wu, Z.; You, Z.; Yi, X.; Ni, K.; Guo, S.; Ruan, J. Effects of organic substitution for synthetic N fertilizer on soil bacterial diversity and community composition: A 10-year field trial in a tea plantation. *Agric. Ecosyst. Environ.* **2018**, *268*, 124–132. [CrossRef]
19. Gu, S.; Hu, Q.; Cheng, Y.; Bai, L.; Liu, Z.; Xiao, W.; Gong, Z.; Wu, Y.; Feng, K.; Deng, Y.; et al. Application of organic fertilizer improves microbial community diversity and alters microbial network structure in tea (*Camellia sinensis*) plantation soils. *Soil Tillage Res.* **2019**, *195*, 104356. [CrossRef]
20. Li, Y.C.; Li, Z.; Li, Z.W.; Jiang, Y.H.; Weng, B.Q.; Lin, W.X. Variations of rhizosphere bacterial communities in tea (*Camellia sinensis* L.) continuous cropping soil by high-throughput pyrosequencing approach. *J. Appl. Microbiol.* **2016**, *121*, 787–799. [CrossRef] [PubMed]
21. Ruan, J.Y.; Gerendas, J.; Härdter, R.; Sattelmacher, B. Effect of Nitrogen Form and Root-zone pH on Growth and Nitrogen Uptake of Tea (*Camellia sinensis*) Plants. *Ann. Bot.* **2007**, *99*, 301–310. [CrossRef] [PubMed]
22. Yan, P.; Wu, L.; Wang, D.; Fu, J.; Shen, C.; Li, X.; Zhang, L.; Zhang, L.; Fan, L.; Wenyan, H. Soil acidification in Chinese tea plantations. *Sci. Total Environ.* **2020**, *715*, 136963. [CrossRef]
23. Arafat, Y.; Tayyab, M.; Khan, M.U.; Chen, T.; Amjad, H.; Awais, S.; Lin, X.; Lin, W.; Lin, S. Long-Term Monoculture Negatively Regulates Fungal Community Composition and Abundance of Tea Orchards. *Agronomy* **2019**, *9*, 466. [CrossRef]
24. Ji, L.; Ni, K.; Wu, Z.; Zhang, J.; Yi, X.; Yang, X.; Ling, N.; You, Z.; Guo, S.; Ruan, J. Effect of organic substitution rates on soil quality and fungal community composition in a tea plantation with long-term fertilization. *Biol. Fertil. Soils* **2020**, *56*, 633–646. [CrossRef]
25. Cheng, J.; Lee, X.; Tang, Y.-J.; Zhang, Q. Long-term effects of biochar amendment on rhizosphere and bulk soil microbial communities in a karst region, southwest China. *Appl. Soil Ecol.* **2019**, *140*, 126–134. [CrossRef]
26. Cooper, J.; Greenberg, I.; Ludwig, B.; Hippich, L.; Fischer, D.; Glaser, B.; Kaiser, M. Effect of biochar and compost on soil properties and organic matter in aggregate size fractions under field conditions. *Agric. Ecosyst. Environ.* **2020**, *295*, 106882. [CrossRef]
27. Nguyen, T.T.N.; Wallace, H.M.; Xu, C.-Y.; Zwieten, L.; Weng, Z.H.; Xu, Z.; Che, R.; Tahmasbian, I.; Hu, H.-W.; Bai, S.H. The effects of short term, long term and reapplication of biochar on soil bacteria. *Sci. Total Environ.* **2018**, *636*, 142–151. [CrossRef] [PubMed]
28. Naeem, I.; Masood, N.; Turan, V.; Iqbal, M. Prospective usage of magnesium potassium phosphate cement combined with Bougainvillea alba derived biochar to reduce Pb bioavailability in soil and its uptake by *Spinacia oleracea* L. *Ecotoxicol. Environ. Saf.* **2021**, *208*, 111723. [CrossRef]

29. Zhou, Z.; Gao, T.; Zhu, Q.; Yan, T.; Li, D.; Xue, J.; Wu, Y. Increases in bacterial community network complexity induced by biochar-based fertilizer amendments to karst calcareous soil. *Geoderma* **2019**, *337*, 691–700. [CrossRef]
30. Hussain, M.; Farooq, M.; Nawaz, A.; Al-Sadi, A.M.; Solaiman, Z.M.; Alghamdi, S.S.; Ammara, U.; Ok, Y.S.; Siddique, K.H.M. Biochar for crop production: Potential benefits and risks. *J. Soils Sediments* **2017**, *17*, 685–716. [CrossRef]
31. Zhou, Z.; Gao, T.; Van Zwieten, L.; Zhu, Q.; Yan, T.; Xue, J.; Wu, Y. Soil Microbial Community Structure Shifts Induced by Biochar and Biochar-Based Fertilizer Amendment to Karst Calcareous Soil. *Soil Sci. Soc. Am. J.* **2019**, *83*, 398–408. [CrossRef]
32. Zhang, M.; Riaz, M.; Zhang, L.; Xia, H.; El-desouki, Z.; Jiang, C. Response of fungal communities in different soils to biochar and chemical fertilizers under simulated rainfall conditions. *Sci. Total Environ.* **2019**, *691*, 654–663. [CrossRef]
33. Oladele, S.O.; Adeyemo, A.J.; Awodun, M.A. Influence of rice husk biochar and inorganic fertilizer on soil nutrients availability and rain-fed rice yield in two contrasting soils. *Geoderma* **2019**, *336*, 1–11. [CrossRef]
34. Faloye, O.T.; Ajayi, A.E.; Alatise, M.O.; Ewulo, B.S.; Horn, R. Nutrient uptake, maximum yield production, and economic return of maize under deficit irrigation with biochar and inorganic fertiliser amendments. *Biochar* **2019**, *1*, 375–388. [CrossRef]
35. Rafael, R.B.A.; FernÁNdez-Marcos, M.L.; Cocco, S.; Ruello, M.L.; Fornasier, F.; Corti, G. Benefits of Biochars and NPK Fertilizers for Soil Quality and Growth of Cowpea (*Vigna unguiculata* L. Walp.) in an Acid Arenosol. *Pedosphere* **2019**, *29*, 311–333. [CrossRef]
36. Yan, S.; Zhao, J.; Ren, T.; Liu, G. Correlation between soil microbial communities and tobacco aroma in the presence of different fertilizers. *Ind. Crops Prod.* **2020**, *151*, 112454. [CrossRef]
37. Saha, A.; Basak, B.B.; Gajbhiye, N.A.; Kalariya, K.A.; Manivel, P. Sustainable fertilization through co-application of biochar and chemical fertilizers improves yield, quality of Andrographis paniculata and soil health. *Ind. Crops Prod.* **2019**, *140*, 111607. [CrossRef]
38. Lustosa Filho, J.F.; Carneiro, J.S.d.S.; Barbosa, C.F.; de Lima, K.P.; Leite, A.d.A.; Melo, L.C.A. Aging of biochar-based fertilizers in soil: Effects on phosphorus pools and availability to Urochloabrizantha grass. *Sci. Total Environ.* **2020**, *709*, 136028. [CrossRef]
39. Sun, R.; Dsouza, M.; Gilbert, J.A.; Guo, X.; Wang, D.; Guo, Z.; Ni, Y.; Chu, H. Fungal community composition in soils subjected to long-term chemical fertilization is most influenced by the type of organic matter. *Environ. Microbiol.* **2016**, *18*, 5137–5150. [CrossRef]
40. Soil Survey Staff. *Keys to Soil Taxonomy*, 12th ed.; USDA-Natural Resources Conservation Service: Washington, DC, USA, 2014.
41. Shi, X.Z.; Yu, D.S.; Xu, S.X.; Warner, E.D.; Wang, H.J.; Sun, W.X.; Zhao, Y.C.; Gong, Z.T. Cross-reference for relating Genetic Soil Classification of China with WRB at different scales. *Geoderma* **2010**, *155*, 344–350. [CrossRef]
42. Bao, S.D. *Analysis of Soil Agricultural Chemistry*, 3rd ed.; China Agriculture Press: Beijing, China, 2005; pp. 263–271. (In Chinese)
43. Grant, A. Growing Tea from Seed—Tips for Germinating Tea Seeds. [Online] Gardening Know How. 2019. Available online: https://www.gardeningknowhow.com/edible/herbs/tea-plant/growing-tea-from-seed.htm (accessed on 18 November 2023).
44. Ni, Z.; Kim, E.-D.; Chen, Z.J. Chlorophyll and starch assays. Protocol Exchange. *Protoc. Exch.* **2009**, *10*, 1038.
45. Bass, A.M.; Bird, M.I.; Kay, G.; Muirhead, B. Soil properties, greenhouse gas emissions and crop yield under compost, biochar and co-composted biochar in two tropical agronomic systems. *Sci. Total Environ.* **2016**, *550*, 459–470. [CrossRef]
46. Kamau, S.; Karanja, N.K.; Ayuke, F.O.; Lehmann, J. Short-term influence of biochar and fertilizer-biochar blends on soil nutrients, fauna and maize growth. *Biol. Fertil. Soils* **2019**, *55*, 661–673. [CrossRef]
47. Haider, G.; Steffens, D.; Moser, G.; Müller, C.; Kammann, C.I. Biochar reduced nitrate leaching and improved soil moisture content without yield improvements in a four-year field study. *Agric. Ecosyst. Environ.* **2017**, *237*, 80–94. [CrossRef]
48. Helliwell, R. Effect of biochar on plant growth. *Arboric. J.* **2015**, *37*, 238–242. [CrossRef]
49. Berihun, T.; Tolosa, S.; Tadele, M.; Kebede, F. Effect of Biochar Application on Growth of Garden Pea (*Pisum sativum* L.) in Acidic Soils of Bule Woreda Gedeo Zone Southern Ethiopia. *Int. J. Agron.* **2017**, *2017*, 6827323. [CrossRef]
50. Wilson, K. *Biochar for Forest Restoration in Western United States*; Wilson Biochar Associates, South Umpqua Rural Community Partnership (SURCP): Tiller, OR, USA, 2015.
51. Khaliq, A.; Abbasi, M.K.; Hussain, T. Effects of integrated use of organic and inorganic nutrient sources with effective microorganisms (EM) on seed cotton yield in Pakistan. *Bioresour. Technol.* **2006**, *97*, 967–972. [CrossRef] [PubMed]
52. Ramana, S.; Biswas, A.K.; Singh, A.B.; Yadava, R.B.R. Relative efficacy of different distillery effluents on growth, nitrogen fixation and yield of groundnut. *Bioresour. Technol.* **2002**, *81*, 117–121. [CrossRef]
53. Hao, X.; Chang, C. Effect of 25 annual cattle manure applications on soluble and exchangeable cations in soil. *Soil Sci.* **2002**, *167*, 126–134. [CrossRef]
54. Ashihara, H.; Deng, W.-W.; Mullen, W.; Crozier, A. Distribution and biosynthesis of flavan-3-ols in *Camellia sinensis* seedlings and expression of genes encoding biosynthetic enzymes. *Phytochemistry* **2010**, *71*, 559–566. [CrossRef]
55. Bagchi, A.; Swain, D.; Mitra, A. Comparative assessment of organic and inorganic tea leaf extract feeding on anxiety behaviour status of colchicine-induced rat model of Alzheimer's disease. *Inflammopharmacology* **2022**, *30*, 609–620. [CrossRef] [PubMed]
56. Stafford, H.A. *Pathway to Proanthocyanindins (Condensed Tannins), Flavan−3−ols, and Unsubstituted Flavans*; CRC Press: New York, NY, USA, 1990; pp. 63–99.
57. Wellmann, F.; Griesser, M.; Schwab, W.; Martens, S.; Eisenreich, W.; Matern, U.; Lukacin, R. Anthocyanidin synthase from Gerbera hybridacatalyzes the conversion of (+)-catechin to cyanidin and a novel procyanidin. *FEBS Lett.* **2006**, *580*, 1642–1648. [CrossRef] [PubMed]
58. Dong, C.; Li, F.; Yang, T.; Feng, L.; Zhang, S.; Li, F.; Li, W.; Xu, G.; Bao, S.; Wan, X.; et al. Theanine transporters identified in tea plants (*Camellia sinensis* L.). *Plant J.* **2020**, *101*, 57–70. [CrossRef] [PubMed]

59. Li, F.; Dong, C.; Yang, T.; Ma, J.; Zhang, S.; Wei, C.; Wan, X.; Zhang, Z. Seasonal Theanine Accumulation and Related Gene Expression in the Roots and Leaf Buds of Tea Plants (*Camellia sinensis* L.). *Front. Plant Sci.* **2019**, *10*, 1397. [CrossRef]
60. Petruccelli, R.; Bonetti, A.; Traversi, M.; Faraloni, C.; Valagussa, M.; Pozzi, A. Influence of biochar application on nutritional quality of tomato (*Lycopersicon esculentum*). *Crop Pasture Sci.* **2015**, *66*, 747–755. [CrossRef]
61. Xu, L.; Yi, M.; Yi, H.; Guo, E.; Zhang, A. Manure and mineral fertilization change enzyme activity and bacterial community in millet rhizosphere soils. *World J. Microbiol. Biotechnol.* **2017**, *34*, 8. [CrossRef] [PubMed]
62. Kropp, B.R.; Albee-Scott, S. Inocybetauensis, a new species from the Samoan Archipelago with biogeographic evidence for a Paleotropical origin. *Fungal Biol.* **2010**, *114*, 790–796. [CrossRef]
63. Yang, Y.; Wang, F.; Wan, Q.; Ruan, J. Transcriptome analysis using RNA-Seq revealed the effects of nitrogen form on major secondary metabolite biosynthesis in tea (*Camellia sinensis*) plants. *Acta Physiol. Plant.* **2018**, *40*, 127. [CrossRef]
64. Ruan, J.Y.; Gerendas, J.; Härdter, R.; Sattelmacher, B. Effect of root zone pH and form and concentration of nitrogen on accumulation of quality-related components in green tea. *J. Sci. Food Agric.* **2007**, *87*, 1505–1516. [CrossRef]
65. Huang, H.; Yao, Q.; Xia, E.-H.; Gao, L.-Z. Metabolomics and Transcriptomics Analyses Reveal Nitrogen Influences on the Accumulation of Flavonoids and Amino Acids in Young Shoots of Tea Plant (*Camellia sinensis* L.) Associated with Tea Flavor. *J. Agric. Food Chem.* **2018**, *66*, 9828–9838. [CrossRef]
66. Zhu, Y.; Luo, Y.; Wang, P.; Zhao, M.; Li, L.; Hu, X.; Chen, F. Simultaneous determination of free amino acids in Pu-erh tea and their changes during fermentation. *Food Chem.* **2016**, *194*, 643–649. [CrossRef] [PubMed]
67. Chen, L.; Chen, Q.; Zhang, Z.; Wan, X. A novel colorimetric determination of free amino acids content in tea infusions with 2,4-dinitrofluorobenzene. *J. Food Compos. Anal.* **2009**, *22*, 137–141. [CrossRef]
68. Thippeswamy, R.; Gouda, K.; Rao, D.; Martin, A.; Gowda, R. Determination of Theanine in Commercial Tea by Liquid Chromatography with Fluorescence and Diode Array Ultraviolet Detection. *J. Agric. Food Chem.* **2006**, *54*, 7014–7019. [CrossRef]
69. Yi, Y.; Xing, J.; Wan, M.; Yu, L.; Lu, Y.; Jian, Y. Effect of Cu on microstructure, crystallography and mechanical properties in Fe-B-C-Cu alloys. *Mater. Sci. Eng. A* **2017**, *708*, 274–284. [CrossRef]
70. Ghosh, P.; Kumar, A.; Bandyopadhyay, K.; Manna, M.; Mandal, K.G.; Misra, A.K.; Hati, K. Comparative effectiveness of cattle manure, poultry manure, phosphocompost and fertilizer-NPK on three cropping systems in vertisols of semi-arid tropics. II. Dry matter yield, nodulation, chlorophyll content and enzyme activity. *Bioresour. Technol.* **2004**, *95*, 85–93. [CrossRef] [PubMed]
71. Amujoyegbe, B.J.; Opabode, J.T.; Olayinka, A. Effect of Organic and Inorganic Fertilizer on Yield and Chlorophyll Content of Maize (*Zea mays* L.) and Sorghum *Sorghum bicolor* (L.) Moench. *Afr. J. Biotechnol.* **2007**, *6*, 1869–1873.
72. Nayek, S.; Choudhury, I.; Haque; Nishika, J.; Roy, S. Spectrophotometric Analysis of Chlorophylls and Carotenoids from Commonly Grown Fern Species by Using Various Extracting Solvents. *Res. J. Chem. Sci.* **2014**, *4*, 2231–2606.
73. Wang, L.F.; Park, S.C.; Chung, J.O.; Baik, J.H.; Park, S.K. The Compounds Contributing to the Greenness of Green Tea. *J. Food Sci.* **2004**, *69*, S301–S305. [CrossRef]
74. Baba, R.; Kumazawa, K. Characterization of the potent odorants contributing to the characteristic aroma of Chinese green tea infusions by aroma extract dilution analysis. *J. Agric. Food Chem.* **2014**, *62*, 8308–8313. [CrossRef]
75. Yang, W.; Li, C.; Wang, S.; Zhou, B.; Mao, Y.; Rensing, C.; Xing, S. Influence of biochar and biochar-based fertilizer on yield, quality of tea and microbial community in an acid tea orchard soil. *Appl. Soil Ecol.* **2021**, *166*, 104005. [CrossRef]
76. Xie, S.; Yang, F.; Feng, H.; Yu, Z.; Liu, C.; Wei, C.; Liang, T. Organic fertilizer reduced carbon and nitrogen in runoff and buffered soil acidification in tea plantations: Evidence in nutrient contents and isotope fractionations. *Sci. Total Environ.* **2021**, *762*, 143059. [CrossRef] [PubMed]
77. Yan, P.; Shen, C.; Fan, L.; Li, X.; Zhang, L.; Zhang, L.; Han, W. Tea planting affects soil acidification and nitrogen and phosphorus distribution in soil. *Agric. Ecosyst. Environ.* **2018**, *254*, 20–25. [CrossRef]
78. Ruan, J.; Zhang, F.; Wong, M.H. Effect of nitrogen form and phosphorus source on the growth, nutrient uptake and rhizosphere soil property of *Camellia sinensis* L. *Plant Soi.* **2000**, *223*, 65–73. [CrossRef]
79. Wang, L.; Butterly, C.R.; Chen, Q.; Mu, Z.; Wang, X.; Xi, Y.; Zhang, J.; Xiao, X. Surface Amendments Can Ameliorate Subsoil Acidity in Tea Garden Soils of High-Rainfall Environments. *Pedosphere* **2016**, *26*, 180–191. [CrossRef]
80. Dai, Z.; Zhang, X.; Tang, C.; Muhammad, N.; Wu, J.; Brookes, P.C.; Xu, J. Potential role of biochars in decreasing soil acidification—A critical review. *Sci. Total Environ.* **2017**, *581–582*, 601–611. [CrossRef] [PubMed]
81. Zhang, M.; Riaz, M.; Zhang, L.; El-desouki, Z.; Jiang, C. Biochar Induces Changes to Basic Soil Properties and Bacterial Communities of Different Soils to Varying Degrees at 25 mm Rainfall: More Effective on Acidic Soils. *Front. Microbiol.* **2019**, *10*, 1321. [CrossRef] [PubMed]
82. Ibrahim, M.M.; Tong, C.; Hu, K.; Zhou, B.; Xing, S.; Mao, Y. Biochar-fertilizer interaction modifies N-sorption, enzyme activities and microbial functional abundance regulating nitrogen retention in rhizosphere soil. *Sci. Total Environ.* **2020**, *739*, 140065. [CrossRef]
83. Mahmood, F.; Khan, I.; Ashraf, U.; Shahzad, T.; Hussain, S.; Shahid, M.; Abid, M.; Ullah, S. Effects of organic and inorganic manures on maize and their residual impact on soil physico-chemical properties. *J. Soil Sci. Plant Nutr.* **2017**, *17*, 22–32. [CrossRef]
84. Birkhofer, K.; Bezemer, T.M.; Bloem, J.; Bonkowski, M.; Christensen, S.; Dubois, D.; Ekelund, F.; Fließbach, A.; Gunst, L.; Hedlund, K.; et al. Long-term organic farming fosters below and aboveground biota: Implications for soil quality, biological control and productivity. *Soil Biol. Biochem.* **2008**, *40*, 2297–2308. [CrossRef]

85. Adeniyan, O.N.; Ojo, A.O.; Akinbode; Adediran, J.A. Comparative study of different organic manures and NPK fertilizer for improvement of soil chemical properties and dry matter yield of maize in two different soils. *J. Soil Sci. Environ. Manag.* **2011**, *2*, 9–13.
86. Kimetu, J.; Lehmann, J. Stability and Stabilisation of Biochar and Green Manure in Soil with Different Organic Carbon Contents. *Soil Res.* **2010**, *48*, 577–585. [CrossRef]
87. Shi, R.; Hong, Z.; Li, J.; Jiang, J.; Baquy, M.A.-A.; Xu, R.-K.; Qian, W. Mechanisms for Increasing the pH Buffering Capacity of an Acidic Ultisol by Crop Residue-Derived Biochars. *J. Agric. Food Chem.* **2017**, *65*, 8111–8119. [CrossRef] [PubMed]
88. Busari, M.A.; Salako, F.K.; Tuniz, C. Stable isotope technique in the evaluation of tillage and fertilizer effects on soil carbon and nitrogen sequestration and water use efficiency. *Eur. J. Agron.* **2016**, *73*, 98–106. [CrossRef]
89. Chen, Y.; Zhang, X.; He, H.; Xie, H.; Yan, Y.; Zhu, P.; Ren, J.; Wang, L. Carbon and nitrogen pools in different aggregates of a Chinese Mollisol as influenced by long-term fertilization. *J. Soils Sediments* **2010**, *10*, 1018–1026. [CrossRef]
90. Shahzad, T.; Chenu, C.; Genet, P.; Barot, S.; Perveen, N.; Mougin, C.; Fontaine, S. Contribution of exudates, arbuscular mycorrhizal fungi and litter depositions to the rhizosphere priming effect induced by grassland species. *Soil Biol. Biochem.* **2015**, *80*, 146–155. [CrossRef]
91. Adediran, J.A.; Taiwo, L.B.; Akande, M.O.; Sobulo, R.A.; Idowu, O.J. Application of Organic and Inorganic Fertilizer for Sustainable Maize and Cowpea Yields in Nigeria. *J. Plant Nutr.* **2005**, *27*, 1163–1181. [CrossRef]

plants

MDPI

Article

Nitrogen Application Timing and Levels Affect the Fate and Budget of Fertilizer Nitrogen in the Apple–Soil System

Fen Wang [1], Chaoran Wang [2], Binghao Yang [1], Xinyu Luo [1], Gaowei Qi [1], Fajin Ji [1], Xinkai Guo [1], Tao Yang [1], Xuehui Zhao [1], Ming Li [1], Qianqian Jiang [1], Ling Peng [3] and Hui Cao [1,*]

1 School of Advanced Agricultural Sciences, Weifang University, Weifang 261061, China; fenwang@wfu.edu.cn (F.W.); 19709656512@163.com (B.Y.); luowill1221@163.com (X.L.); 13676368263@163.com (G.Q.); 17306368595@163.com (F.J.); 18063274689@163.com (X.G.); 15253068797@163.com (T.Y.)

2 Agriculture & Forestry Technology College, Weifang Vocational College, Weifang 261061, China

3 Shandong Key Laboratory of Eco-Environmental Science for Yellow River Delta, Shandong University of Aeronautics, Binzhou 256600, China

* Correspondence: hui5232@163.com

Abstract: This study aimed to determine the effects of the nitrogen (N) application period and level on the fate of fertilizer N and the contribution of N absorption and translocation to apple organ N. Two N application periods (labeled by the ^{15}N tracer technique in spring and summer, represented by SP and SU, respectively) and three N levels (N0, MN, and HN) were used to determine the physiological indexes and aboveground, root, and soil ^{15}N content of 4-year-old dwarf ('Red Fuji'/M9T337) and arborized ('Red Fuji'/*Malus hupehensis* Rehd.) apple trees. The results showed that HN led to shoot overgrowth, which was not conducive to the growth of the apple root system (root length, root tips, root surface area, and root volume) or the improvement of root activity. The contribution of soil N to apple organ N accounted for more than 50%, and the contribution of N application in summer to fruit N was higher than that in spring. Under HN treatment, the proportion of soil N absorbed by trees decreased, while that of fertilizer N increased; however, the highest proportion was still less than 50%, so apple trees were highly dependent on soil N. Under MN treatment, fertilizer N residue was similar to soil N consumption, and soil N fertility maintained a basic balance. Under HN treatment, fertilizer N residue was significantly higher than soil N consumption, indicating that excessive N application increased fertilizer N residue in the soil. Overall, the ^{15}N utilization rate of arborized trees (17.33–22.38%) was higher than that of dwarf trees (12.89–16.91%). A total of 12.89–22.38% of fertilizer ^{15}N was absorbed by trees, 30.37–35.41% of fertilizer ^{15}N remained in the soil, and 44.65–54.46% of fertilizer ^{15}N was lost. The ^{15}N utilization rate and ^{15}N residual rate of summer N application were higher than those of spring N application, and the ^{15}N loss rate was lower than that of spring N application. High microbial biomass N (MBN) may be one of the reasons for the high N utilization rate and the low loss rate of N application in summer.

Keywords: apple; ^{15}N labeling; N fate; N use efficiency; microbial biomass N

1. Introduction

China holds the top position globally in both apple cultivation area and apple yield [1]. Nitrogen (N) fertilizer plays a decisive role in increasing yield in the barren orchard soil of China. To ensure large fruit and high yield, the amount of N fertilizer used in the main apple-producing areas in China continues to increase, which far exceeds the demand of the trees [2,3]. Excessive unused N accumulates in the soil, which can easily cause deep leaching and surface runoff loss [4,5].

After N fertilizer is applied to the orchard, it is absorbed by the trees through the roots system to meet the needs of growth and development; however, it exists in the form of inorganic N or in combination with organic N (such as microbial biomass N (MBN)) in the soil profile. Other N is lost to the environment from the plant–soil system through processes such as ammonia volatilization, nitrification/denitrification, and leaching [6]. The fate of fertilizer N is influenced by crop type, soil type, N application rate, fertilization method, and fertilization period, which vary greatly [7,8]. Ge et al. [9] showed that the soil residual rate and loss rate of fertilizer N in 2-year-old apple orchards were 19.38–31.28% and 21.50–29.13%, respectively.

For the fertilization period, due to the high activity of N transformation (including ammonia volatilization, nitrification/denitrification, and leaching), separate fertilization generally has a higher N use efficiency than one-time fertilization. Researchers have found that the utilization rate of topdressing N is significantly higher than that of base fertilizer, while the loss rate is lower than that of base fertilizer [8,10,11]. Most studies only focus on N use efficiency and do not consider the residue of fertilizer N in the soil—that is, the compensation effect on the soil N pool. In the period of N application in apples, trees are fertilized in spring and summer [12]. Furthermore, the two fertilizer application periods corresponded to the two peak periods of apple root growth. However, differences in the direction of N fertilizer application and the absorption of apple trees in different periods have rarely been reported.

The ^{15}N isotopic tracer technique has been used to monitor the budget of N fertilizer and is still widely used to evaluate fertilization management [13–15]. The ^{15}N tracer method can quantify the source of N absorbed by the plant—that is, the amount of N from fertilizer and soil. If the amount of N from the soil is greater than the residual fertilizer N, it means that the residual fertilizer N cannot maintain the balance of the soil N pool. In contrast, residual fertilizer N can maintain the balance of the soil N pool. Therefore, the ^{15}N tracer method can better reflect the compensation effect of fertilization on the soil N pool.

Therefore, the present study used ^{15}N isotope tracing technology to determine the effects of N supply levels and N application periods (spring and summer) on N absorption, utilization, residue, and loss in apples. This can further clarify the relationship between fertilizer N, soil N, and tree N, improve N management in apple production, improve N application efficiency, and reduce negative impacts on the environment, which is of great significance for improving N conservation and efficiency in apple production.

2. Results

2.1. Biomass

Different levels of N application had different effects on the biomass of apple organs (Figure 1 and Table S1). Overall, the total biomass of arborized trees was higher than that of dwarf trees. With the increase in the N application level, the root biomass first increased and then decreased. The root biomass of the HN treatment was lowest, at 13.98% ('Red Fuji'/M9T337) and 16.44% ('Red Fuji'/*M. hupehensis* Rehd.) lower than the N0 treatment ($p > 0.05$). With the increase in the N application level, the biomass of annual branch, leaf, and fruit increased gradually and was highest under the HN treatment. HN was not conducive to root growth and led to shoot overgrowth.

2.2. Annual Branch Length, Leaf Area, and Chlorophyll Content

Figure 2 and Table S2 show the effects of N application in spring and summer on annual branch length, leaf area, and chlorophyll content of 'Red Fuji'/M9T337 and 'Red Fuji'/*M. hupehensis* Rehd. under different N levels. Over time, the annual branch length, leaf area, and chlorophyll content of 'Red Fuji'/M9T337 and 'Red Fuji'/*M. hupehensis* Rehd. increased gradually. There was no significant difference in the new annual branch length in May under the N treatment, but it reached a significant level after July. In October, the growth rate of the new annual branch length of arborized trees was higher than that of dwarf trees. Compared with N0, MN and HN treatments increased the annual branch length and leaf area of 'Red Fuji'/M9T337 and 'Red Fuji'/*M. hupehensis* Rehd. to different degrees. In different periods, the chlorophyll content of the MN and HN treatments was higher than that of N0, although there was no significant difference in May. Over time, the chlorophyll content of 'Red Fuji'/M9T337 and 'Red Fuji'/*M. hupehensis* Rehd. showed an increase with the increase in N application level. HN had the most significant effect on annual branch length.

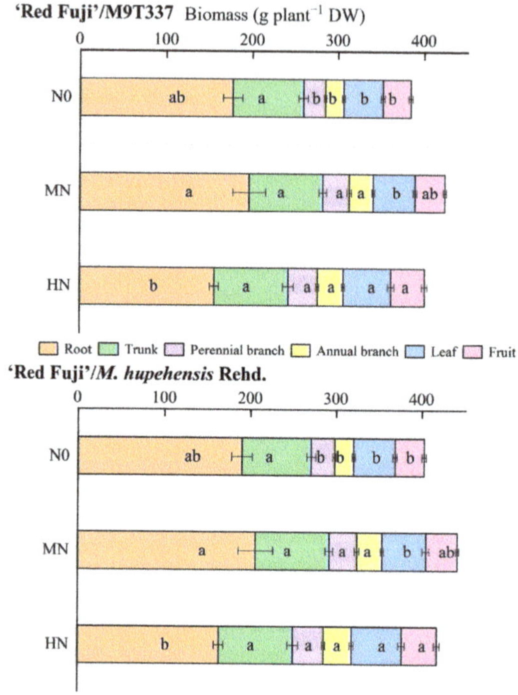

Figure 1. The biomass of apple organs. DW means dry weight. Error bar indicates the standard deviation of three replications. Different letters indicate statistically significant differences ($p < 0.05$).

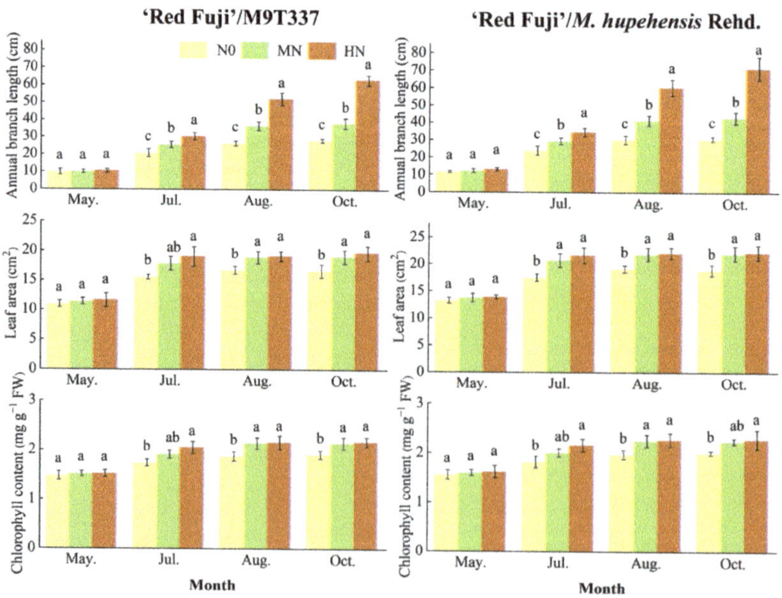

Figure 2. Annual branch length, leaf area, and chlorophyll content of apples. Error bar indicates the standard deviation of three replications. Different letters indicate statistically significant differences ($p < 0.05$).

2.3. Root Activity and Root Morphology

As shown in Figure 3 and Table S3, the effect of N application levels on apple root activity differed. Sampling in different periods showed that compared with N0, the root activity of 'Red Fuji'/M9T337 under MN was the highest, while HN decreased the root activity to different degrees, although the difference was not significant in April. Early April, early June, and early October correspond to the three peak periods of apple root growth (spring, summer, and autumn). The root activity of 'Red Fuji'/M9T337 in early June and early October was relatively high (47.73–65.20 μg h^{-1} g^{-1}), followed by early April (40.81–45.78 μg h^{-1} g^{-1}), while the root activity in May and July was relatively low (29.88–48.98 μg h^{-1} g^{-1}). 'Red Fuji'/*M. hupehensis* Rehd. followed the same trend.

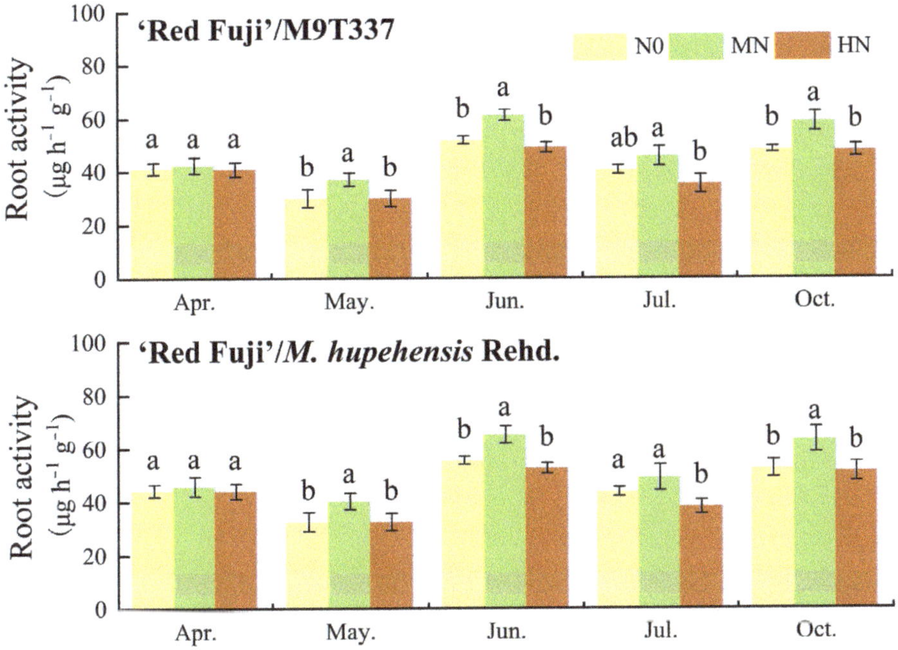

Figure 3. Root activity of apple under different N levels at different stages. Error bar indicates the standard deviation of three replications. Different letters indicate statistically significant differences ($p < 0.05$).

N level had a significant effect on root morphology at the mature stage (Figures 4 and S1, and Table S3). With the increase in N level, the root length, root tips, root surface area, and root volume of 'Red Fuji'/M9T337 increased first and then decreased, and were highest under the MN treatment. Compared with N0, the root length, root tip number, root surface area, and root volume of the MN treatment increased by 138.31, 95.95, 35.85, and 84.88%, respectively, and the differences all reached a significant level ($p < 0.05$). However, the root morphology of HN increased compared with N0 but decreased significantly compared with MN. 'Red Fuji'/*M. hupehensis* Rehd. showed the same trend. The results showed that the HN treatment decreased root activity and was not conducive to root growth.

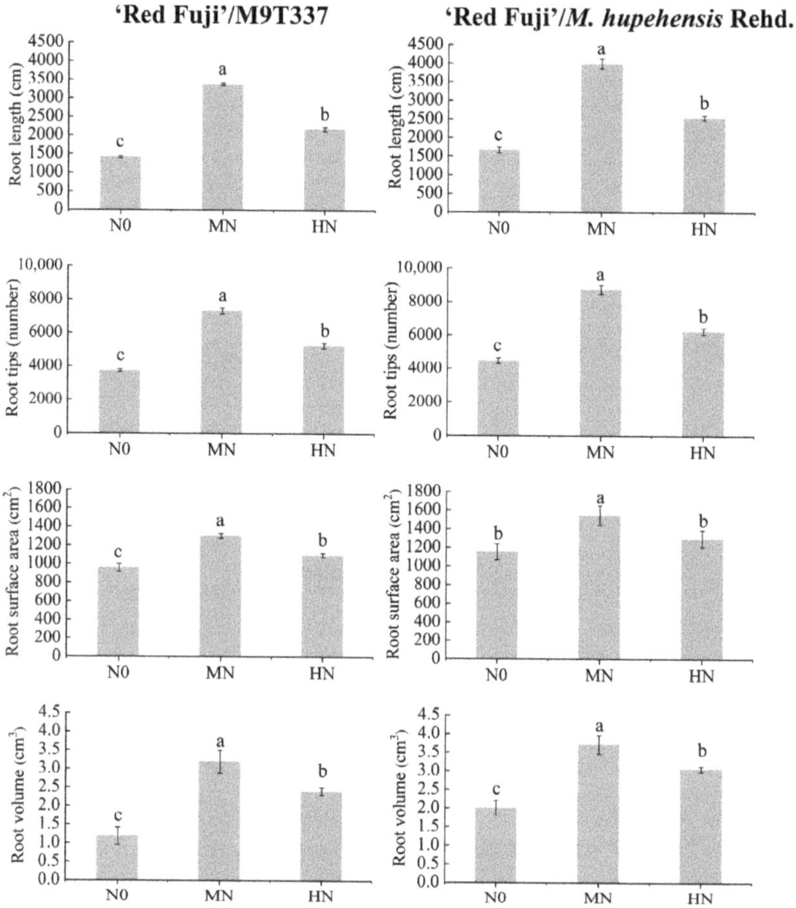

Figure 4. Root morphology of apples under different N levels during maturity stage. Error bar indicates the standard deviation of three replications. Different letters indicate statistically significant differences ($p < 0.05$).

2.4. Apple Organ N Derived from Fertilizer (%Ndff) and Soil (%Ndfs)

The study of 'Red Fuji'/M9T337 showed that the contribution of soil N to apple organ N (52.74–74.63%) was significantly higher than that of spring N application (10.43–22.39%) and summer N application (11.71–25.20%) (Figure 5 and Table S4). Under the MN treatment, fertilizer N contributed the most to fruit N (42.13%); spring N and summer N contributed 18.56% and 23.57%, respectively; fertilizer N contributed the least to trunk N (22.14%), and spring N and summer N contributed 10.43% and 11.71%, respectively. In contrast, soil N contributed the most to trunk N (77.86%) and the least to fruit N (57.87%). The HN treatment increased the contribution of spring and summer N applications and decreased the contribution of soil N to each organ. A comparison of different N application periods showed that a higher proportion of N applied in spring was transferred to annual branches (15.01–18.69%) and leaves (18.95–22.39%), while a higher proportion of N applied in summer was transferred to apple fruit (23.57–25.20%). 'Red Fuji'/*M. hupehensis* Rehd. showed the same trend. The contribution of soil N to apple organ N was more than 50%, and the contribution of summer N application to fruit N was higher than that of spring N application.

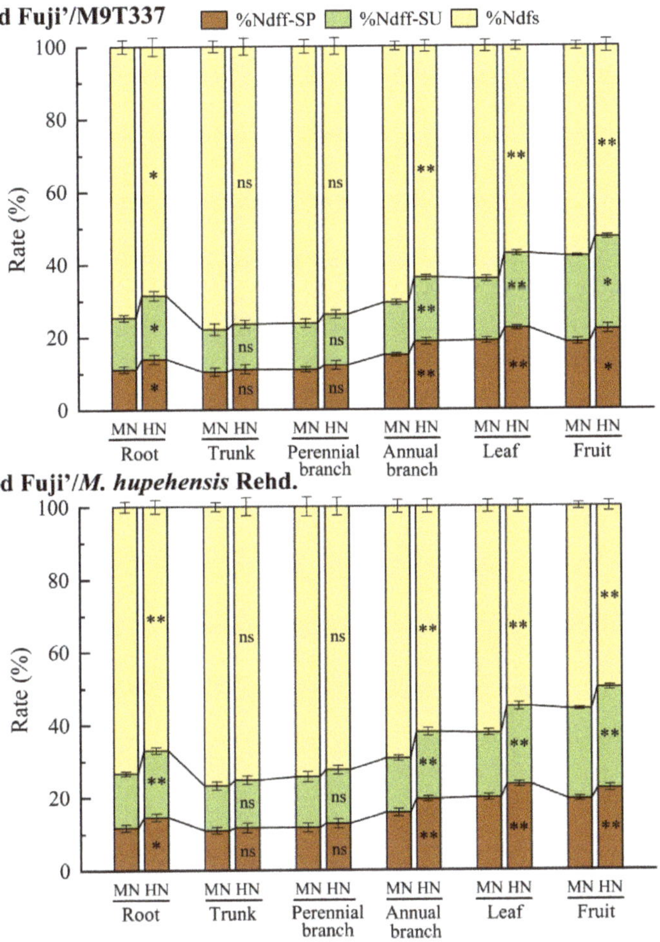

Figure 5. Contribution of spring N application (%Ndff-SP), summer N application (%Ndff-SU), and soil N (%Ndfs) to apple organ N. SP and SU represent N application in spring and summer, respectively. Error bar indicates the standard deviation of the three replications. * represents $p < 0.05$, ** represents $p < 0.01$, ns represents not significant.

2.5. ^{15}N Residue

Figure 6 shows the residual ^{15}N abundance of soil in the root layer (0–60 cm) at different stages after N fertilizer application in the spring and summer. 'Red Fuji'/M9T337 and 'Red Fuji'/*M. hupehensis* Rehd. showed the same trend. In May (one month after applying N fertilizer in spring), the ^{15}N abundance of the 0–40 cm soil layer was higher than that of the 40–60 cm soil layer; the abundance in each soil layer was higher under SP-HN treatment than under SP-MN. In July (one month after applying N fertilizer in summer), the ^{15}N abundance of spring N application treatment (SP-MN and SP-HN) was highest in the 20–40 cm soil layer, which was significantly higher than that of the 0–20 cm and 40–60 cm soil layers. The trends in August and October were similar. With the deepening of the soil layer, the abundance of ^{15}N increased gradually, and under the same amount of N application, the abundance of residual ^{15}N in each soil layer was higher than that in spring. With the extension of time, the N application in spring and summer showed a downward trend, and the residue of N fertilizer applied in summer was higher than that in spring at the mature stage.

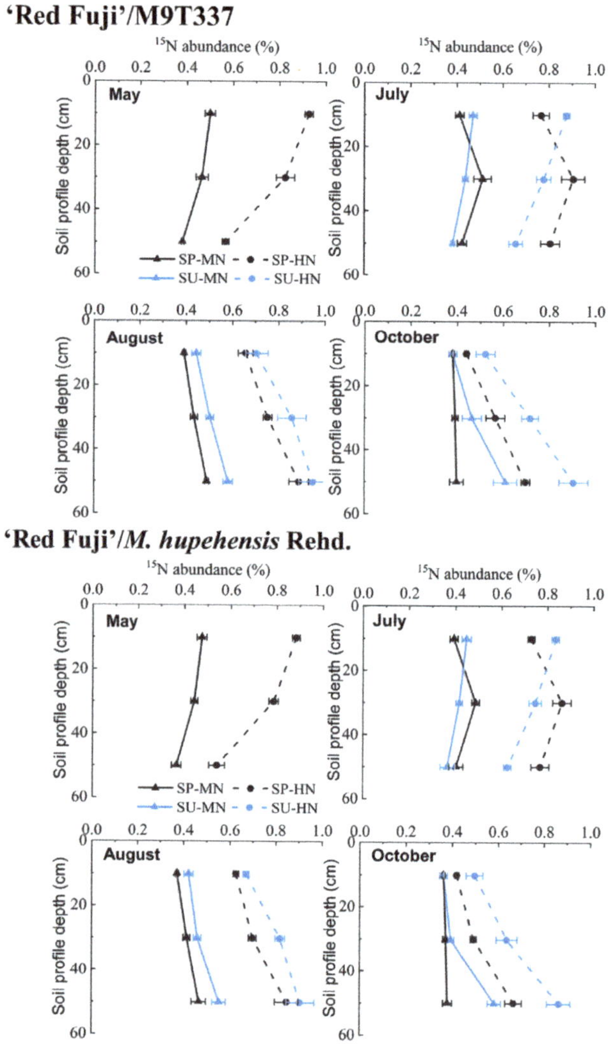

Figure 6. ^{15}N residue abundance in different soil layers under different treatments. SP and SU represent N application in spring and summer, respectively. Error bar indicates the standard deviation of the three replications.

2.6. Utilization, Residue, and Loss of ^{15}*N*

Figure 7 and Table S5 show the fate of fertilizer N from spring and summer N applications at different N levels. The study of 'Red Fuji'/M9T337 showed that the ^{15}N utilization rate of spring N application was lower than that of summer N application under MN and HN treatments. The total ^{15}N utilization rate of the HN treatment was significantly lower than that of the MN treatment (decreased by 31.26%, $p < 0.01$), and the ^{15}N utilization rate of spring N application and summer N application significantly decreased by 38.20 and 26.25%, respectively. Under the MN and HN treatments, the ^{15}N residual rate of spring N application was lower than that of summer N application, and the ^{15}N residual rate of the

HN treatment was significantly lower than that of the MN ($p < 0.05$). The residual rates of spring and summer N application decreased by 12.91 and 5.17%, respectively. Under MN and HN treatments, the ^{15}N loss rate of the spring N application was higher than that of the summer N application, and the ^{15}N loss rate of the HN treatment was significantly higher than that of the MN treatment (increased by 14.24%, $p < 0.01$). The ^{15}N loss rates of spring N application and summer N application increased by 14.33 and 14.13%, respectively. 'Red Fuji'/*M. hupehensis* Rehd. also showed the same trend, and the ^{15}N utilization rate (SP and SU) under different N levels was higher than that of 'Red Fuji'/M9T337, while the residual rate and loss rate were lower than that of 'Red Fuji'/M9T337. The ^{15}N utilization rate and soil residual rate were higher and the loss rate was lower in summer, and the HN treatment significantly decreased the fertilizer ^{15}N utilization rate and residual rate and increased the ^{15}N loss rate.

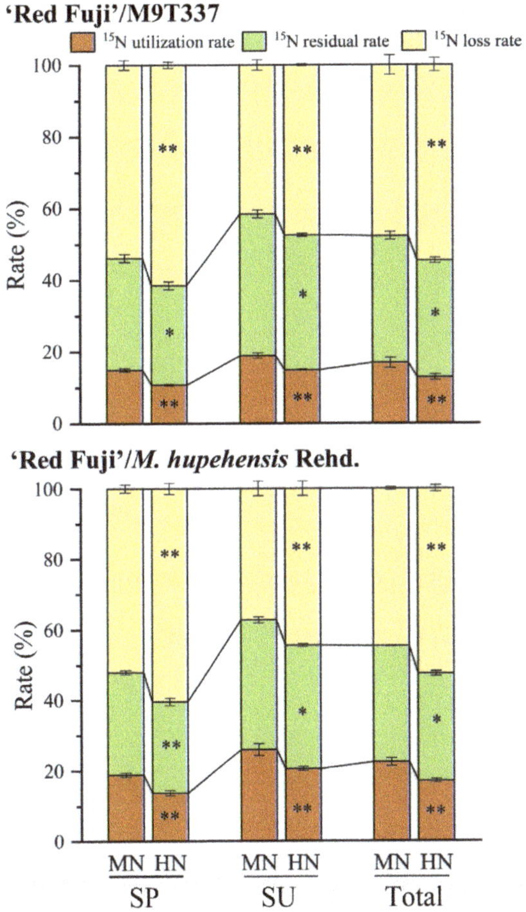

Figure 7. Utilization rate, residual rate, and loss rate of fertilizer ^{15}N under different N application methods. SP and SU represent ^{15}N application in spring and summer, respectively. Error bar indicates the standard deviation of the three replications. * represents $p < 0.05$, ** represents $p < 0.01$.

Figure 8 and Table S6 show the utilization, residue, and loss of fertilizer N under different N application levels and periods. Under MN and HN treatments, fertilizer N showed the following trend: loss > residual > absorption. The ^{15}N uptake and residual amount of N applied in summer were higher than those applied in spring, and the ^{15}N loss of N applied in spring was higher than that applied in summer. The HN treatment significantly increased the total absorption, total residue, and total loss of fertilizer N for 'Red Fuji'/*M. hupehensis* Rehd. Residual fertilizer N in the soil was an important supplement to the consumption of the soil N pool. Under the MN treatment, the residual fertilizer N was similar to soil N consumption (Table 1), and soil N fertility maintained a basic balance. Under the HN treatment, the residual fertilizer N was significantly higher than soil N consumption, indicating that excessive N application increased the fertilizer N residue in the soil.

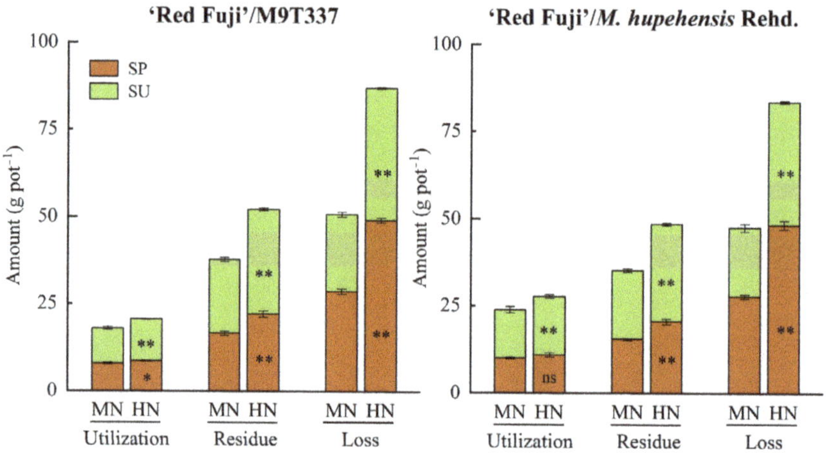

Figure 8. Utilization, residue, and loss of fertilizer N under different N application methods. SP and SU represent N application in spring and summer, respectively. Error bar indicates the standard deviation of the three replications. * represents $p < 0.05$, ** represents $p < 0.01$, ns represents not significant.

Table 1. The amount of N absorbed by apple trees from fertilizer sources and soil sources.

Types of Trees	Treatment	Total N Uptake by Trees (g Plant^{-1})	N Absorbed from Fertilizer (g Plant^{-1})			N Absorbed from Soil (g Plant^{-1})
			Absorption of Spring N Application	Absorption of Summer N Application	Total	
'Red Fuji'/M9T337	MN	55.03	7.97	10.07	18.04	36.99
	HN	64.15 ***	8.65 *	11.96 **	20.62 **	43.53 **
'Red Fuji'/*M. hupehensis* Rehd.	MN	59.27	10.06	13.81	23.87	35.40
	HN	70.56 **	10.99 *	16.73 **	27.72 **	42.84 *

Note: * represents $p < 0.05$, ** represents $p < 0.01$, *** represents $p < 0.001$.

2.7. MBN

Figure 9 and Table S7 show the MBN-SP and MBN-SU contents in the 0–60 cm soil layer under different treatments. Overall, the MBN content at 20–40 cm (135.15–203.40 mg pot^{-1}) was higher than that at 0–20 cm (106.66–166.99 mg pot^{-1}) and 40–60 cm (88.35–140.93 mg pot^{-1}). The MBN content of HN in each soil layer was higher than that of MN, while the difference was not significant ($p > 0.05$). The MBN-SU content in each soil layer was significantly higher than that of MBN-SP ($p < 0.01$). Compared with MBN-SP, MBN-SU significantly increased by 35.46–38.74% (0–20 cm), 33.17–42.22% (20–40 cm), and 37.24–47.30% (40–60 cm), respectively.

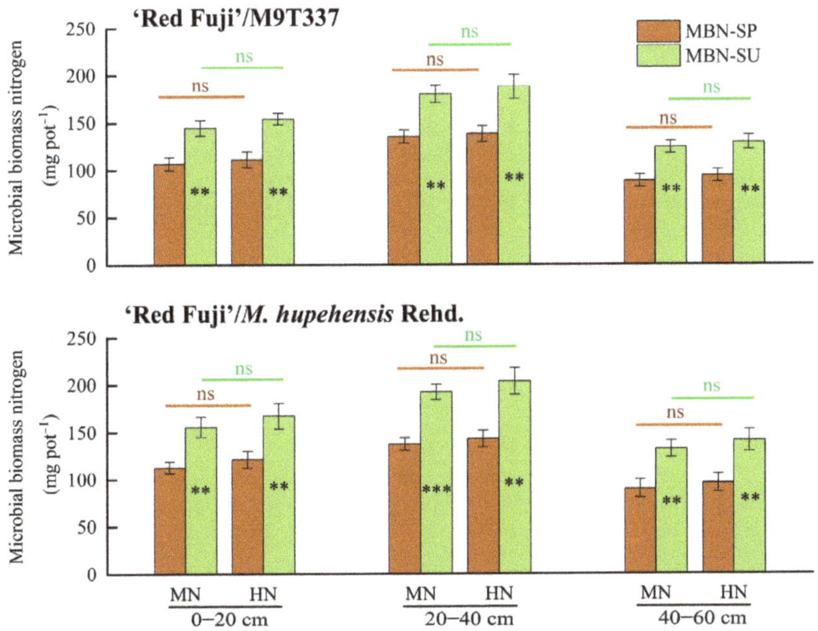

Figure 9. MBN-SP and MBN-SU contents in the 0–60 cm soil layer under different treatments. SP and SU represent N application in spring and summer, respectively. Error bar indicates the standard deviation of three replications. ** represents $p < 0.01$, *** represents $p < 0.001$, ns represents not significant.

3. Discussion

Higher root activity values indicate stronger root metabolic activities and nutrient absorption capacities [16]. Our results showed that root activity was inhibited under N0 and HN treatments, while the inhibitory effect was more obvious under HN than under N0 conditions. This is consistent with the study of Chen et al. [17] in pear seedlings.

The N absorbed by plants mainly comes from soil and fertilizer N. Stevens et al. [18] found that Ndfs accounted for 54–83% of the total N assimilated by maize, with the percentage decreasing with the long-term N application rate increasing, and explained that the increase in fertilization history reduced the absolute N uptake from soil sources. Yang et al. [11] showed that the N absorbed by summer maize mainly came from the soil, and the proportion of fertilizer N was less than 44%. In the present study, the contribution of soil N to apple organ N accounted for more than half. A comparison of different N application periods showed that a higher proportion of N applied in spring was transferred to annual branches and leaves, while a higher proportion of N applied in summer was transferred to apple fruit. Under the high N treatment, the proportion of soil N absorbed by trees decreased, while that of fertilizer N increased, but the highest proportion was still less than 50%, which shows that apple trees are highly dependent on soil N.

After applying N fertilizer into the soil, its utilization rate is 30–50%, and this varies greatly with different fertilization methods, soil properties, and management methods [19,20]. The timing and level of N application affect the N fate. The present study found that the N fertilizer utilization rate of spring N application ranged from 10.81 to 14.95%, while that of summer N application ranged from 14.96 to 18.88%. HN treatment significantly reduced the utilization and residual rate of fertilizer N and increased the N loss rate of fertilizer, consistent with Wang et al. [3]. This study also found that the ^{15}N utilization rate of N application in summer was higher than that in spring due to strong root growth and more new roots in spring, but the growth period of new roots is relatively short. In summer, due to high temperatures, high humidity, a large leaf area in fruit trees, and high root vitality, root utilization efficiency after N application is higher. In addition, this study found that the N fertilizer utilization rate of arborized trees was higher than that of dwarf trees because the root morphology indicators (root length, root tips, root surface area, and root volume) and root activity of arborized trees were higher than those of dwarf trees. Thus, the root system of arborized trees has a higher absorption and utilization rate of fertilizer N. Webster [21] concluded that the decrease in absorption capacity was related to the smaller root systems in apple dwarf rootstocks and to the graft union, which had very convoluted xylem vessels that act as filters, affecting the balance of different solutes reaching its scion.

The portion of N not absorbed by the plants remains in the soil. Zhu [22] summarized that the residual rate of fertilizer N during crop harvests in China is generally 15–30%. The N residue rate of fertilizer in this experiment ranged from 25.78 to 39.56%. With the increase in the N application rate, the N residual amount increased, while the N residual rate decreased. This is consistent with Wang et al. [3]. The residual fertilizer N in the soil is an important supplement to the soil N pool. In this study, the ^{15}N tracer method better reflected the compensation effect of N application in the soil N pool. Under MN treatment, fertilizer N residue was similar to soil N consumption, and soil N fertility maintained a basic balance. Under HN treatment, fertilizer N residue was significantly higher than soil N consumption, indicating that excessive N application increased fertilizer N residue in the soil. The abundance of ^{15}N in the soil reflected the residual fertilizer N in each soil layer. The results showed that with the extension of time, both spring and summer N applications showed a downward trend of migration. The ^{15}N residual abundance of summer N application in each soil layer was higher than that of spring N application, and the ^{15}N residual abundance of high N treatment was higher than that of medium N treatment, indicating that excessive N application could aggravate the accumulation of fertilizer N in the soil.

MBN is closely related to soil N supply, which can be used as both an N source and N sink to regulate soil N supply through microbial immobilization and mineralization. Zhao et al. [23] showed that MBN was positively correlated with wheat biomass and N uptake, which could be used as an index of soil N supplying capacity. At the same time, microbial fixation of inorganic N was also an effective way to reduce N loss. Our research found that MBN-SP and MBN-SU were significantly correlated with N uptake by trees in SP and SU, respectively ($p < 0.01$), and significantly negatively correlated with N loss in SP and SU ($p < 0.01$). Therefore, high MBN-SU may be one of the reasons for high N use efficiency and low loss rate in SU. In this study, it was found that MBN content did not increase significantly under HN (Figure 9). The reason is that the application of high N decreases the C/N in soil, which is not conducive to the fixation of N by microorganisms. In addition, the limited carrying capacity of soil microorganisms under high N does not significantly increase MBN [24].

Under the conditions of continuous precipitation or massive irrigation, excessive nitrate N accumulates in the soil, which can be easily leached from the root zone to the deep soil layer, threatening the safety of shallow groundwater [25]. Other studies have shown that residual N in the soil is still available to future crops [11,26]. Therefore, in the late management of apple orchards, the water and fertilizer status should be reasonably regulated. The amount of N fertilizer should be reasonably controlled and the residual N resources in the root layer soil should be excavated to exert the aftereffect of residual N. However, it is necessary to reasonably regulate the water content of the root layer to prevent nitrate leaching [27].

The input of N fertilizer supplements the soil N pool, which is beneficial to improve soil fertility, but there are some problems, such as waste of resources and environmental pollution. Studies of winter wheat and summer maize in the North China Plain by Jia et al. [28] and Yang et al. [11] pointed out that when the amount of N fertilizer exceeded the crop demand, N loss increased sharply. Wang et al. [3] showed that the loss amount and rate of

N fertilizer increased with an increase in the N application level. The same conclusion was drawn in the present study, and the trend was as follows: loss amount of N fertilizer > soil residual N > N absorbed by trees under medium and high N levels. Therefore, in fruit tree production, N fertilizer application should not only ensure the growth needs of trees and achieve efficient use of fertilizer N but also maintain the balance of the soil N pools and reduce N loss. This study showed that in the medium N treatment, the N utilization rate was higher and the N loss rate was lower. Under high N treatment, excessive N led to tree overgrowth, lower N utilization rate, and higher N loss rate, causing lower agricultural production efficiency and increasing the risk of environmental pollution.

4. Materials and Methods

4.1. Experimental Site and Materials

This experiment was performed from March 2022 to October 2022 at the Weifang University Modern Horticulture Research Institute located in Weifang, Shandong Province, China. The climate is semi-humid. The tested soil was collected from the surface layer of the apple orchard (0–20 cm) at the experimental station. Table S8 shows the average results of air-dried soil properties.

The test materials were 4-year-old apple trees. The variety was commercially important apple cultivar 'Red Fuji', which was grafted to dwarf rootstock M9T337 ('Red Fuji'/M9T337) and arborized rootstock Malus hupehensis Rehd. ('Red Fuji'/*M. hupehensis* Rehd.), respectively.

4.2. Experimental Design and Sampling

In an open space, pot experiments were carried out in pots with an inner diameter of 50 cm and height of 60 cm. The soil bulk density at the depth of 0−20, 20−40, and 40−60 cm was 1.09, 1.21, and 1.32 g cm^{-3}, respectively, and the wet soil weight per pot was 142 kg. An apple tree was planted in each pot. The experiment was a two-factor design with 3 replicates. The application of N fertilizer in different periods was factor A, and there were two N fertilization periods: spring fertilization (SP) and summer fertilization (SU). The amount of N application was factor B, with no N application (N0), medium N application (MN, 0.21 g N kg^{-1} soil), and high N application (HN, 0.42 g N kg^{-1} soil) [29]. Of the total amount of N applied, 50% was applied in spring (March 20), and the remaining 50% was applied in summer (1 June). Each plant was treated with 0 (N0), 65 g (MN), or 130 g (HN) of urea. A total of 6.5 (MN)/13 (HN) g of normal urea applied per plant was replaced in each treatment with 6.5/13 g of ^{15}N-urea ($CO(^{15}NH_2)_2$, 10.22% abundance, Shanghai Research Institute of Chemical Industry) in spring or summer, namely ^{15}N-SP+^{14}N-SU and ^{14}NH-SP+^{15}N-SU, respectively. Calcium superphosphate (0.11 g kg^{-1} P_2O_5) and potassium sulfate (0.22 g kg^{-1} K_2O) was applied as the base fertilizer to each plant. The fertilizers used and their composition percentages are shown in Table S9. The method of fertilization included digging a circular trench (radius of 30 cm, width and depth of 10 cm) around each tree. The field management measures such as diseases and insect pests in each growth period were consistent with the local conventional model. The soil moisture content was measured regularly using a soil moisture meter (Handi-TRASE), and when the soil moisture content was below the lower limit (60%), drip irrigation was performed.

Twenty-four trees from two rootstocks were treated with each N level in each N application period. Among these, three trees of each rootstock were sampled and analyzed before treatment, and the N content of the plant was used as the basic value to calculate the N uptake of the trees. Annual branch length, leaf area, chlorophyll content, and soil ^{15}N residues were measured in May, July, August, and October, respectively, and root activity was measured in April, May, June, July, and October (Figure 10).

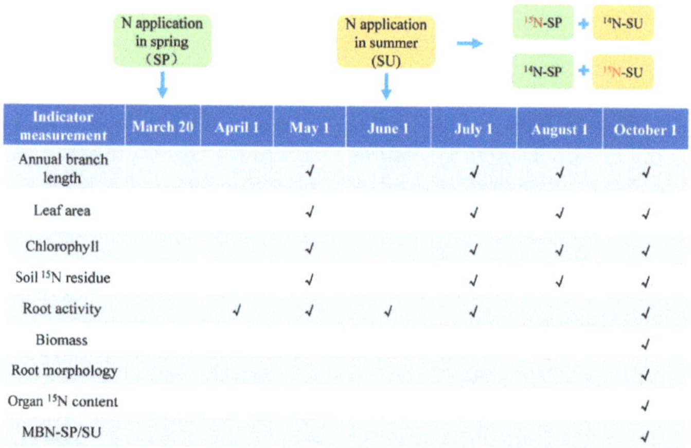

Figure 10. Processing and sampling time.

Soil collection was carried out at 6 sampling sites where each pot was uniformly distributed. In the vertical direction of each sampling site, soil samples at depths of 0–20, 20–40, and 40–60 cm were collected, and 6 soil samples in each layer were evenly mixed into one replicate. After collection, the soil samples were transferred to the laboratory for ^{15}N abundance determination. On October 1st, all plants were destructively sampled to determine the N content, ^{15}N abundance, MBN, and root morphology.

4.3. Annual Branch Length, Leaf Area, and Chlorophyll Content

Ten annual branches and leaves were randomly selected from each tree to determine the annual branch length, leaf area, and chlorophyll content. The average of the 10 measured values for each tree was considered as one replicate, with a total of 3 replicates. The annual branch length was measured using a meter ruler. The leaf area was measured using a leaf area meter (YMJ-B, Zhejiang Top Yunong Technology Co., Ltd., Hangzhou, China). The chlorophyll concentration (total chlorophyll, chlorophyll a, and chlorophyll b) was calculated according to Li et al. [30].

4.4. Root Activity and Root Morphology

Root activity was measured using the triphenyltetrazolium chloride (TTC) method, and the reduction strength of tetrazolium (μg h^{-1} g^{-1} FW) was used to express root activity [31]. WinRhizo software (WinRHIZO Pro 2017a, Regent Instruments Canada) was used to analyze root morphological index [32].

4.5. ^{15}N Abundance

Soil ^{15}N-MBN was determined by chloroform fumigation extraction as described by Zhang et al. [13]. MBN was calculated as the difference in N content between fumigated and unfumigated samples divided by a conversion coefficient of 0.45 [13,33]. The ^{15}N abundance of fumigated and unfumigated solution was determined as for inorganic ^{15}N by DELTAV advantage isotope ratio mass spectrometer (Thermo Fisher Scientific Inc., Waltham, MA, USA) according to Zhang et al. [13].

Apple plants were divided into fruits, leaves, annual branches, perennial branches, trunk, and roots. The samples were heated at 105 °C for 30 min and dried at 80 °C, followed by homogenization and filtration with a 0.25 mm mesh screen. The N content was determined using the Kjeldahl method [34]. The ^{15}N abundance of the plants and soil was determined using a DELTAV advantage isotope ratio mass spectrometer [6].

The calculation formulas are as follows:

$$\%\text{Ndff} = \frac{\text{amount of fertilizer }^{15}\text{N absorbed by trees (g)}}{\text{amount of N absorbed by trees (g)}} \times 100\% \quad (1)$$

where %Ndff represents the percentage of N absorbed by tree from fertilizer.

$$\%\text{Ndfs} = 100\% - \%\text{Ndff} \quad (2)$$

where %Ndfs represents the percentage of N absorbed by tree from soil.

$$\text{Ndff }(\%) = \frac{\text{abundance of }^{15}\text{N in plant natural abundance of }^{15}\text{N}}{\text{abundance of }^{15}\text{N in fertilizer-natural abundance of }^{15}\text{N}} \times 100\% \quad (3)$$

$$^{15}\text{N utilization rate }(\%) = \frac{\text{Ndff} \times \text{totalNoforgans (g)}}{^{15}\text{N fertilization (g)}} \times 100\% \quad (4)$$

$$^{15}\text{N residual rate }(\%) = \frac{^{15}\text{Nresidue in soil (g)}}{^{15}\text{Nfertilization (g)}} \times 100\% \quad (5)$$

$$^{15}\text{N loss rate }(\%) = 100\% - {}^{15}\text{N utilization rate }(\%) - {}^{15}\text{N residual rate }(\%) \quad (6)$$

4.6. Statistical Analysis

All graphs were plotted using Origin 8.0 (Northhampton, MA, USA). Data were analyzed with SPSS 20.0 (IBM Corporation, Armonk, NY, USA). The significance of differences between two and three treatments was analyzed using a *t*-test (Student's *t*-test) and one-way factorial analysis of variance (ANOVA), respectively.

5. Conclusions

i: Soil N contributed more than half of the N of various apple organs, and N application in summer contributed more to fruit N than N application in spring.

ii: Under MN treatment, the fertilizer N residue was similar to soil N consumption, and soil N fertility maintained a basic balance. Under HN treatment, the fertilizer N residue was significantly higher than the soil N consumption, indicating that excessive N application intensifies the fertilizer N residue in the soil.

iii: The ^{15}N utilization rate and ^{15}N residual rate of summer N application were higher than those of spring N application, and the ^{15}N loss rate was lower than that of spring N application. High MBN may be one of the reasons for high N utilization rate and low loss rate of N application in summer. The loss rate of N fertilizer was high (44.65–54.46%), and measures such as appropriately increasing the proportion of N application in summer, slow-release fertilizers, and water–fertilizer integration should be taken to reduce fertilizer N loss.

Supplementary Materials: The following supporting information can be downloaded at https://www.mdpi.com/article/10.3390/plants13060813/s1, Figure S1: photos of dwarf rootstock M9T337 ('Red Fuji'/M9T337) and arborized rootstock *Malus hupehensis* Rehd. ('Red Fuji'/*M. hupehensis* Rehd.); Table S1: ANOVA analysis for the effects of studied factors (N rate and rootstock) on the biomass of apple organs. Significant ($p < 0.05$) values are highlighted using bold type; Table S2: ANOVA analysis for the effects of studied factors (N rate and rootstock) on annual branch length, leaf area, and chlorophyll content of apples. Significant ($p < 0.05$) values are highlighted using bold type; Table S3: ANOVA analysis for the effects of studied factors (N rate, DMPP application, and year) on root activity and root morphology. Significant ($p < 0.05$) values are highlighted using bold type; Table S4: ANOVA analysis for the effects of studied factors (N rate and rootstock) on %Ndff-SP, %Ndff-SU, and %Ndfs. SP and SU represent ^{15}N application in spring and summer, respectively. Significant ($p < 0.05$) values are highlighted using bold type; Table S5: ANOVA analysis for the effects of studied factors (N rate and rootstock) on utilization rate, residual rate, and loss rate of fertilizer ^{15}N. SP and SU represent ^{15}N application in spring and summer, respectively. Significant ($p < 0.05$) values are highlighted using bold type; Table S6: ANOVA analysis for the effects of studied factors (N rate and rootstock) on utilization, residue, and loss of fertilizer N. SP and SU represent N application in spring and summer, respectively. Significant ($p < 0.05$) values are highlighted using bold type; Table S7: ANOVA analysis for the effects of studied factors (N rate and rootstock) on MBN-SP and MBN-SU. SP and SU represent ^{15}N application in spring and summer, respectively. Significant ($p < 0.05$) values are highlighted using bold type; Table S8: chemical properties of experimental soil; Table S9: nutrient content of chemical fertilizer.

Author Contributions: Conceptualization, F.W. and H.C.; methodology, C.W.; software, B.Y.; validation, G.Q., F.J., and X.G.; formal analysis, X.L. and T.Y.; investigation, X.Z.; resources, M.L.; data curation, Q.J.; writing—original draft preparation, F.W.; writing—review and editing, F.W. and H.C.; visualization, L.P.; supervision, F.W. and C.W. All authors have read and agreed to the published version of the manuscript.

Funding: This research was supported by the National Natural Science Foundation of China (32072518 and 32302471), the Natural Science Foundation of Shandong Province (ZR2023QC020 and ZR2020QC168), and Weifang University Doctor Initiation Fund Project (2022BS15).

Data Availability Statement: All data are included in the main text and Supplementary Materials.

Acknowledgments: The authors would like to thank Zengwen Liang, Guobin Xin, and Youzhong Liang for their help in apple cultivation. We thank LetPub (www.letpub.com (accessed on 1 October 2023)) for its linguistic assistance during the preparation of this manuscript.

Conflicts of Interest: The authors declare no conflicts of interest.

References

1. FAOSTAT. Statistical Database of the Food and Agricultural Organization of the United Nations. 2022. Available online: https://www.faostat.fao.org (accessed on 10 December 2023).
2. Zhao, Z.P.; Yan, S.; Liu, F.; Wang, X.Y.; Tong, Y.A. Analysis of nitrogen inputs and soil nitrogen loading in different kinds of orchards in Shaanxi Province. *Acta Oncol.* **2014**, *34*, 5642–5649. [CrossRef]
3. Wang, F.; Ge, S.; Lyu, M.; Liu, J.; Li, M.; Jiang, Y.; Xu, X.; Xing, Y.; Cao, H.; Zhu, Z.; et al. DMPP reduces nitrogen fertilizer application rate, improves fruit quality, and reduces environmental cost of intensive apple production in China. *Sci. Total Environ.* **2022**, *802*, 149813. [CrossRef]
4. Raese, J.T.; Drake, S.R.; Curry, E.A. Nitrogen fertilizer influences fruit quality, soil nutrients and cover crops, leaf color and nitrogen content, biennial bearing and cold hardiness of 'golden delicious'. *J. Plant Nutr.* **2007**, *30*, 1585–1604. [CrossRef]
5. Ju, X. Direct pathway of nitrate produced from surplus nitrogen inputs to the hydrosphere. *Proc. Natl. Acad. Sci. USA* **2014**, *111*, E416. [CrossRef]
6. Vinzent, B.; Fuß, R.; Maidl, F.X.; Hülsbergen, K.J. N_2O emissions and nitrogen dynamics of winter rapeseed fertilized with different N forms and a nitrification inhibitor. *Agric. Ecosyst. Environ.* **2018**, *259*, 86–97. [CrossRef]
7. Ting, L.; Huang, Y.; Song, X.; Deng, O.; Zhou, W.; Luo, L.; Tang, X. Biological nitrification inhibitor co-application with urease inhibitor or biochar yield different synergistic interaction effects on NH_3 volatilization, N leaching, and N use efficiency in a calcareous soil under rice cropping. *Environ. Pollut.* **2022**, *293*, 118499. [CrossRef]
8. López-Bellido, L.; López-Bellido, R.J.; López-Bellido, F.J. Fertilizer nitrogen efficiency in durum wheat under rainfed mediterranean conditions: Effect of split application. *Agron. J.* **2006**, *98*, 55–62. [CrossRef]
9. Ge, S.; Jiang, Y.; Wei, S.; Fang, J. Nitrogen balance under different nitrogen application rates in young apple orchards. *Plant Nutr. Fert. Sci.* **2011**, *17*, 949–955. [CrossRef]
10. Ju, X.; Liu, X.; Pan, J.; Zhang, F. Fate of ^{15}N-labeled urea under a winter wheat-summer maize rotation on the North China Plain. *Pedosphere* **2007**, *17*, 52–61. [CrossRef]
11. Yang, Y.; Wang, X.; Dai, K.; Jia, S.; Meng, C.; Zhao, Q.; Zhang, X.; Zhang, D.; Feng, Z.; Sun, Y.; et al. Fate of labeled urea-^{15}N as basal and topdressing applications in an irrigated wheat–maize rotation system in North China Plain: II summer maize. *Nutr. Cycl. Agroecosyst.* **2011**, *90*, 379–389. [CrossRef]
12. Zhang, D.; Li, Q.; Chu, B.; Zou, Y. Effects of nitrogen application levels on apple growth, yield and quality in different periods. *Northern Horti.* **2022**, *17*, 42–50.
13. Zhang, J.; Zhang, L.; Qiu, S. Biochar amendment benefits ^{15}N fertilizer retention and rhizosphere N enrichment in a maize-soil system. *Geoderma* **2022**, *412*, 115713. [CrossRef]
14. Jiang, Z.; Cui, H.; Lin, J.; Mo, C.; Liu, Y.; Li, Y.; Yang, J. Incorporation of spring maize into paddy fields increased nitrogen uptake in late rice by enhancing soil nitrogen supply: Field ^{15}N tracer studies. *Field Crop. Res.* **2023**, *299*, 108990. [CrossRef]
15. Li, X.; Wang, Y.; Feng, G.; Xu, Z.; Meng, F.; Gao, Q. Differential fertilizer nitrogen fates in maize cropping system among three soil textures based on ^{15}N. *Field Crop. Res.* **2023**, *291*, 108780. [CrossRef]
16. Halusková, L.; Valentovicová, K.; Huttová, J.; Mistrík, I.; Tamás, L. Effect of abiotic stresses on glutathione peroxidase and glutathione S-transferase activity in barley root tips. *Plant Physiol. Biochem.* **2009**, *47*, 1069–1074. [CrossRef]

17. Chen, G.; Wang, L.; Fabricel, M.R.; Tian, Y.; Qi, K.; Chen, Q. Physiological and nutritional responses of pear seedlings to nitrate concentrations. *Front. Plant Sci.* **2018**, *9*, 1679. [CrossRef] [PubMed]
18. Stevens, W.B.; Hoeft, R.G.; Mulvaney, R.L. Fate of nitrogen-15 in a long-term nitrogen rate study: I. interactions with soil nitrogen. *Agron. J.* **2005**, *97*, 1037–1045. [CrossRef]
19. Ladha, J.K.; Pathak, H.; Krupnik, T.J.; Six, J.; Kessel, C.V. Efficiency of fertilizer nitrogen in cereal production: Retrospects and prospects. *Adv. Agron.* **2005**, *87*, 85–156. [CrossRef]
20. Sun, C.X.; Hao, L.; Wang, D.; Li, C.; Zhang, C.; Chen, X.; Fu, J.; Zhang, Y.L. Nitrogen utilisation and metabolism in maize (*Zea mays* L.) plants under different rates of biochar addition and nitrogen input conditions. *Plant Biol.* **2019**, *21*, 882–890. [CrossRef]
21. Webster, A.D. Vigour mechanisms in dwarfing rootstocks for temperate fruit trees. *Acta Hortic.* **2004**, *658*, 29–41. [CrossRef]
22. Zhu, Z.L. Nitrogen management and food production and environment. *Acta Pedol. Sin.* **2022**, *39*, 3–11.
23. Zhao, J.; Yu, Z.; Li, Y.; Wang, X. Effects of nitrogen application rate on soil inorganic nitrogen distribution microbial biomass nitrogen content and yield of wheat. *Plant Nutr. Fertil. Sci.* **2006**, *12*, 466–472.
24. Wang, C.; Lu, X.; Mori, T.; Mao, Q.; Zhou, K.; Zhou, G.; Nie, Y.; Mo, J. Responses of soil microbial community to continuous experimental nitrogen additions for 13 years in a nitrogen-rich tropical forest. *Soil Biol. Biochem.* **2018**, *121*, 103–112. [CrossRef]
25. Mo, X.; Peng, H.; Xin, J.; Wang, S. Analysis of urea nitrogen leaching under high-intensity rainfall using HYDRUS-1D. *J. Environ. Manag.* **2022**, *312*, 114900. [CrossRef]
26. Wang, X.; Wang, Z.; Li, H.; Wang, R.; Li, S. Dynamics and availability to crops of residual fertilizer nitrogen in upland soil. *Acta Pedol. Sin.* **2016**, *53*, 1202–1212. [CrossRef]
27. Cheng, Q.; Ma, J.; Ren, R.; Zheng, L.; Guo, X.; Sun, X. Effects of fertilization management under wspi on soil nitrogen distribution and nitrogen absorption in apple orchard in loess plateau. *Agronomy* **2020**, *10*, 1386. [CrossRef]
28. Jia, S.; Wang, X.; Yang, Y.; Dai, K.; Meng, C.; Zhao, Q.; Zhang, X.; Zhang, D.; Feng, Z.; Sun, Y.; et al. Fate of labeled urea-^{15}N as basal and topdressing applications in an irrigated wheat–maize rotation system in North China Plain: I winter wheat. *Nutr. Cycl. Agroecosyst.* **2011**, *90*, 331–346. [CrossRef]
29. Wang, F.; Ge, S.; Xu, X.; Xing, Y.; Du, X.; Zhang, X.; Lv, M.; Liu, J.; Zhu, Z.; Jiang, Y. Multiomics analysis reveals new insights into the apple fruit quality decline under high nitrogen conditions. *J. Agric. Food Chem.* **2021**, *69*, 5559–5572. [CrossRef] [PubMed]
30. Li, J.; Liu, J.; Wang, G.; Cha, J.Y.; Li, G.; Chen, S.; Li, Z.; Guo, J.; Zhang, C.; Yang, Y.; et al. A chaperone function of NO CATALASE ACTIVITY1 is required to maintain catalase activity and for multiple stress responses in Arabidopsis. *Plant Cell* **2015**, *27*, 908–925. [CrossRef] [PubMed]
31. Liang, H.; Liu, Q.; Gu, J.; Gong, J.; Guan, C.; Lepo, J.E.; Rong, X.; Song, H.; Zhang, Z. V-ATPase and V-PPase at the tonoplast affect NO_3^- content in *Brassica Napus* by controlling distribution of NO_3^- between the cytoplasm and vacuole. *J. Plant Growth Regul.* **2014**, *34*, 22–34. [CrossRef]
32. Liu, C.; Zhou, G.; Qin, H.; Guan, Y.; Wang, T.; Ni, W.; Xie, H.; Xing, Y.; Tian, G.; Lyu, M.; et al. Metabolomics combined with physiology and transcriptomics reveal key metabolic pathway responses in apple plants exposure to different selenium concentrations. *J. Hazard. Mater.* **2024**, *464*, 132953. [CrossRef] [PubMed]
33. Tuo, Y.; Wang, Z.; Zheng, Y. Effect of water and fertilizer regulation on the soil microbial biomass carbon and nitrogen, enzyme activity, and saponin content of *Panax notoginseng*. *Agric. Water Manag.* **2023**, *278*, 108145. [CrossRef]
34. Liu, L.; Xiao, W.; Li, L. Effect of exogenously applied molybdenum on its absorption and nitrate metabolism in strawberry seedlings. *Plant Physiol. Biochem.* **2017**, *115*, 200–211. [CrossRef] [PubMed]

plants

MDPI

Article

N_2 Fixation, N Transfer, and Land Equivalent Ratio (LER) in Grain Legume–Wheat Intercropping: Impact of N Supply and Plant Density

Sebastian Salinas-Roco [1], Amanda Morales-González [1], Soledad Espinoza [2], Ricardo Pérez-Díaz [3], Basilio Carrasco [3], Alejandro del Pozo [4] and Ricardo A. Cabeza [1,*]

[1] Laboratory of Plant Nutrition, Department of Crop Sciences, Faculty of Agricultural Sciences, University of Talca, Talca 3460000, Chile; sebastian.salinas@utalca.cl (S.S.-R.); amanda.morales@utalca.cl (A.M.-G.)

[2] Centro Regional de Investigación Quilamapu, Instituto de Investigaciones Agropecuarias, Chillán 3780000, Chile; soledad.espinoza@inia.cl

[3] Centro de Estudios en Alimentos Procesados (CEAP), Talca 3480094, Chile; rperez@ceap.cl (R.P.-D.); bcarrasco@ceap.cl (B.C.)

[4] Plant Phenomics Center, Faculty of Agricultural Sciences, University of Talca, Talca 3460000, Chile; adelpozo@utalca.cl

* Correspondence: rcabeza@utalca.cl

Abstract: Intercropping legumes with cereals can lead to increased overall yield and optimize the utilization of resources such as water and nutrients, thus enhancing agricultural efficiency. Legumes possess the unique ability to acquire nitrogen (N) through both N_2 fixation and from the available N in the soil. However, soil N can diminish the N_2 fixation capacity of legumes. It is postulated that in intercropping, legumes uptake N mainly through N_2 fixation, leaving more soil N available for cereals. The latter, in turn, has larger root systems, allowing it to explore greater soil volume and absorb more N, mitigating its adverse effects on N_2 fixation in legumes. The goal of this study was to evaluate how the supply of N affects the intercropping of faba beans (*Vicia faba* L.) and peas (*Pisum sativum* L.) with wheat under varying plant densities and N levels. We measured photosynthetic traits, biomass production, the proportion of N derived from air (%Ndfa) in the shoot of the legumes, the N transferred to the wheat, and the land equivalent ratio (LER). The results revealed a positive correlation between soil N levels and the CO_2 assimilation rate (An), chlorophyll content, and N balance index (NBI) in wheat. However, no significant effect was observed in legumes as soil N levels increased. Transpiration (E) increased in wheat intercropped with legumes, while stomatal conductance (g_s) increased with N addition in all crops. Water use efficiency (WUE) decreased in faba beans intercropped with wheat as N increased, but it showed no significant change in wheat or peas. The shoot dry matter of wheat increased with the addition of N; however, the two legume species showed no significant changes. N addition reduced the %Ndfa of both legume species, especially in monoculture, with peas being more sensitive than faba beans. The intercropping of wheat alleviated N_2 fixation inhibition, especially at high wheat density and increased N transfer to wheat, particularly with peas. The LER was higher in the intercropping treatments, especially under limited N conditions. It is concluded that in the intercropping of wheat with legumes, the N_2 fixation inhibition caused by soil N is effectively reduced, as well as there being a significant N transfer from the legume to the wheat, with both process contributing to increase LER.

Keywords: legumes; cereal; intercropping; N_2 fixation; N transfer

1. Introduction

Legumes in symbiosis with rhizobia have the unique ability to fix di-nitrogen (N_2) from the atmosphere for their own metabolic processes. Thanks to this, the use of legumes in agricultural systems reduces the negative impact on the agroecosystem caused by the use of synthetic nitrogen (N) fertilizers [1,2]. Furthermore, legumes, and especially pulses, are a valuable source of protein for human nutrition [3,4] and animal feed [5]. Legumes can also be a good alternative to use in combination with other crops (intercrop) to take advantage of complementary processes [6,7]. This is because growth-promoting resources, such as light, water, and nutrients, are used more efficiently in intercropping [8,9], leading to higher productivity compared to monocrop systems [10]. Furthermore, the N_2 fixation capacity of legumes is significantly enhanced in intercropping systems with other species, thus increasing soil N levels and improved yields of the companion crops, particularly cereals [11–14].

Indeed, early studies of peas (*Pisum sativum* L.)–barley (*Hordeum vulgare* L.) intercropping showed strong competition for soil N by the cereal, leading to a notable increase in the proportion of N derived from air (%Ndfa) in peas compared to the monocrop system [15,16]. Also, in faba beans (*Vicia faba* L.)–barley intercropping, the %Ndfa increased on average from 74 to 92% depending on the cropping intensity [17]. Furthermore, the meta-analysis performed by Rodriguez et al. [18], showed that the proportion of %Ndfa increased in average from 66 to 76%, in various grain legumes intercropped with cereals compared to legumes cropped on their own. Apparently, the rise in the %Ndfa in legumes cultivated in intercropping systems correlates with the decline in available soil N, which is attributed to an increased N uptake by the cereal, which subsequently diminishes the inhibition of N_2 fixation [11,16,19,20].

In legume–cereal intercropping, the latter take advantage of the N transfer facilitated by the N_2 fixation process carried out by legumes [6]. This N transfer is achieved through the release of N compounds by the legume roots, through a process called rhizodeposition, which proceeds in two ways: (1) the senescence, death, and decomposition of roots/nodules; and (2) the exudation of compounds from the roots, which contain assimilable N [6,21]. Another mechanism by which N transfer occurs is through arbuscular mycorrhizal networks, which spread and connect the root system of legumes and cereals, especially under limited N conditions [22,23]. The N transfer is variable, depending on plant growth conditions. For instance, the N transfer from soybeans to sorghum ranged from 32 to 58% [24], while Chu et al. [25], showed that N transfer from peanut to rice in intercrop was between 6.2 and 12.2%, depending on the N applied as fertilizer. Similarly, Gungaabayar et al. [26] reported N transfer in an intercrop of peas with cereals between 17 and 43%.

Therefore, legume–cereal intercropping is an effective strategy to increase the N availability for the cereal and also to promote the N_2 fixation of the legumes [27–30]. However, the benefit of N_2 fixation is affected by N fertilization by reducing the nodule number and their specific activity in legumes cultivated in monoculture [31–33]. The study of Xiao et al. [14] in faba bean–wheat (*Triticum aestivum* L.) intercropping showed that the N fertilization can be reduced by 5 to 15%, but still increasing wheat yield by 16 to 30%. But the impact of N fertilization on N_2 fixation and the nodulation of grain legumes grown in intercropping systems with cereals remains unclear.

Intercropping systems offer the advantage of more efficient land use, measured by the land equivalent ratio (LER). LER compares the yield of intercropped species to the yield of the same species grown in monoculture on the same area [34]. A value of one indicates no difference, while values above one indicate intercropping outperforms monoculture, and values below one indicate lower yields due to competition [35,36]. Studies have shown LER values above one in legume–cereal intercropping systems [37,38]. For instance, corn intercropped with peanuts, soybeans, and mungbeans had LER values of 1.66, 1.60, and 1.48, respectively [39]. Similarly, pea–barley intercropping showed LER values of 1.14 and 1.10, respectively, indicating 14% and 10% more efficiency compared to monoculture, with the intercrop performing better without additional N fertilizer [40].

In the current study, it was hypothesized that the inhibition of N_2 fixation induced by the presence of N in the soil could potentially be alleviated through intercropping. This hypothesis is based on the premise that N uptake by the cereal would reduce soil N levels, thus promoting N_2 fixation by the legume. The objectives of this research were to assess the impact of N supply on the intercropping of peas and faba beans with wheat across two wheat plant densities and three N levels, focusing on: leaf gas exchange and pigment content of the three species, biomass production, the %Ndfa in the legumes, the N transfer to the cereal, and the land equivalent ratio (LER) for the intercropping treatments.

2. Results

2.1. Gas Exchange Parameters

The N level did not affect the net assimilation of CO_2 (A_n) of legumes in monocrop or in intercrop with wheat, but had positives effects on wheat growing in monocrop or intercropped with faba beans or peas (Figure 1a,b). A_n was slightly higher in legumes intercropped with wheat than legumes cultivated alone (Figure 1c). The plant density had no effects on A_n in legumes (Figure 1c). Similarly, the N increased transpiration (E) in wheat (Figure 1d), but not in legumes (Figure 1d). When wheat was intercropped with legumes, E increased, particularly in intercropping with faba beans. For faba beans, E decreased as the N increased and was higher at the plant density of 1:2.3 than at 1:1 (Figure 1f). Conversely, in peas, E nearly doubled at both plant densities compared to sole cultivation, despite N having no effect on E (Figure 1f). The stomatal conductance (g_s) followed a similar pattern to A_n; increased in wheat with increasing N level, both in monoculture and when intercropped with legumes (Figure 1g,h), whereas legumes showed no response to N addition (Figure 1g). Interestingly, when intercropped with wheat, g_s doubled at both plant densities compared to sole cultivation, despite N level showing no significant effect (Figure 1i). Intercellular CO_2 concentration (C_i) remained unchanged with varying N level in both legumes and wheat in monoculture (Figure 2a). Water use efficiency (WUE) exhibited minimal change with N, with a slight increase observed in wheat (Figure 2d).

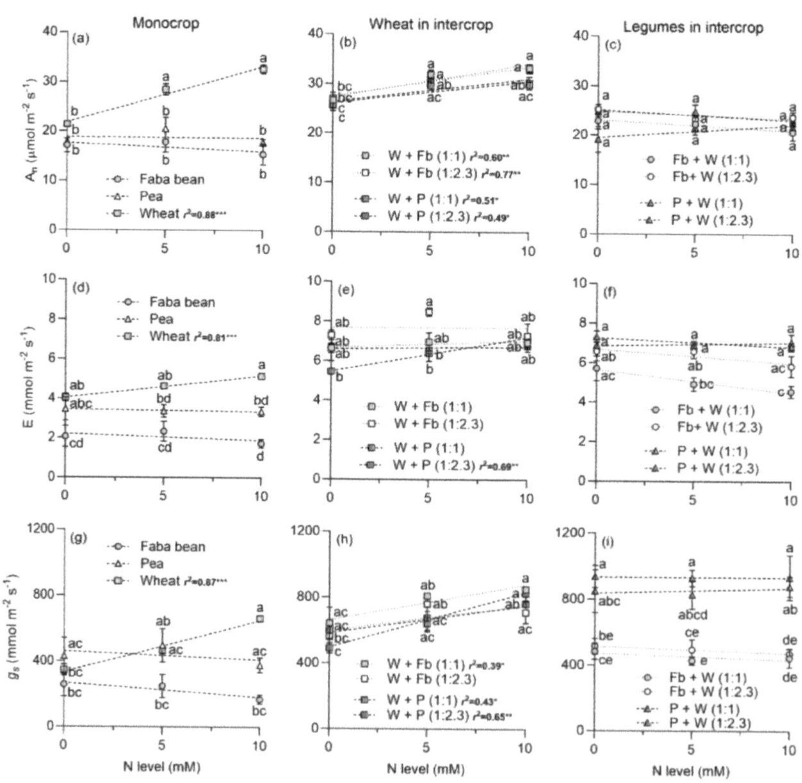

Figure 1. Responses of leaf gas exchange parameters to N supply of faba beans, peas, and wheat growing in monocrop and intercropped at two plant densities: 1:1 and 1:2.3, respectively. A_n is the

net assimilation of CO_2 (**a–c**); E is the transpiration (**d–f**) and g_s is the stomatal conductance (**g–i**). N levels were zero (N available in the soil), 5, and 10 mM applied in the form of NH_4NO_3. Symbols represent the mean and bars the standard error (n = 4 for each N level). r^2 is the coefficient of determination for linear regressions and asterisks indicate statistical significance at p-values: * < 0.05, ** < 0.001, and *** < 0.0001, respectively. Small letters indicate differences between treatments according to two-way ANOVA and Tukey tests ($p < 0.05$).

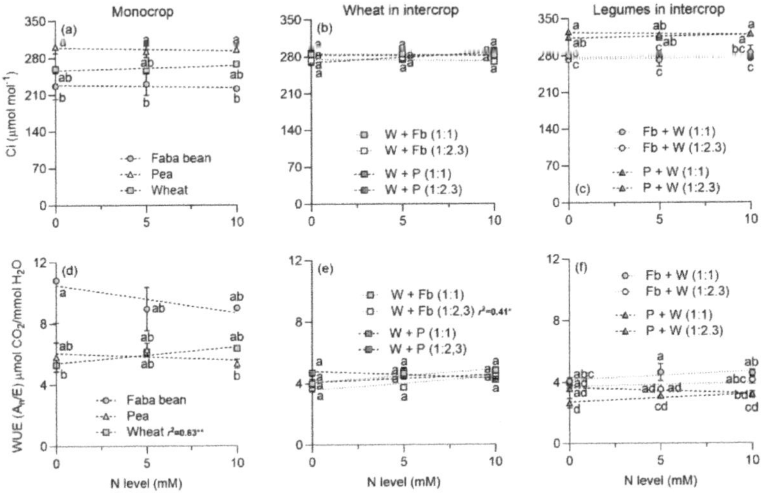

Figure 2. Responses of intercellular carbon (C_i) and water use efficiency (WUE) to the N level of faba beans, peas, and wheat in monocrop and intercropped at two plant densities: 1:1 and 1:2.3, respectively. C_i is the intercellular CO_2 concentration (**a–c**) and WUE is the water use efficiency (**d–f**). N levels were zero (N available in the soil), 5, and 10 mM applied in the form of NH_4NO_3. Symbols represent the mean and bars the standard error (n = 4 for each N level). r^2 is the coefficient of determination for linear regressions and asterisks indicate statistical significance at p-values: * < 0.05, and ** < 0.001, respectively. Small letters indicate differences between treatments according to two-way ANOVA and Tukey tests ($p < 0.05$).

2.2. Photosynthetic Pigments

Photosynthetic pigments were analyzed at 59, 66, and 74 days after sowing (DASs). Since pigment contents were similar at the three measurement dates, data obtained at 59 DASs are presented in Figure 3 (see Supplemental Material Figures S1 and S2 for pigments at 66 and 74 DAS). For wheat cropped alone or intercropped, the chlorophyll content increased with N level, while for legumes there was no significant effect (Figure 3a–c). Flavonoids decreased in wheat as the N increased, either cropped alone or intercropped with legumes (Figure 3d,e). In addition, N did not alter the flavonoid content in legumes (Figure 3f). When it comes to the nitrogen balance index (NBI), this increased in wheat as the N level increased, either in monocrop or intercropped (Figure 3g,h). However, it remained constant in peas, both in monocrop and intercropped. Interestingly, NBI increased as the N level increased for faba beans intercropped with wheat at both plant densities tested (Figure 3i).

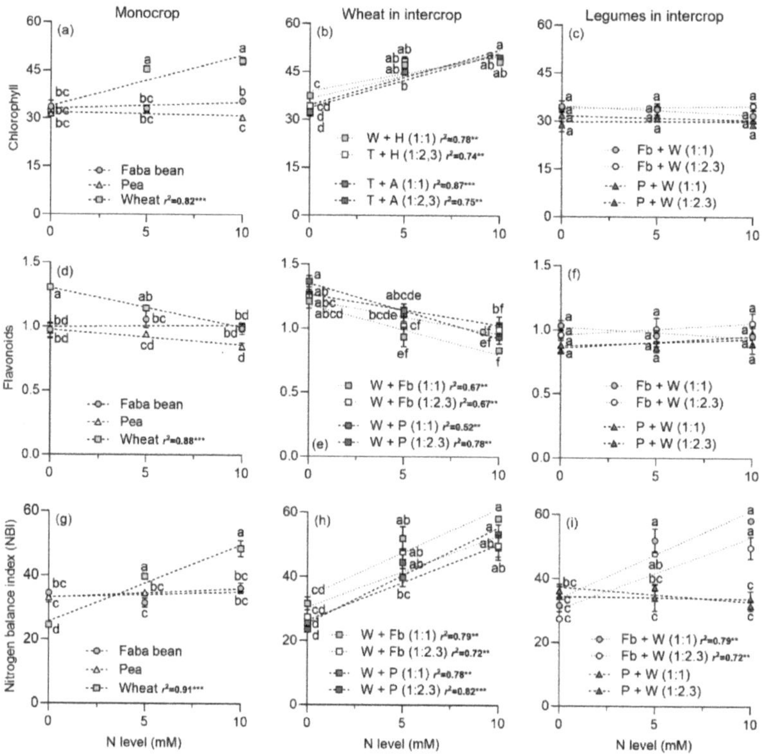

Figure 3. Responses of leaf chlorophyll (**a–c**), flavonoids (**d–f**) and nitrogen balance index (NBI) (**g–i**) at 59 DAS to N level in faba beans, peas, and wheat growing in monocrop and intercropped at two plant densities: 1:1 and 1:2.3, respectively. N levels were zero (N available in the soil), 5, and 10 mM applied in the form of NH_4NO_3. Symbols represent the mean and bars the standard error (n = 4 for each N level). r^2 is the coefficient of determination for linear regressions and asterisks indicate statistical significance at p-values: ** < 0.001, and *** < 0.0001, respectively. Small letters indicate differences between treatments according to two-way ANOVA and Tukey tests ($p < 0.05$).

2.3. Dry Matter and Land Equivalent Ratio (LER)

In monocrops, the shoot dry matter (DM) of faba beans was higher than wheat or peas, at any N level (Figure 4a). For the faba bean–wheat intercrop, the shoot DM was higher at the 1:1 plant density compared to the 1:2.3 (Figure 4b); the slope of the relationship between DM yield and N level was significant at the 1:2.3 plant density. For the pea–wheat intercrop, the shoot DM was higher compared to the monocrop, especially at 10 mM of N (Figure 4c). For this combination, the factors plant density, N level, and their interaction were significant (Table 1).

For faba beans, the nodule biomass per plant slightly decreased as N increased in monocrop and intercrop with wheat (Figure 5a); the nodule biomass (nodule density) was lower in the monocrop compared to the intercrop. For peas in monocrop, nodule biomass decreased in response to the added N (Figure 5b). However, there was no variation in the nodule biomass across the N levels for peas intercropped with wheat, except for peas in monocrop at the high N dose.

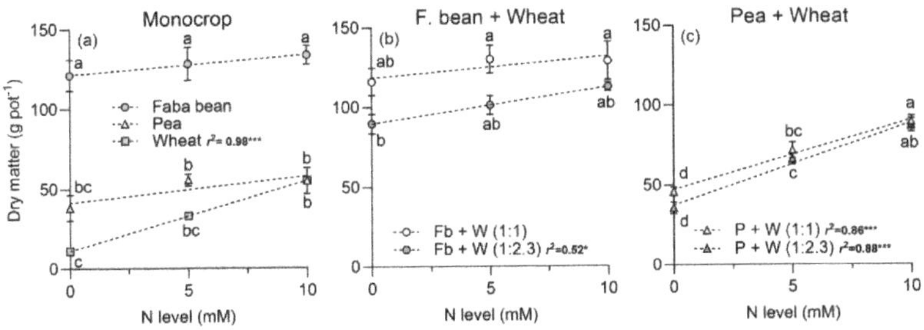

Figure 4. Shoot dry matter per pot in (**a**) faba beans, peas, and wheat in monocrop; (**b**) total dry matter produced by the faba bean–wheat intercrop, and (**c**) total dry matter produced by the pea–wheat intercrop, at two plant densities, 1:1 and 1:2.3 and three N levels, respectively. N levels were 0 (N available in the soil), 5, and 10 mM applied in the form of NH_4NO_3. Symbols represent the mean and bars the standard error (n = 4 for each N supply). r^2 is the coefficient of determination for linear regressions and asterisks indicate statistical significance at p-values: * < 0.05 and *** < 0.0001, respectively. Small letters indicate differences among treatments according to two-way ANOVA and Tukey tests (p < 0.05).

Table 1. Two-way ANOVA results for the effects of plant density, N dose, and their interactions in DM produced per area, DM of nodules, and LER.

Factors	Variable	*df*	F-Value	*p*-Value
A: Plant density	*V. faba* shoot DM	2	10.20	<0.001 ***
B: N dose		2	3.11	0.06
A × *B*		4	0.23	0.92
A: Plant density	*P. sativum* shoot DM	2	11.17	<0.001 ***
B: N dose		2	40.36	<0.001 ***
A × *B*		4	3.23	0.03 *
A: Plant density	*V. faba* nodule DM	2	6.19	0.006 **
B: N dose		2	1.44	0.25
A × *B*		4	0.39	0.81
A: Plant density	*P. sativum* nodule DM	2	3.08	0.12
B: N dose		2	0.95	0.40
A × *B*		4	2.03	0.12
A: Plant density	LER for *V. faba*	1	0.94	0.35
B: N dose		2	38.3	<0.001 ***
A × *B*		2	0.09	0.91
A: Plant density	LER for *P. sativum*	1	0.93	0.35
B: N dose		2	2.16	0.14
A × *B*		2	0.12	0.88

Asterisks indicate statistical significance at * < 0.05, ** < 0.01, and *** < 0.001.

The LER decreased as the N level increased for faba beans intercropped with wheat independent of plant density (Figure 6a; Table 1). The relative contribution of wheat was higher when it was intercropped with peas than with faba beans (Figure 6c,d). As the N level increased, the relative contribution of wheat decreased when it was intercropped with faba beans (Figure 6c); although its contribution increased when it was intercropped with peas at 1:1 plant density; otherwise, the wheat contribution was maintained at the different N levels when density increased (Figure 6d).

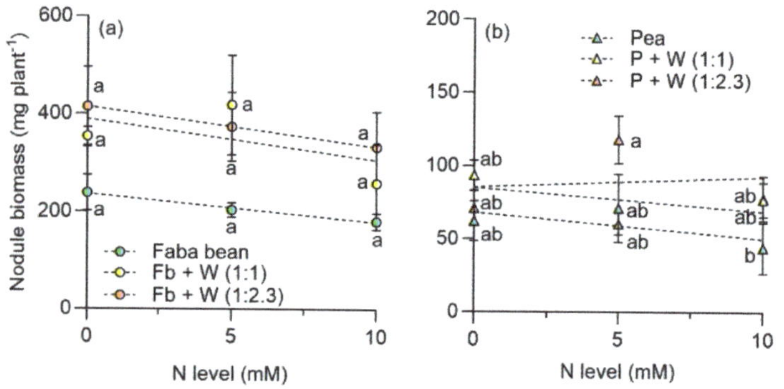

Figure 5. Nodule biomass per plant in (**a**) faba beans in monocrop and intercropped with wheat and (**b**) peas in monocrop and intercropped with wheat at two plant densities, 1:1 and 1:2.3 and at three N levels, respectively. N levels were zero (N available in the soil), 5, and 10 mM applied in the form of NH_4NO_3. Symbols represent the mean and bars the standard error ($n = 4$ for each N level). Small letters indicate differences among treatments according to two-way ANOVA and Tukey tests ($p < 0.05$).

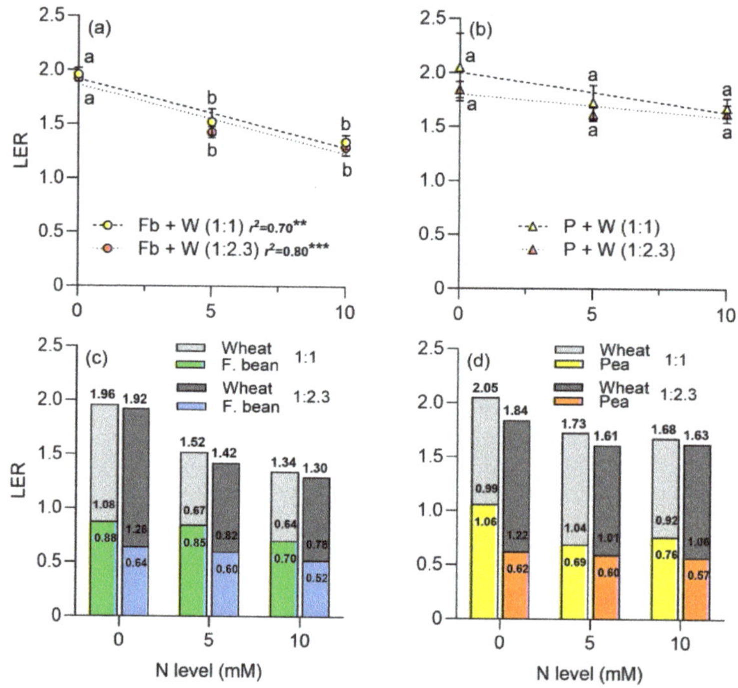

Figure 6. Land equivalent ratio (LER) of (**a**) faba bean–wheat intercrop, (**b**) pea–wheat intercrop, at two plant densities, 1:1 and 1:2.3 and at three N levels. The relative contribution of faba beans and wheat (**c**) and peas and wheat (**d**) in the LER are also shown. N levels were zero (N available in the soil), 5, and 10 mM applied in the form of NH_4NO_3. Symbols represent the mean and bars the standard error ($n = 4$ for each N level). In (**a**,**b**), r^2 is the coefficient of determination for linear regression and asterisks indicate statistical significance at p-values: ** < 0.001 and *** < 0.0001, respectively. In (**c**,**d**), the numbers above the bars are the LER and the numbers inside the bars are the relative contribution of wheat and legumes to the LER. Small letters in (**a**,**b**) indicate differences between treatments according to two-way ANOVA and Tukey tests ($p < 0.05$).

2.4. Proportion of Nitrogen Derived from Air (%Ndfa), N Transfer, and Equivalent N Uptake Ratio (LER_N)

The %Ndfa was above 80% in both legumes when there was no N addition. For faba beans, the %Ndfa decreased as the N level increased, and no differences were observed between intercropped or monocropped treatments (Figure 7a). For peas, the intercropping with wheat reduced the inhibition of %Ndfa due to N addition (Figure 7b). The maximum reduction of the %Ndfa by N was lower in faba beans (12%) than in peas (40%). In the pea–wheat intercrop the %Ndfa at the highest N level was maintained at 77.4 and 82% at 1.1 and 1:2.3 plant density, respectively (Figure 7b). The effects of plant density and N dose were significant in faba beans and peas, and also the plant density × N level interaction was significant for peas (Table 2).

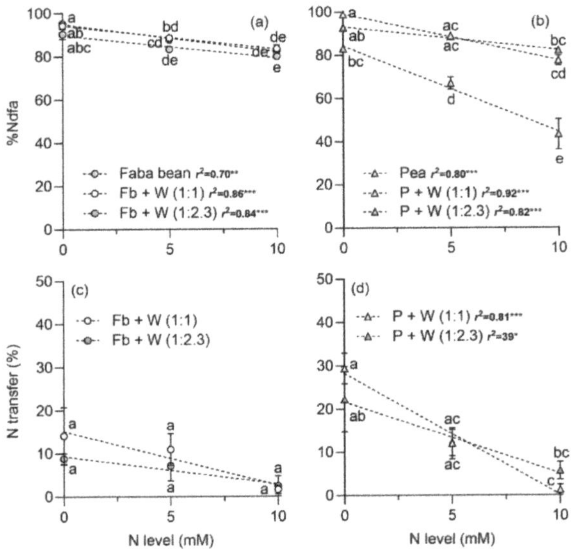

Figure 7. Proportion of N derived from air (%Ndfa) in (**a**) faba beans in monocrop and intercrop with wheat and (**b**) peas in monocrop and intercrop with wheat. The N transferred from (**c**) faba beans to wheat and (**d**) peas to wheat at two plant densities, 1:1 and 1:2.3, and at three N levels, respectively. N levels were zero (N available in the soil), 5, and 10 mM applied in the form of NH_4NO_3. Symbols represent the mean and bars the standard error (n = 4 for each N level). r^2 is the coefficient of determination for linear regressions and asterisks indicate statistical significance at p-values: * < 0.05, ** < 0.001, and *** < 0.0001, respectively. Small letters indicate differences among treatments according to two-way ANOVA and Tukey tests (p < 0.05).

Nitrogen transfer from legumes to wheat was drastically reduced as the N level increased (Figure 7c,d; Table 2). In the faba bean–wheat intercrop at the 1:1 plant density, the N transfer to the wheat was of 14.1%, 10.9%, and 1.6% at 0, 5, and 10 mM of N, respectively. In contrast, at the 1:2.3 plant density, N transfer decreased from 8.8% to 7.2% and 2.4% at the 0, 5, and 10 mM of N rates, respectively (Figure 7c). In the pea–wheat intercrop at the 1:1 plant density there was a maximum transfer of 29.4%, 12.0%, and 1.2% at 0, 5, and 10 mM of N, respectively. At the 1:2.3 plant density, the N transfer decreased from 22.3 to 12.2 and to 5.7% at 0, 5, and 10 mM of N, respectively (Figure 7d).

For faba beans, the land equivalent ratio for N uptake (LER_N) decreased with N supply, especially at the 1:1 plant density (Figure 8; Table 2). At the 1:1 plant density, a maximum value of 2.3 and a minimum of 1.5 were recorded at 0 and 10 mM of N, respectively (Figure 8a). For the pea–wheat intercrop, the LER_N values were similar at both plant densities and were not affected by N supply (Figure 8b).

Table 2. Two-way ANOVA results for the effects of plant density, N dose, and their interactions in %Ndfa, %N transferred, and LER_N in different crop treatments.

Factors	Variable	*df*	F-Value	*p*-Value
A: Plant density	%Ndfa *V. faba*	2	9.96	<0.001 ***
B: N level		2	57.5	<0.001 ***
A × *B*		4	0.32	0.86
A: Plant density	%Ndfa *P. sativum*	2	72.3	<0.00 ***
B: N level		2	56.3	<0.001 ***
A × *B*		4	7.07	<0.001 ***
A: Plant density	%N transferred *V. faba*	1	0.87	0.36
B: N level		2	3.74	0.04 *
A × *B*		4	0.38	0.68
A: Plant density	%N transferred *P. sativum*	1	0.06	0.80
B: N level		2	15.8	<0.001 ***
A × *B*		4	1.05	0.37
A: Plant density	LER_N *V. faba*	1	2.22	0.15
B: N level		2	8.02	0.004 **
A × *B*		2	2.60	0.10
A: Plant density	LER_N *P. sativum*	1	0.008	0.93
B: N level		2	0.93	0.41
A × *B*		2	0.03	0.96

Asterisks indicate statistical significance at *p* * < 0.05, ** < 0.01, and *** < 0.001.

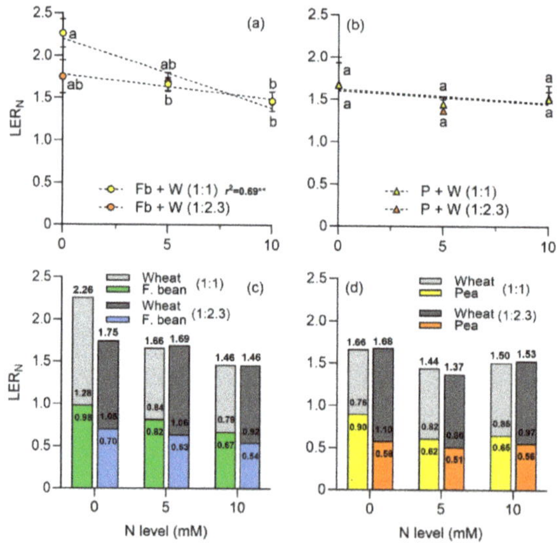

Figure 8. Land equivalent ratio for N (LER_N) of (**a**) faba bean–wheat intercrop, (**b**) pea–wheat intercrop, at two plant densities, 1:1 and 1:2.3 and at three N levels. The relative contribution of faba beans and wheat (**c**) and peas and wheat (**d**) in the LER_N are also shown. N levels were zero (N available in the soil), 5, and 10 mM applied in the form of NH_4NO_3. Symbols represent the mean and bar the standard error (n = 4 for each N level). In (**a**,**b**), r^2 is the coefficient of determination for linear regression and asterisks indicate statistical significance at *p*-value lower than ** < 0.001. In (**c**,**d**), the numbers above the bars are the LER_N and the numbers inside the bars are the relative contribution of wheat and legumes to the LER_N. Small letters in (**a**,**b**) indicate differences between treatments according to two-way ANOVA and Tukey tests ($p < 0.05$).

In relation to the relative contribution of each species, in the intercropping of faba beans and wheat, a higher contribution of wheat was observed with respect to faba beans when there was no N addition, especially at the 1:1 plant density (Figure 8c). In the pea–wheat mixture, the lowest N level also presented the highest ratio in favor of peas at the 1:1 plant density (Figure 8d). However, this value also decreased as N increased. On the other hand, at the 1:2.3 plant density, as the N level increased, the relative contribution of wheat underwent less variation than at the 1:1 plant density, particularly in the intercropping of faba beans with wheat (Figure 8c).

3. Discussion

3.1. Overall Aspects

Intercropping legumes with cereals enhances complementarity processes [7]. The addition of N to the intercropping systems predominantly benefits wheat, fostering increased photosynthesis and the accumulation of photosynthetic pigments in both monoculture or intercropped with *P. sativum* or *V. faba* (Figures 1–3). Notably, legumes showed no significant response to N addition, indicating that N_2 fixation adequately supplied the N required under the experimental conditions. Furthermore, the addition of N boosted DM yield in wheat, and it had no positive effect on legumes (Figure 4). The DM yield is reflected in the LER, which declined as N addition and wheat density increased, with higher values observed in intercrops without N supply (Figure 6). The increased N uptake by wheat helped to reduce the inhibition of N_2 fixation caused by the N addition, especially of *P. sativum* (Figure 7a,b). Alternatively, legumes use N_2 fixation for their own metabolism, thereby increasing the availability of N for wheat and the transfer of N compounds via rhizodeposition [6] (Figure 7c,d). The higher %Ndfa found in legumes without N addition, suggests that N_2 fixation and N transfer to the cereal are more efficient under low N conditions. In the following sections, we will describe these effects in detail.

3.2. Leaf Photosynthetic Traits

The increase in leaf gas exchange (A_n, E and g_s) of wheat growing in monoculture or intercrop is consistent with the increase observed in the leaf chlorophyll content (Figures 1 and 2). The positive relationship between leaf N or chlorophyll content and A_n or g_s has been well stablished in different species, including wheat and faba beans [41–43]. The higher A_n and g_s of wheat growing in intercropping compared to monoculture, particularly with no N addition, can be explained by the lower competition for resources in the intercrop. The lack of response of A_n and g_s of legumes to N addition was probably a consequence of the leaf N status of legumes (NBI), especially for peas which showed no change at the different N levels (Figure 3). This indicates that N_2 fixation can sustain the growth of legumes without extra N.

Photosynthetic pigments are useful for estimating the level of N in plants and are directly related to biomass production and photosynthesis [44,45]. Other investigations on legumes intercropped with non-legumes have demonstrated an increase in wheat chlorophyll content compared to monoculture, particularly combined with N supplementation, e.g., Tosti and Guiducci [46] in wheat–faba bean; and Suryapani et al. [47] in lentil–wheat. Experiments with soybeans reinforce the idea that intercropping with a legume increases the amount of chlorophyll in both species [39], even if the companion crop is not a cereal, e.g., mint or tea [39,48,49].

The slightly higher NBI of wheat in intercrop compared to being cropped alone can be associated with a higher N availability in soil and the N transferred from the intercropped legumes. Flavonoids exhibit antioxidant characteristics and play a role in safeguarding cells during periods of stress, such as N deficiency. In conditions of N sufficiency, plants prioritize chlorophyll synthesis, whereas during N deficit, they employ flavonoid production as a protective measure [50]. In the three measurements performed on wheat, flavonoid content decreased as the N level increased (Figure 3 and Supplemental Figure S2). This contrasted with the results obtained in legumes, where the concentration of flavonoids did not vary with the amount of N applied. Therefore, our data indicate that cereals have a higher susceptibility to stress when they lack N to meet their demand, leading to higher flavonoid production. In contrast, legumes are less likely to suffer from N deficiency stress because they can meet their requirements through N_2 fixation.

3.3. Dry Matter Production and Land Equivalent Ratio (LER)

The height of the crop causes interspecific competition for light, benefiting if it is taller, in this case, for wheat [51,52]. Furthermore, plants in intercrop have a higher radiation use efficiency compared to those in monocrop [53–55]. The lower biomass production of peas intercropped with wheat compared to faba beans in intercrop suggests that the latter has more complementary physiological and morphological characteristics, resulting in a more efficient use of resources. Faba beans having an upright growth habit with branching stems similar to cereals enables them to compete better with wheat than peas, which have prostrate growth [45].

The positive effect of N in wheat shoot DM was probably related with its more extensive root system and greater soil exploration capacity, which allows it to take up more N, mainly during the vegetative stage [16,29,56]. On the other hand, legumes, both in monoculture and intercropping, did not significantly increase DM when N was added. These results suggest that increased competition from cereals in intercropping forces legumes to supply their N_2 fixation to meet their demand [10,29,40]. Previous studies supported these results, confirming that cereals are strong competitors for soil resources, especially for N in mixtures of cereals and legumes [11,18].

In relation to the LER, various authors have reported LER values higher than one in intercrop of legumes and cereals, including faba beans [14,57,58] and peas [59–61]. This indicates a higher resource use efficiency in intercrop systems compared to monocrops. In the present work, LER decreased as the proportion of wheat plants increased in the intercrop, as reported by Dhima et al. [62] for faba beans intercropped with oat. The higher LER value without N addition suggests that the intercropping of legumes and cereals is especially advantageous in low-input agriculture, especially when N availability is lower [63–65]. Regarding the relative contribution of the species to the LER, the shoot DM obtained by wheat represents the high proportion of the LER index [10,63]. Conversely, when the intercrop involves herbaceous grass species such as ryegrass, the legume has a higher relative contribution to the LER [66,67]. Therefore, this pot experiment is aligned with the idea that, under N limited conditions, legume–cereal mixtures can be more efficient in terms of equivalent land use.

3.4. N_2 Fixation, N Transfer, and Land Equivalent Ratio for N (LER_N)

The results revealed that the %Ndfa of faba beans was less affected by N level than peas, when cultivated as monocrops, as reported previously by Guinet et al. [68]. The nodule biomass was much higher in faba beans, in both monocrop and intercrop (Figure 5). A study by Liu et al. [69] showed that the number of nodules in faba beans slightly decreases as the N level increases, reaffirming the idea that intercropping can inhibit the negative effect of N on nodulation; in the case of peas, nodule biomass was not affected by N increase, although the %Ndfa was significantly reduced (Figure 7). This can be explained by a reduction of specific nodule activity which reduces the amount of N fixed by each nodule.

Previous studies indicate that the intercropping of legumes with cereals improves the nodulation capacity, increasing the number and weight of nodules, especially when there is a higher density of cereal in the mixture [70]. Our results show that, for faba beans, the nodule biomass increased in intercropping with wheat plants. Similar results were reported by Li et al. [71], who recorded higher nodule biomass in intercropping than in monoculture for faba beans. In the case of peas, the nodule biomass was similar in the monocrop and intercrop; however, Hu et al. [72] reported higher nodule biomass in relation to the pea cropped alone. The higher nodule biomass in intercropping systems could be related to a decrease in the soil available N due to the cereal N uptake, which results in the production of nodules to compensate for N uptake by the legume.

The intercropping of wheat and faba beans or peas improved the efficiency of N_2 fixation and attenuated its inhibition to soil N, particularly benefiting the more sensitive species—i.e., peas (Figure 7b). This can be explained by the fact that cereals have a root system with a greater capacity to take up N compared to legumes, presenting a greater interspecific competition and promoting the process of N_2 fixation by legumes [27,73,74]. Furthermore, the N transferred to wheat was higher when intercropped with peas than faba beans (Figure 7c,d), suggesting that peas have a higher rate of decomposition and renewal of roots and nodules, which upon completing their life cycle releases N compounds that are mineralized by microorganisms, which is known as rhizodeposition [6,75]. The above could also be related to the precocity of peas compared to faba beans, which could also improve N transfer. In addition, the N application significantly decreased N transfer, particularly in peas, which may be due to the inhibition of N_2 fixation caused by N application [76,77]. Alternatively, it cannot be ruled out that the addition of N led to a dilution of N from rhizodeposition that affects N transfer. Interestingly, when the intercrop has the same plant density of cereals and legumes, the N transfer was higher than with less wheat plants. This suggests a high N availability for wheat due to lower intraspecific competition.

With respect to the LER_N, the results support previous findings showing values higher than one in cereal and legume intercrops [29,78], indicating higher efficiency of intercropping over monocrop. Furthermore, in the intercrop of faba beans with wheat at a 1:1 plant density, LER_N was higher with no N addition. This coincides with the results of Jiao et al. [79] and Wang et al. [77], who pointed out that N supply decreased the LER_N. This could be explained by the fact that legumes stop N_2 fixation and increase competition for N uptake, decreasing intercrop efficiency. In addition to the above, Zhu et al. [80] reported that LER_N decreased significantly in legume and cereal intercrops when N, P, and water were more available. This indicates that, at lower resource availability, a positive interspecific interaction is promoted, mainly by stimulating N_2 fixation in the absence of available soil N [80,81]. The results obtained confirm a higher N use efficiency in intercrop systems with legumes and cereals, in contrast to monocrops. This is largely due to the complementary use of the mineral N present in the soil and the N_2 fixed by legumes [76,78].

4. Materials and Methods

4.1. Plant Growth Conditions and Experimental Design

The effects of intercropping on the inhibition of N_2 fixation by exogenous N addition were evaluated in a pot experiment using the faba bean var. 'Super Agua Dulce' and the pea var. 'Utrillo' intercropped with soft wheat var. 'Pandora', at three levels of N. The experiment was carried out using 10 L plastic pots filled with 12 kg of a 1:1 soil–sand mixture based on weight, which were placed in a laterally open greenhouse (only the roof-covered the plants to shield them from the rain) at the Phenomics Center of the Talca University, Chile, during the 2021 season (Supplemental Figure S3). The soil used was an Inceptisol with a pH of 6.3 (pH–soil to water ratio of 1:2.5) and 1.0, 20, and 196 mg kg^{-1} of N, P, and K, respectively, 1.2% of soil organic matter (SOM), and 0.03 dS m^{-1} of electric conductivity.

Before sowing, seeds of the three species were washed with distilled water, selected according to shape and size and germinated in plastic clamshell containers with sterilized vermiculite and subsequently irrigated with 200 mL of distilled water and placed in a growth chamber (Pitec® BIOREF-38L) at 23 °C with a 12 h light/12 h dark photoperiod. After 3–6 days, the most homogeneous seedlings were selected and transplanted at a depth of 3 cm. In all treatments (monoculture and intercrop) ten seedlings were transplanted per pot. In intercrop treatments, seedlings were transplanted at two densities: 1:1 = 5 legumes and 5 cereals, and 1:2.3 = 3 legumes and 7 cereals. After transplanting, seedlings of legumes were inoculated with 4 mL/plant of a stationary *Rhizobium leguminosarum* bv. viciae YEM culture (local strains 373-3007-Su303 for peas; and strain 1400 for faba beans), with an approximate cell density of 10^9 mL^{-1} to promote nodulation. Treatments were completely randomized with four replicates and three N levels applied as ammonium nitrate (NH_4NO_3): 0, 5, and 10 mM. To achieve an optimal P supply, in addition to the P present in the soil-sand mixture (1:1 ratio), 200 mg of P per kg of soil mixture was added in the form of potassium phosphate (K_2HPO_4). All the plants were fertilized with a nutrient solution composed of 0.7 mM of K_2SO_4, 0.5 mM of $MgSO_4$, 0.8 mM of $CaCl_2$, 4.0 μM of H_3BO_3, 0.1 μM of Na_2MoO_4, 1.0 μM of $ZnSO_4$, 2.0 μM of $MnCl_2$, 0.2 μM of $CoCl_2$, 1.0 μM of $CuCl_2$, and 1.0 μM of FeNaEDTA. The nutrient solution was applied at a frequency of every four days, with each application consisting of a volume of 500 mL. In the case of N, it was applied alongside the nutrient solution according to the respective treatment, 0, 5, or 10 mM, but only once a week and starting three weeks after transplanting. The soil water content was maintained at 75% of the maximum water-holding capacity.

The experimental design considered two factors: the N dose and plant density in the intercropping system, using four replicates for all treatments, except for the treatment of faba beans intercropped with wheat at a plant density of 1:1 and 10 mM of N, where 3 replicates were recorded. Analyses of variance (two-way ANOVA) were performed for all measured variables after testing for normality and homogeneity of variances using Shapiro–Wilk and Levene's tests, respectively. Plots, regressions, and their parameters were calculated using GraphPad Prism® version 10.0.

4.2. Gas Exchange and Photosynthetic Pigment Content

Leaf gas exchange was determined using a CIRAS-2 gas analyzer. Measurements were taken between 11:00 and 14:00 h, during the peak of photosynthetic activity. The parameters evaluated were net assimilation of CO_2 (A_n), transpiration (E), stomatal conductance (g_s), intercellular CO_2 concentration (C_i), and water use efficiency (WUE), the latter calculated as a ratio of A_n–E. The conditions under which this measurement was carried out were with an intensity of photosynthetic active radiation (PAR) from 1500 to 2100 μmol $m^{-2}s^{-1}$, leaf temperature between 23 and 25 °C, and an airflow rate of 248 mL/min. In the case of the legumes, measurements were made in the middle of the plant, considering the central part of a leaf in good condition. For the wheat, measurements were made on the flag leaf. Measurements were taken 63 days after sowing (DAS), with the legumes at an advanced stage of flowering and the wheat at boot stage. The parameters were measured three times in each leaf and in triplicate for each treatment.

The content of photosynthetic pigments (chlorophyll, flavonoids, and the nitrogen balance index [NBI]) was determined using a portable chlorophyll meter (DUALEX®) on the same leaves used previously for the measurement of gas exchange, but three measurements were made on different dates. The first measurement was taken at 59 DAS, at which time the legumes were in full flowering and the wheat was in the boot stage. The second measurement was taken at 66 DAS when the legumes were in pod formation and the wheat was at anthesis. Finally, the third measurement was made at 74 DAS when the legumes and wheat were in the grain-filling stage. Three plants per treatment, one leaf per plant, and three measurements on the same leaf were evaluated.

4.3. Determination of Plant and Nodule Dry Matter

The dry matter (DM) of shoots and nodules was evaluated at the grain filling stage. All plants were harvested at 77 DASs. Shoots were separated from the roots and dried in an air oven at 65 °C for 48 h until reaching a constant weight. The roots were carefully separated from the soil, washed, and stored at −80 °C. Then, the nodules were separated from the roots manually with steel tweezers. Subsequently, the nodules were dried at 65 °C for 48 h for DM determination. Shoot DM was finely milled and stored in paper bags for subsequent chemical analysis.

4.4. Determining the Proportion of N Derived from Air (%Ndfa)

The N concentration and ^{15}N natural abundance (expressed as ‰ $\delta^{15}N$ relative to the ^{15}N composition of atmospheric N_2) were determined using an elemental analyzer and an isotope ratio mass spectrometer at the Laboratory of Applied Chemistry and Physics, Ghent University, Belgium. The natural abundance method of ^{15}N, to measure the fixation of N_2, is based on the difference that exists between the abundance of ^{15}N from atmospheric N_2 ($\delta^{15}N = 0$‰) and the N present in the soil ($\delta^{15}N$ generally > 0‰) [82]. The percentage of N derived from the air (%Ndfa) in legumes was calculated by comparing natural abundance of ^{15}N of the legume ($\delta^{15}N_{leg}$) with that of the reference plants ($\delta^{15}N_{ref}$), using Equation (1) [83]:

$$\%Ndfa = 100 \times \left[\frac{\delta^{15}N_{ref} - \delta^{15}N_{leg}}{\delta^{15}N_{ref} - \beta} \right] \quad (1)$$

where, the reference plant provides the natural abundance of ^{15}N that comes from the soil and the β value represents the ^{15}N abundance in legumes that rely solely on N_2 fixation for growth [82]. With this method, the greater the difference in the natural abundance of ^{15}N between the reference plants and the legumes, the greater the N_2 fixation, and the closer the $\delta^{15}N_{leg}$ is to the β value. The reference plants used were wheat (*Triticum aestivum* L.), oat (*Avena sativa* L.), sunflower (*Helianthus annuus* L.), and quinoa (*Chenopodium quinoa* Willd.). These plants were grown under the same experimental conditions as the legumes, and with the addition of 5 mM of N. The shoots of the reference plants were harvested, dried, weighed, and milled in the same way as described above for the legumes. The reference plants were harvested manually at the phenological stage of anthesis for wheat and oat or at full flowering for quinoa and sunflower.

4.5. Determination of the β Value

Legumes have different N_2 fixing capacities, which is reflected in the β value, which indicates the amount of N_2 fixed in the absence of N in the substrate [82,83]. Because the β value was not determined in the absence of N, a sensitivity analysis was performed, with a range of β values obtained from the literature and serving to evaluate the effect of N on the inhibition of N_2 fixation (Supplemental Figure S4).

4.6. N Transfer from Legume to Cereal

To measure N transfer from legumes to cereals, the $\delta^{15}N$ present in the cereal in monocrop was compared with the $\delta^{15}N$ of the cereal intercropped with the legumes. The %Ndfa transferred from legume to cereal ($\%N_t$) was calculated by Equation (2):

$$\%N_t = 100 \times \left[1 - \frac{\delta^{15}N_{Cereal\ IC}}{\delta^{15}N_{Cereal\ MC}} \right] \quad (2)$$

where $\delta^{15}N_{Cereal\ IC}$ represents the $\delta^{15}N$ in the cereal intercropped with legumes and $\delta^{15}N_{Cereal\ MC}$ in the monocrop.

4.7. Land Equivalent Ratio (LER)

To calculate the land use efficiency, the land equivalent ratio (LER) was used, which compares the yield achieved by intercropping two species with the yield obtained in a monocrop. The LER was calculated according to Equation (3) proposed by Mead and Willey [34]:

$$LER = (SB_{Leg\ IC}/SB_{Leg\ MC}) + (SB_{Cereal\ IC}/SB_{Cereal\ MC}) \quad (3)$$

where, $SB_{Leg\ IC}$ is the legume shoot biomass in intercropping, $SB_{Leg\ MC}$ is the legume shoot biomass in monocrop, $SB_{Cereal\ IC}$ is the cereal shoot biomass in intercropping and $SB_{Cereal\ MC}$ is the cereal shoot biomass in monocrop. In addition, to calculate the equivalent N uptake ratio (LER_N), the same Equation (3) was used, but with the values of N content accumulated in the shoot.

5. Conclusions

The intercrop of legumes with wheat can improve N use efficiency, since cereals, being more competitive, decrease the inhibition of N_2 fixation caused by the presence of N in the soil. In addition, peas were found to be more sensitive to the presence of N than faba beans, so the latter could have a better complementarity with wheat. However, peas can transfer up to 30% of the N fixed, so this is also a species that should be considered in intercropping systems. In addition, when there is a lower amount of N in the soil, N_2 fixation and N transfer are promoted, which directly benefits the performance of wheat, increasing its N content, photosynthetic parameters, and LER and LER_N indexes.

Therefore, this study provides a deeper understanding of the dynamics in legume–cereal intercropping. The role of N in competition, N fixation and transfer, and photosynthesis, highlighting the complexity of interspecific interactions and how they can influence agricultural productivity and the efficient use of N. These findings have important implications for the planning of sustainable cropping systems and the optimization of food production.

Supplementary Materials: The following supporting information can be downloaded at: https://www.mdpi.com/article/10.3390/plants13070991/s1, Figure S1: Responses of leaf chlorophyll (a–c), flavonoids (d–f), and nitrogen balance index (NBI) (g–i) at 66 DAS to N level in faba beans, peas, and wheat grown in monocrop and intercropped at two plant densities: 1:1 and 1:2.3, respectively. N levels were zero (N available in the soil), 5, and 10 mM applied in the form of NH_4NO_3. Symbols represent the mean and bars the standard error (n = 4 for each N level). r^2 is the coefficient of determination for linear regressions and asterisks indicate statistical significance at p-values: * < 0.05, ** < 0.001, and *** < 0.0001, respectively. Small letters indicate differences between treatments according to two-way ANOVA and Tukey tests ($p < 0.05$); Figure S2: Responses of leaf chlorophyll (a–c), flavonoids (d–f), and the nitrogen balance index (NBI) (g–i) at 74 DAS to N level in faba beans, peas, and wheat grown in monocrop and intercropped at two plant densities: 1:1 and 1:2.3, respectively. N levels were zero (N available in the soil), 5, and 10 mM applied in the form of NH_4NO_3. Symbols represent the mean and bars the standard error (n = 4 for each N level). r^2 is the coefficient of determination for linear regressions and asterisks indicate statistical significance at p-values: * < 0.05, ** < 0.001, and *** < 0.0001, respectively. Small letters indicate differences between treatments according to two-way ANOVA and Tukey tests ($p < 0.05$); Figure S3. Daily mean air temperature, air humidity, and hours of daylight (photoperiod [P]) during the growth period of faba beans, peas, and wheat grown in monocrop and intercropped at two plant densities: 1:1 and 1:2.3, respectively. Plants were grown during the springtime in the southern hemisphere (September to November 2021); Figure S4. Sensitivity analysis for β values for *V. faba* (a–c) and *P. sativum* (d–f) in monocrop and intercropped with *T. aestivum* at two plant densities (1:1 and 1:2.3) and three N levels, respectively. N levels were zero (N available in the soil), 5, and 10 mM applied in the form of NH_4NO_3. A range of β values were tested as recorded in the literature. The β values resulted in an average of −0.49 and −0.61, for peas and faba beans, respectively. In the case of faba beans, the β value proposed by the literature correctly fitted the linear regression (the green line in Figure S4a–c), since it never exceeded 100% of %Ndfa. However, the β value used for the peas did not fit well, thus a range of β values were analyzed in a sensitivity analysis and the one that best fitted the linear regression was β = −1.3 (the green line in Figure S4d–f). Independently of the β value used, it was observed that the %Ndfa decreases as N increases.

Author Contributions: Conceptualization, R.A.C. and A.d.P.; methodology, R.P.-D., S.E., B.C. and R.A.C.; formal analysis, S.S.-R. and R.A.C.; investigation, R.P.-D., S.S.-R., B.C., S.E. and A.M.-G.; resources, A.M.-G. and S.S.-R.; data curation, S.S.-R.; visualization, S.S.-R. and R.A.C.; writing—original draft preparation, S.S.-R. and R.A.C.; writing—review and editing, R.A.C. and A.d.P.; supervision, R.A.C. and A.d.P.; project administration, R.A.C.; funding acquisition, R.A.C. and A.d.P. All authors have read and agreed to the published version of the manuscript.

Funding: This work and the APC were funded by ANID Chile FONDECYT, grant number 1201050.

Data Availability Statement: The data presented in this study are available on request from the corresponding author.

Acknowledgments: R.A.C. thanks Millennium Nucleus for the Development of Super Adaptable Plants (MN-SAP), Project N° NCN2021_010, and the Office of Research Administration of the University of Talca. B.C. thanks ANID Fortalecimiento Científico de los Centros Regionales, Project N° R20F0001.

Conflicts of Interest: The authors declare no conflicts of interest.

References

1. Vance, C.P. Legume Symbiotic Nitrogen Fixation: Agronomic Aspects. In *The Rhizobiaceae: Molecular Biology of Model Plant-Associated Bacteria*; Spaink, H.P., Kondorosi, A., Hooykaas, P.J.J., Eds.; Springer: Dordrecht, The Netherlands, 1998; pp. 509–530, ISBN 978-94-011-5060-6.
2. Gan, Y.; Liang, C.; Chai, Q.; Lemke, R.L.; Campbell, C.A.; Zentner, R.P. Improving Farming Practices Reduces the Carbon Footprint of Spring Wheat Production. *Nat. Commun.* **2014**, *5*, 5012. [CrossRef] [PubMed]
3. Foyer, C.H.; Lam, H.-M.; Nguyen, H.T.; Siddique, K.H.M.; Varshney, R.K.; Colmer, T.D.; Cowling, W.; Bramley, H.; Mori, T.A.; Hodgson, J.M.; et al. Neglecting Legumes Has Compromised Human Health and Sustainable Food Production. *Nat. Plants* **2016**, *2*, 16112. [CrossRef] [PubMed]
4. Didinger, C.; Thompson, H.J. The Role of Pulses in Improving Human Health: A Review. *Legume Sci.* **2022**, *4*, e147. [CrossRef]
5. Gdala, J. Composition, Properties, and Nutritive Value of Dietary Fibre of Legume Seeds. A Review. *J. Anim. Feed. Sci.* **1998**, *7*, 131–150. [CrossRef]
6. Fustec, J.; Lesuffleur, F.; Mahieu, S.; Cliquet, J.-B. Nitrogen Rhizodeposition of Legumes. A Review. *Agron. Sustain. Dev.* **2010**, *30*, 57–66. [CrossRef]
7. Monti, M.; Pellicanò, A.; Pristeri, A.; Badagliacca, G.; Preiti, G.; Gelsomino, A. Cereal/Grain Legume Intercropping in Rotation with Durum Wheat in Crop/Livestock Production Systems for Mediterranean Farming System. *Field Crops Res.* **2019**, *240*, 23–33. [CrossRef]
8. Lithourgidis, A.S.; Dhima, K.V.; Vasilakoglou, I.B.; Dordas, C.A.; Yiakoulaki, M.D. Sustainable Production of Barley and Wheat by Intercropping Common Vetch. *Agron. Sustain. Dev.* **2007**, *27*, 95–99. [CrossRef]
9. Stomph, T.; Dordas, C.; Baranger, A.; de Rijk, J.; Dong, B.; Evers, J.; Gu, C.; Li, L.; Simon, J.; Jensen, E.S.; et al. Chapter One-Designing Intercrops for High Yield, Yield Stability and Efficient Use of Resources: Are There Principles? In *Advances in Agronomy*; Sparks, D.L., Ed.; Academic Press: Cambridge, MA, USA, 2020; Volume 160, pp. 1–50.
10. Bedoussac, L.; Journet, E.-P.; Hauggaard-Nielsen, H.; Naudin, C.; Corre-Hellou, G.; Jensen, E.S.; Prieur, L.; Justes, E. Ecological Principles Underlying the Increase of Productivity Achieved by Cereal-Grain Legume Intercrops in Organic Farming. A Review. *Agron. Sustain. Dev.* **2015**, *35*, 911–935. [CrossRef]
11. Yu, Y.; Stomph, T.-J.; Makowski, D.; Zhang, L.; van der Werf, W. A Meta-Analysis of Relative Crop Yields in Cereal/Legume Mixtures Suggests Options for Management. *Field Crop Res.* **2016**, *198*, 269–279. [CrossRef]
12. Yu, L.; Tang, Y.; Wang, Z.; Gou, Y.; Wang, J. Nitrogen-Cycling Genes and Rhizosphere Microbial Community with Reduced Nitrogen Application in Maize/Soybean Strip Intercropping. *Nutr. Cycl. Agroecosyst.* **2019**, *113*, 35–49. [CrossRef]
13. Kermah, M.; Franke, A.C.; Adjei-Nsiah, S.; Ahiabor, B.D.K.; Abaidoo, R.C.; Giller, K.E. Maize-Grain Legume Intercropping for Enhanced Resource Use Efficiency and Crop Productivity in the Guinea Savanna of Northern Ghana. *Field Crops Res.* **2017**, *213*, 38–50. [CrossRef] [PubMed]
14. Xiao, J.; Yin, X.; Ren, J.; Zhang, M.; Tang, L.; Zheng, Y. Complementation Drives Higher Growth Rate and Yield of Wheat and Saves Nitrogen Fertilizer in Wheat and Faba Bean Intercropping. *Field Crop Res.* **2018**, *221*, 119–129. [CrossRef]
15. Izaurralde, R.C.; McGill, W.B.; Juma, N.G. Nitrogen Fixation Efficiency, Interspecies N Transfer, and Root Growth in Barley-Field Pea Intercrop on a Black Chernozemic Soil. *Biol. Fertil. Soils* **1992**, *13*, 11–16. [CrossRef]
16. Jensen, E.S. Grain Yield, Symbiotic N_2 Fixation and Interspecific Competition for Inorganic N in Pea-Barley Intercrops. *Plant Soil* **1996**, *182*, 25–38. [CrossRef]
17. Danso, S.K.A.; Zapata, F.; Hardarson, G.; Fried, M. Nitrogen Fixation in Fababeans as Affected by Plant Population Density in Sole or Intercropped Systems with Barley. *Soil Biol. Biochem.* **1987**, *19*, 411–415. [CrossRef]
18. Rodriguez, C.; Carlsson, G.; Englund, J.-E.; Flöhr, A.; Pelzer, E.; Jeuffroy, M.-H.; Makowski, D.; Jensen, E.S. Grain Legume-Cereal Intercropping Enhances the Use of Soil-Derived and Biologically Fixed Nitrogen in Temperate Agroecosystems. A Meta-Analysis. *Eur. J. Agron.* **2020**, *118*, 126077. [CrossRef]
19. Hauggaard-Nielsen, H.; Jensen, E.S. Evaluating Pea and Barley Cultivars for Complementarity in Intercropping at Different Levels of Soil N Availability. *Field Crops Res.* **2001**, *72*, 185–196. [CrossRef]
20. Andersen, M.K.; Hauggaard-Nielsen, H.; Ambus, P.; Jensen, E.S. Biomass Production, Symbiotic Nitrogen Fixation and Inorganic N Use in Dual and Tri-Component Annual Intercrops. *Plant Soil* **2005**, *266*, 273–287. [CrossRef]
21. Nguyen, C. Rhizodeposition of Organic C by Plant: Mechanisms and Controls. In *Sustainable Agriculture*; Lichtfouse, E., Navarrete, M., Debaeke, P., Véronique, S., Alberola, C., Eds.; Springer: Dordrecht, The Netherlands, 2009; pp. 97–123, ISBN 978-90-481-2666-8.
22. Moyer-Henry, K.A.; Burton, J.W.; Israel, D.W.; Rufty, T.W. Nitrogen Transfer between Plants: A ^{15}N Natural Abundance Study with Crop and Weed Species. *Plant Soil* **2006**, *282*, 7–20. [CrossRef]
23. Zhang, T.; Yu, L.; Shao, Y.; Wang, J. Root and Hyphal Interactions Influence N Transfer by Arbuscular Mycorrhizal Fungi in Soybean/Maize Intercropping Systems. *Fungal Ecol.* **2023**, *64*, 101240. [CrossRef]
24. Fujita, K.; Ogata, S.; Matsumoto, K.; Masuda, T.; Ofosu-Budu, G.K.; Kuwata, K. Nitrogen Transfer and Dry Matter Production in Soybean and Sorghum Mixed Cropping System at Different Population Densities. *Soil Sci. Plant Nutr.* **1990**, *36*, 233–241. [CrossRef]
25. Chu, G.X.; Shen, Q.R.; Cao, J.L. Nitrogen Fixation and N Transfer from Peanut to Rice Cultivated in Aerobic Soil in an Intercropping System and Its Effect on Soil N Fertility. *Plant Soil* **2004**, *263*, 17–27. [CrossRef]

26. Gungaabayar, A.; Jha, A.; Warkentin, T.; Knight, D.; Penner, G.; Biligetu, B. Forage Yield and Biological Nitrogen Fixation of Pea–Cereal Intercrops for Hay Production. *Agron. J.* **2023**, *115*, 607–619. [CrossRef]
27. Corre-Hellou, G.; Fustec, J.; Crozat, Y. Interspecific Competition for Soil N and Its Interaction with N2 Fixation, Leaf Expansion and Crop Growth in Pea–Barley Intercrops. *Plant Soil* **2006**, *282*, 195–208. [CrossRef]
28. Neumann, A.; Schmidtke, K.; Rauber, R. Effects of Crop Density and Tillage System on Grain Yield and N Uptake from Soil and Atmosphere of Sole and Intercropped Pea and Oat. *Field Crops Res.* **2007**, *100*, 285–293. [CrossRef]
29. Hauggaard-Nielsen, H.; Gooding, M.; Ambus, P.; Corre-Hellou, G.; Crozat, Y.; Dahlmann, C.; Dibet, A.; von Fragstein, P.; Pristeri, A.; Monti, M.; et al. Pea–Barley Intercropping for Efficient Symbiotic N_2-Fixation, Soil N Acquisition and Use of Other Nutrients in European Organic Cropping Systems. *Field Crop Res.* **2009**, *113*, 64–71. [CrossRef]
30. Bedoussac, L.; Journet, E.-P.; Hauggaard-Nielsen, H.; Naudin, C.; Corre-Hellou, G.; Prieur, L.; Jensen, E.S.; Justes, E. Eco-Functional Intensification by Cereal-Grain Legume Intercropping in Organic Farming Systems for Increased Yields, Reduced Weeds and Improved Grain Protein Concentration. In *Organic Farming, Prototype for Sustainable Agricultures*; Springer: Dordrecht, The Netherlands, 2014; pp. 47–63, ISBN 978-94-007-7927-3.
31. Voisin, A.-S.; Salon, C.; Munier-Jolain, N.G.; Ney, B. Quantitative Effects of Soil Nitrate, Growth Potential and Phenology on Symbiotic Nitrogen Fixation of Pea (*Pisum sativum* L.). *Plant Soil* **2002**, *243*, 31–42. [CrossRef]
32. Naudin, C.; Corre-Hellou, G.; Voisin, A.-S.; Oury, V.; Salon, C.; Crozat, Y.; Jeuffroy, M.-H. Inhibition and Recovery of Symbiotic N_2 Fixation by Peas (*Pisum sativum* L.) in Response to Short-Term Nitrate Exposure. *Plant Soil* **2011**, *346*, 275–287. [CrossRef]
33. Zhao, F.; Sun, Z.; Feng, L.; Zhang, Y.; Feng, C.; Bai, W.; Zheng, J.; Zhang, Z.; Yang, N.; Cai, Q.; et al. Biological N Fixation but Not Mineral N Fertilization Enhances the Accumulation of N in Peanut Soil in Maize/Peanut Intercropping System. *J. Agric. Food Res.* **2022**, *10*, 100365. [CrossRef]
34. Mead, R.; Willey, R.W. The Concept of a 'Land Equivalent Ratio' and Advantages in Yields from Intercropping. *Exp. Agric.* **1980**, *16*, 217–228. [CrossRef]
35. Vandermeer, J.H. *The Ecology of Intercropping*; Cambridge University Press: Cambridge, UK, 1989; ISBN 978-0-521-34689-4.
36. Fetene, M.; Feten, M. Intra- and Inter-Specific Competition between Seedlings of Acacia Etbaica and a Perennial Grass (*Hyparrhenia hirta*). *J. Arid. Environ.* **2003**, *55*, 441–451. [CrossRef]
37. Kaci, G.; Ouaret, W.; Rahmoune, B. Wheat-Faba Bean Intercrops Improve Plant Nutrition, Yield, and Availability of Nitrogen (N) and Phosphorus (P) in Soil. *Agron. Res.* **2022**, *20*, 603–616. [CrossRef]
38. Tavoletti, S.; Merletti, A. A Comprehensive Approach to Evaluate Durum Wheat–Faba Bean Mixed Crop Performance. *Front. Plant Sci.* **2022**, *13*, 733116. [CrossRef] [PubMed]
39. Polthanee, A.; Trelo-ges, V. Growth, Yield and Land Use Efficiency of Corn and Legumes Grown under Intercropping Systems. *Plant Prod. Sci.* **2003**, *6*, 139–146. [CrossRef]
40. Cowden, R.J.; Shah, A.N.; Lehmann, L.M.; Kiær, L.P.; Henriksen, C.B.; Ghaley, B.B. Nitrogen Fertilizer Effects on Pea–Barley Intercrop Productivity Compared to Sole Crops in Denmark. *Sustainability* **2020**, *12*, 9335. [CrossRef]
41. Del Pozo, A.; Garnier, E.; Aronson, J. Contrasted Nitrogen Utilization in Annual C3 Grass and Legume Crops: Physiological Explorations and Ecological Considerations. *Acta Oecol.* **2000**, *21*, 79–89. [CrossRef]
42. Del Pozo, A.; Pérez, P.; Gutiérrez, D.; Alonso, A.; Morcuende, R.; Martínez-Carrasco, R. Gas Exchange Acclimation to Elevated CO_2 in Upper-Sunlit and Lower-Shaded Canopy Leaves in Relation to Nitrogen Acquisition and Partitioning in Wheat Grown in Field Chambers. *Environ. Exp. Bot.* **2007**, *59*, 371–380. [CrossRef]
43. Yu-zheng, Z.; Han-qing, Z.; Ping, L.; Dong-sheng, Z.; Xing-yu, H.; Zhi-qiang, G. Leaf Nitrogen Have a Better Relationship with Photosynthesis Performance across Wheat Species under Elevated CO_2 and Drought. *Plant Physiol. Biochem.* **2021**, *166*, 964–973. [CrossRef] [PubMed]
44. Dordas, C.A.; Lithourgidis, A.S.; Matsi, T.; Barbayiannis, N. Application of Liquid Cattle Manure and Inorganic Fertilizers Affect Dry Matter, Nitrogen Accumulation, and Partitioning in Maize. *Nutr. Cycl. Agroecosyst.* **2008**, *80*, 283–296. [CrossRef]
45. Lithourgidis, A.S.; Dordas, C.A. Forage Yield, Growth Rate, and Nitrogen Uptake of Faba Bean Intercrops with Wheat, Barley, and Rye in Three Seeding Ratios. *Crop Sci.* **2010**, *50*, 2148–2158. [CrossRef]
46. Tosti, G.; Guiducci, M. Durum Wheat–Faba Bean Temporary Intercropping: Effects on Nitrogen Supply and Wheat Quality. *Eur. J. Agron.* **2010**, *33*, 157–165. [CrossRef]
47. Suryapani, S.; Umar, S.; Malik, A.A.; Ahmad, A. Symbiotic Nitrogen Fixation by Lentil Improves Biochemical Characteristics and Yield of Intercropped Wheat Under Low Fertilizer Input. *J. Crop Improv.* **2013**, *27*, 53–66. [CrossRef]
48. Amani Machiani, M.; Javanmard, A.; Morshedloo, M.R.; Maggi, F. Evaluation of Competition, Essential Oil Quality and Quantity of Peppermint Intercropped with Soybean. *Ind. Crops Prod.* **2018**, *111*, 743–754. [CrossRef]
49. Duan, Y.; Shen, J.; Zhang, X.; Wen, B.; Ma, Y.; Wang, Y.; Fang, W.; Zhu, X. Effects of Soybean–Tea Intercropping on Soil-Available Nutrients and Tea Quality. *Acta Physiol. Plant* **2019**, *41*, 140. [CrossRef]
50. Cartelat, A.; Cerovic, Z.G.; Goulas, Y.; Meyer, S.; Lelarge, C.; Prioul, J.-L.; Barbottin, A.; Jeuffroy, M.-H.; Gate, P.; Agati, G.; et al. Optically Assessed Contents of Leaf Polyphenolics and Chlorophyll as Indicators of Nitrogen Deficiency in Wheat (*Triticum aestivum* L.). *Field Crops Res.* **2005**, *91*, 35–49. [CrossRef]
51. Wang, R.; Sun, Z.; Bai, W.; Wang, E.; Wang, Q.; Zhang, D.; Zhang, Y.; Yang, N.; Liu, Y.; Nie, J.; et al. Canopy Heterogeneity with Border-Row Proportion Affects Light Interception and Use Efficiency in Maize/Peanut Strip Intercropping. *Field Crops Res.* **2021**, *271*, 108239. [CrossRef]

52. Pellicanò, A.; Romeo, M.; Pristeri, A.; Preiti, G.; Monti, M. Cereal-Pea Intercrops to Improve Sustainability in Bioethanol Production. *Agron. Sustain. Dev.* **2015**, *35*, 827–835. [CrossRef]
53. Szumigalski, A.R.; Van Acker, R.C. Land Equivalent Ratios, Light Interception, and Water Use in Annual Intercrops in the Presence or Absence of In-Crop Herbicides. *Agron. J.* **2008**, *100*, 1145–1154. [CrossRef]
54. Du, X.; Chen, B.; Shen, T.; Zhang, Y.; Zhou, Z. Effect of Cropping System on Radiation Use Efficiency in Double-Cropped Wheat–Cotton. *Field Crops Res.* **2015**, *170*, 21–31. [CrossRef]
55. Liu, X.; Rahman, T.; Song, C.; Yang, F.; Su, B.; Cui, L.; Bu, W.; Yang, W. Relationships among Light Distribution, Radiation Use Efficiency and Land Equivalent Ratio in Maize-Soybean Strip Intercropping. *Field Crops Res.* **2018**, *224*, 91–101. [CrossRef]
56. Jensen, E.S.; Carlsson, G.; Hauggaard-Nielsen, H. Intercropping of Grain Legumes and Cereals Improves the Use of Soil N Resources and Reduces the Requirement for Synthetic Fertilizer N: A Global-Scale Analysis. *Agron. Sustain. Dev.* **2020**, *40*, 5. [CrossRef]
57. Boutagayout, A.; Belmalha, S.; Nassiri, L.; El Alami, N.; Jiang, Y.; Lahlali, R.; Bouiamrine, E.H. Weed Competition, Land Equivalent Ratio and Yield Potential of Faba Bean (*Vicia faba* L.)-Cereals (*Triticum aestivum* L. and/or *Avena sativa* L.) Intercropping under Low-Input Conditions in Meknes Region, Morocco. *Vegetos* **2023**, 1–14. [CrossRef]
58. Nurgi, N.; Tana, T.; Dechassa, N.; Tesso, B.; Alemayehu, Y. Effect of Spatial Arrangement of Faba Bean Variety Intercropping with Maize on Yield and Yield Components of the Crops. *Heliyon* **2023**, *9*, e16751. [CrossRef]
59. Bedoussac, L.; Justes, E. A Comparison of Commonly Used Indices for Evaluating Species Interactions and Intercrop Efficiency: Application to Durum Wheat–Winter Pea Intercrops. *Field Crops Res.* **2011**, *124*, 25–36. [CrossRef]
60. Sahota, T.S.; Malhi, S.S. Intercropping Barley with Pea for Agronomic and Economic Considerations in Northern Ontario. *Agric. Sci.* **2012**, *3*, 889–895. [CrossRef]
61. Monti, M.; Pellicanò, A.; Santonoceto, C.; Preiti, G.; Pristeri, A. Yield Components and Nitrogen Use in Cereal-Pea Intercrops in Mediterranean Environment. *Field Crop Res.* **2016**, *196*, 379–388. [CrossRef]
62. Dhima, K.V.; Vasilakoglou, I.B.; Keco, R.X.; Dima, A.K.; Paschalidis, K.A.; Gatsis, T.D. Forage Yield and Competition Indices of Faba Bean Intercropped with Oat. *Grass Forage Sci.* **2014**, *69*, 376–383. [CrossRef]
63. Ghaley, B.B.; Hauggaard-Nielsen, H.; Høgh-Jensen, H.; Jensen, E.S. Intercropping of Wheat and Pea as Influenced by Nitrogen Fertilization. *Nutr. Cycl. Agroecosyst.* **2005**, *73*, 201–212. [CrossRef]
64. Brooker, R.W.; Bennett, A.E.; Cong, W.-F.; Daniell, T.J.; George, T.S.; Hallett, P.D.; Hawes, C.; Iannetta, P.P.M.; Jones, H.G.; Karley, A.J.; et al. Improving Intercropping: A Synthesis of Research in Agronomy, Plant Physiology and Ecology. *New Phytol.* **2015**, *206*, 107–117. [CrossRef]
65. Jensen, E.S.; Bedoussac, L.; Carlsson, G.; Journet, E.-P.; Justes, E.; Hauggaard-Nielsen, H. Enhancing Yields in Organic Crop Production by Eco-Functional Intensification. *Sustain. Agric. Res.* **2015**, *4*, 42–50. [CrossRef]
66. Saia, S.; Urso, V.; Amato, G.; Frenda, A.S.; Giambalvo, D.; Ruisi, P.; Di Miceli, G. Mediterranean Forage Legumes Grown Alone or in Mixture with Annual Ryegrass: Biomass Production, N2 Fixation, and Indices of Intercrop Efficiency. *Plant Soil* **2016**, *402*, 395–407. [CrossRef]
67. Bacchi, M.; Monti, M.; Calvi, A.; Lo Presti, E.; Pellicanò, A.; Preiti, G. Forage Potential of Cereal/Legume Intercrops: Agronomic Performances, Yield, Quality Forage and LER in Two Harvesting Times in a Mediterranean Environment. *Agronomy* **2021**, *11*, 121. [CrossRef]
68. Guinet, M.; Nicolardot, B.; Revellin, C.; Durey, V.; Carlsson, G.; Voisin, A.-S. Comparative Effect of Inorganic N on Plant Growth and N_2 Fixation of Ten Legume Crops: Towards a Better Understanding of the Differential Response among Species. *Plant Soil* **2018**, *432*, 207–227. [CrossRef]
69. Liu, Y.; Yin, X.; Xiao, J.; Tang, L.; Zheng, Y. Interactive Influences of Intercropping by Nitrogen on Flavonoid Exudation and Nodulation in Faba Bean. *Sci. Rep.* **2019**, *9*, 4818. [CrossRef] [PubMed]
70. Banik, P.; Sharma, R.C. Yield and Resource Utilization Efficiency in Baby Corn—Legume-Intercropping System in the Eastern Plateau of India. *J. Sustain. Agric.* **2009**, *33*, 379–395. [CrossRef]
71. Li, Y.-Y.; Yu, C.-B.; Cheng, X.; Li, C.-J.; Sun, J.-H.; Zhang, F.-S.; Lambers, H.; Li, L. Intercropping Alleviates the Inhibitory Effect of N Fertilization on Nodulation and Symbiotic N_2 Fixation of Faba Bean. *Plant Soil* **2009**, *323*, 295–308. [CrossRef]
72. Hu, F.; Zhao, C.; Feng, F.; Chai, Q.; Mu, Y.; Zhang, Y. Improving N Management through Intercropping Alleviates the Inhibitory Effect of Mineral N on Nodulation in Pea. *Plant Soil* **2017**, *412*, 235–251. [CrossRef]
73. Hauggaard-Nielsen, H.; Andersen, M.K.; Jørnsgaard, B.; Jensen, E.S. Density and Relative Frequency Effects on Competitive Interactions and Resource Use in Pea–Barley Intercrops. *Field Crop Res.* **2006**, *95*, 256–267. [CrossRef]
74. Zhao, C.; Fan, Z.; Coulter, J.A.; Yin, W.; Hu, F.; Yu, A.; Fan, H.; Chai, Q. High Maize Density Alleviates the Inhibitory Effect of Soil Nitrogen on Intercropped Pea. *Agronomy* **2020**, *10*, 248. [CrossRef]
75. Thilakarathna, M.S.; McElroy, M.S.; Chapagain, T.; Papadopoulos, Y.A.; Raizada, M.N. Belowground Nitrogen Transfer from Legumes to Non-Legumes under Managed Herbaceous Cropping Systems. A Review. *Agron. Sustain. Dev.* **2016**, *36*, 58. [CrossRef]
76. Naudin, C.; Corre-Hellou, G.; Pineau, S.; Crozat, Y.; Jeuffroy, M.-H. The Effect of Various Dynamics of N Availability on Winter Pea–Wheat Intercrops: Crop Growth, N Partitioning and Symbiotic N_2 Fixation. *Field Crop Res.* **2010**, *119*, 2–11. [CrossRef]
77. Wang, X.; Gao, Y.; Zhang, H.; Shao, Z.; Sun, B.; Gao, Q. Enhancement of Rhizosphere Citric Acid and Decrease of NO_3^-/NH_4^+ Ratio by Root Interactions Facilitate N Fixation and Transfer. *Plant Soil* **2020**, *447*, 169–182. [CrossRef]

78. Bedoussac, L.; Justes, E. The Efficiency of a Durum Wheat-Winter Pea Intercrop to Improve Yield and Wheat Grain Protein Concentration Depends on N Availability during Early Growth. *Plant Soil* **2010**, *330*, 19–35. [CrossRef]
79. Jiao, N.; Wang, F.; Ma, C.; Zhang, F.; Jensen, E.S. Interspecific Interactions of Iron and Nitrogen Use in Peanut (*Arachis hypogaea* L.)-Maize (*Zea mays* L.) Intercropping on a Calcareous Soil. *Eur. J. Agron.* **2021**, *128*, 126303. [CrossRef]
80. Zhu, S.-G.; Zhu, H.; Zhou, R.; Zhang, W.; Wang, W.; Zhou, Y.-N.; Wang, B.-Z.; Yang, Y.-M.; Wang, J.; Tao, H.-Y.; et al. Intercrop Overyielding Weakened by High Inputs: Global Meta-Analysis with Experimental Validation. *Agric. Ecosyst. Environ.* **2023**, *342*, 108239. [CrossRef]
81. Li, C.; Hoffland, E.; Kuyper, T.W.; Yu, Y.; Zhang, C.; Li, H.; Zhang, F.; van der Werf, W. Syndromes of Production in Intercropping Impact Yield Gains. *Nat. Plants* **2020**, *6*, 653–660. [CrossRef]
82. Unkovich, M.; Herridge, D.; Peoples, M.; Cadisch, G.; Boddey, B.; Giller, K.; Alves, B.; Chalk, P.M. *Measuring Plant-Associated Nitrogen Fixation in Agricultural Systems*; ACIAR Monograph; Australian Centre for International Agricultural Research: Canberra, ACT, Australia, 2008.
83. Kohl, D.H.; Shearer, G. Isotopic Fractionation Associated with Symbiotic N_2 Fixation and Uptake of NO_3^- by Plants. *Plant Physiol.* **1980**, *66*, 51–56. [CrossRef]

plants

MDPI

Article

Partial Substitution of Nitrogen Fertilizer with Biogas Slurry Increases Rice Yield and Fertilizer Utilization Efficiency, Enhancing Soil Fertility in the Chaohu Lake Basin

Yangting Lu [1], Qian Xiao [1], Sheng Wu [1], Haoqiang Yuan [1], Tingfeng Gao [1], Tianpei Cai [1], Xiaowen Wu [2], Youhua Ma [1] and Xia Liao [1,*]

1 Key Laboratory of Farmland Ecological Conservation and Pollution Prevention and Control in Anhui Province, College of Resources and Environment, Anhui Agricultural University, Hefei 230036, China; luyt0512@163.com (Y.L.)

2 Agricultural Technique Extension Center of Lujiang County in Anhui Province, Hefei 231500, China

* Correspondence: xliao@ahau.edu.cn

Abstract: To investigate the effects of biogas slurry substitution for fertilizer on rice yield, fertilizer utilization efficiency, and soil fertility, a field experiment was conducted on rice–wheat rotation soil in the Chaohu Lake Basin for two consecutive years, with the following six treatments: no fertilization (CK), conventional fertilization (CF), optimized fertilization (OF), biogas slurry replacing 15% of fertilizer (15% OFB), biogas slurry replacing 30% of fertilizer (30% OFB), and biogas slurry replacing 50% of fertilizer (50% OFB). The field experiment results showed that, compared with CF treatment, OF treatment in 2022 and 2023 significantly increased ($p < 0.05$) rice yield, promoted the uptake of nitrogen (N), phosphorus (P), and potassium (K) by grains and straws, improved fertilizer utilization efficiency, and increased the contents of soil organic C (SOC), NH_4^+-N, NO_3^--N, hydrolysable N, and available P. The 15% OFB and 30% OFB treatments significantly increased ($p < 0.05$) rice grain and straw yields compared with CF treatment, and rice grain and straw yields were the highest in the 30% OFB treatment. Compared with CF and OF treatments, 30% OFB treatment significantly increased ($p < 0.05$) the N, P, and K uptake of grains and straws and increased the fertilizer utilization efficiency. Compared with CF treatment, the grain yield of 50% OFB treatment was significantly decreased ($p < 0.05$) in 2022, and there was no significant difference in 2023, which may be because the biogas slurry was applied before planting in 2023 to provide more nutrients for early rice growth. Compared with CF treatment, 30% OFB treatment significantly increased ($p < 0.05$) the contents of SOC, NH_4^+-N, available K, and hydrolysable N. In summary, optimizing N and K topdressing methods can increase rice yield and improve the fertilizer utilization efficiency and soil fertility. The 30% OFB treatment resulted in the highest rice yield, fertilizer utilization efficiency, and improved soil fertility, indicating that biogas slurry replacing 30% of fertilizer was the best application mode for rice in this region.

Keywords: biogas slurry; fertilizer optimization; rice yield; fertilizer utilization efficiency; soil fertility

1. Introduction

Rice is an important food crop in China and plays an important role in ensuring national food security. Due to the high yield of rice, farmers tend to apply a large amount of fertilizer, which has led to some negative consequences, such as low fertilizer utilization efficiency and environmental pollution [1–3]. Currently, China's fertilizer production and consumption are 3.7 times that of the world average, and the fertilizer intensity and application amount per unit area far exceed the world average [1]. The amount of synthetic fertilizer application in China has increased from7.8 $\times$ 10^4 tons in the 1950s to 6.02 $\times$ 10^7 tons, increasing at a rate of 9.8% per year [1]. The amount of fertilizer applied in the central region of Anhui Province ranked first among the four major regions of east, central, west, and northeast China [2]. From 2006 to 2015, the amount of fertilizer applied in Anhui Province increased by 1.4×10^7 tons, with an annual growth rate of 41.41% [2]. Nitrogen (N) is an essential nutrient element for increasing crop yield and improving crop quality [3]. From 2010 to 2015, the average N fertilizer application rate in China was the highest across the world, but the N fertilizer utilization efficiency in the current season was less than 35% [4], resulting in economic losses, biodiversity reduction, and environmental pollution. Therefore, N fertilizer application technology is a difficult problem that needs to be overcome in agriculture at present [5]. A short-term effect of excessive fertilizer application was the increase in crop yield, but the long-term effects include crop yield reducing, crop quality declining, fertilizer utilization efficiency declining, soil acidification, nutrient loss, and environmental pollution [6].

With the development of animal husbandry, the discharge of livestock and poultry excrement is increasing. Without safe treatment before discharging, it is directly discharged into the environment, causing environmental pollution and ultimately endangering human health [7]. Data have shown that in 2020, Lujiang County, which belongs to Hefei City, Anhui Province, raised 1.50×10^5 pigs, 7.68×10^5 laying hens, and 2.44×10^6 meat and poultry animals, with an annual output of 1.08×10^6 tons of livestock and poultry manure [8]. China produces 3.8 billion tons of livestock and poultry manure every year, and its utilization rate is 70% [9]. However, the allocational rate of livestock and poultry manure treatment facilities is only 63% [9]. Thus, it is necessary to further improve the safe treatment and resource utilization of livestock and poultry manure [9]. In the process of livestock manure composting and biogas engineering, the N and P nutrient contents of livestock and poultry manure, were reduced by 22% and 10%, respectively, and the available N and P nutrient supply amounts were 0.9×10^4 and 0.3×10^4 t, respectively [10,11]. Therefore, the rational consumption and resource-oriented utilization of livestock and poultry manure is an urgent problem to be solved.

A biogas slurry is an organic liquid fertilizer produced by the anaerobic fermentation of livestock and poultry manure. It was rich in N, P, and K, as well as microelements, such as iron, calcium, and copper [12]. Partial replacement of chemical fertilizer with biogas slurry can reduce the amount of chemical fertilizer needed, enrich the soil, and enable the resource consumption of livestock and poultry manure [10]. Compared with conventional fertilizer application, replacing 25% of fertilizer with biogas slurry can increase the soil available K and available P contents, while replacing 100% of fertilizer with biogas slurry can significantly increase the soil total N, organic C, available P, and available K contents [13]. In recent years, there have been many studies on the application of biogas slurry as a fertilizer replacement for different crops. However, the effects of different biogas slurry proportions on yield and fertilizer utilization efficiency were inconsistent. Compared with conventional fertilizer application, the replacement of 15% and 30% of fertilizer with biogas slurry can significantly increase wheat yield, while the replacement of 50% of fertilizer can significantly reduce wheat yield [14]. Replacing 40% of fertilizer with biogas slurry can increase the yield of summer corn and winter wheat and improve the fertilizer utilization efficiency [15,16]. However, the replacement of 50% and 75% of fertilizer with biogas slurry increased rice yield [17]. The effects of biogas slurry replacement of fertilizer on different crops and soil nutrients were jointly determined by the biogas slurry content, replacement ratio, soil properties, climate factors, and farming systems [14]. Therefore, research on the optimal ratio of biogas slurry to replace fertilizer should be carried out in specific regions, aiming at providing a theoretical basis for biogas slurry consumption and fertilizer reduction in the region.

In this study, field experiments with biogas slurry as a fertilizer replacement were carried out in a rice–wheat rotation system in the Chaohu Lake Basin, and three ratios of biogas slurry were set to replace fertilizer (15%, 30%, and 50%) to explore the effects of different ratios of biogas slurry on rice yield, fertilizer utilization efficiency, and soil fertility. In two years of field trials, the optimal ratio of biogas slurry as a fertilizer replacement in the Chaohu Lake Basin was determined, providing a scientific basis for the quality and sustainable development of agricultural products under rice–wheat rotation in central Anhui.

2. Materials and Methods

2.1. Study Site

The field experiment was conducted from June 2022 to November 2023 in Guanghan Village, Guohe Town, Lujiang County, Hefei City, Anhui Province (31°29′ N, 117°12′ E), which belongs to the subtropical monsoon climate zone with a mild climate, abundant rainfall, and an annual average temperature of 15.9 °C. The average precipitation was 1262.9 mm. The soil was derived from alluvial sediments, and the physical and chemical properties of the 0–20 cm soil layer were as follows: soil pH, 4.80, bulk density, 1.01 g·cm^{-3}; SOC, 9.79 g·kg^{-1}; TN, 1.11 g·kg^{-1}; NH_4^+-N, 9.24 mg·kg^{-1}; NO_3^--N, 1.56 mg·kg^{-1}; hydrolysable N, 109.67 mg·kg^{-1}; available P, 10.63 mg·kg^{-1}; and available K, 124.67 mg·kg^{-1}.

2.2. Field Experiment

The field experiment began on 15 June 2022, and consisted of six treatments: no fertilization (CK), conventional fertilization (CF), optimized fertilization (OF), biogas slurry replacing 15% of fertilizer (15% OFB), biogas slurry replacing 30% of fertilizer (30% OFB), and biogas slurry replacing 50% of fertilizer (50% OFB). Each treatment was performed in three repetitions. The plots were designed in a completely random block design, with an area of 30 m^2 (5 m × 6 m), which were separated by a ridge covered with plastic film. Irrigation and drainage channels were used to eliminate the mutual interference of water and fertilizer among the plots, and protection lines were set.

The chemical fertilizers tested were urea (N ≥ 46.4%), superphosphate (P_2O_5 ≥ 12%), and potassium chloride (K_2O ≥ 60%). The biogas slurry was obtained from Wugongshan Livestock and Poultry Breeding Co., Ltd., Hefei, China. The nutrient content of the biogas slurry was measured before application. The N, P_2O_5, and K_2O contents were 0.15%, 0.02%, and 0.03%, respectively, before application in the 2022 rice season. In 2023, the N, P_2O_5, and K_2O contents in the biogas slurry were 0.11%, 0.05%, and 0.03%, respectively.

The amount of fertilization in each treatment is shown in Table 1. P fertilizer treatment was applied as a base fertilizer, and N and K fertilizers were divided into base fertilizer and topdressing fertilizer. N fertilizer treatment for conventional fertilization was applied as 50% base fertilizer, 35% tillering fertilizer, and 15% booting fertilizer, and K fertilizer was applied as 50% base fertilizer and 50% booting fertilizer. The optimized fertilization treatment of N fertilizer was applied as 50% base fertilizer, 30% tillering fertilizer, and 20% booting fertilizer, and K fertilizer was applied as 45% base fertilizer, 15% tillering fertilizer, and 40% booting fertilizer. In 2022, the biogas slurry was applied as a tillering fertilizer, and in 2023, the biogas slurry was applied as a base fertilizer. The application of N, P, and K fertilizers was consistent with the optimal fertilization treatment. All fertilizers were distributed manually, and the management measures of the field plot were consistent with local practices.

The rice variety planted was Wankenuo No. 2, and plant space was 30 cm × 14 cm, with two or three seedlings in each hole. The specific application times of base fertilizer, tillering fertilizer, booting fertilizer, and rice harvest time are shown in Table 2.

2.3. Collection and Analysis of Soil and Plant Samples

Before the experiment and after rice harvest, fresh soil samples were collected from the 0–20 cm layer by soil drilling in each plot according to the five-point sampling method. After fully mixing, a portion of the soil samples was stored at 4 °C. The other portion of the soil samples were air-dried and screened with 2 and 0.15 mm sieves. The methods for determining the physical and chemical properties of the soil are shown in Table 3.

Table 1. Fertilizer amount of different treatments in 2022 and 2023 rice seasons.

Years	Treatment	Organic Fertilizer (kg·ha^{-1})	Fertilizer (kg·ha^{-1})			Converted to Pure Nutrients (kg·ha^{-1})		
		Biogas Slurry	Urea	Superphosphate	Potassium Chloride	N	P_2O_5	K_2O
2022	CK	0	0	0	0	0	0	0
	CF	0	485	500	200	225	60	120
	OF	0	485	500	200	225	60	120
	15% OFB	22,500	412	462	189	225	60	120
	30% OFB	45,000	339	425	177	225	60	120
	50% OFB	75,000	242	375	162	225	60	120
2023	CK	0	0	0	0	0	0	0
	CF	0	485	500	200	225	60	120
	OF	0	485	500	200	225	60	120
	15% OFB	30,682	412	372	185	225	60	120
	30% OFB	61,363	339	244	169	225	60	120
	50% OFB	102,272	242	74	149	225	60	120

Table 2. Field management time in 2022 and 2023 rice seasons.

Management Measure	2022	2023
Base fertilizer	15 June 2022	9 June 2023
Booting fertilizer	11 July 2022	13 July 2023
Tillering fertilizer	10 August 2022	17 August 2023
Harvest	2 November 2022	15 November 2023

Table 3. Soil measurement indexes and methods.

Sample	Measurement Indexes	Measurement Methods
Soil	pH	Potentiometry
	SOC	Potassium dichromate oil bath plus heat capacity method
	TN	Kjeldahl method for nitrogen determination
	NH_4^+-N	Indophenol blue colorimetry
	NO_3^--N	Ultraviolet spectrophotometry
	Available P	Sodium bicarbonate extraction-molybdenum antimony resistance spectrophotometry
	Available K	Flame photometer method
	Hydrolysable N	Alkalolytic diffusion method

During the mature stage of rice, straw and grain yields were measured separately in each plot. Plant samples from each plot were collected at multiple points using the double diagonal method to determine plant height, 1000-grain weight, and kernels per spike. The samples were threshed separately according to straw and grain and then crushed for N, P, and K content analysis in grain and straw. The N, P, and K contents in plants were determined using the Kellner method, molybdenum yellow colorimetry, and flame spectrophotometry, respectively.

2.4. Calculations and Statistical Analysis

N (P and K) uptake in rice grain was calculated by the product of rice grain yield and N (P and K) content. N (P and K) uptake of rice straw was calculated by the product of rice straw yield and straw N (P and K) uptake content. The fertilizer utilization efficiency was determined by the difference in N (P and K) uptake between rice grain and straw in fertilized and non-fertilized areas, and the ratio of the N (P and K) application amount.

$$\text{N (P and K)uptake in rice grain (NUG, PUG, KUG, kg·ha}^{-1}\text{)} = GY \times GN(GP, GK) \quad (1)$$

$$\text{N (P and K)uptake of rice straw (NUS, PUS, KUS, kg·ha}^{-1}\text{)} = SY \times SN\ (SP, SK) \quad (2)$$

$$\text{N utilization efficiency (NUE, \%)} = \frac{(NUG + NUS) - (NUG_0 + NUS_0)}{TNA} \times 100 \quad (3)$$

$$\text{P utilization efficiency (PUE, \%)} = \frac{(PUG + PUS) - (PUG_0 + PUS_0)}{TPA} \times 100 \quad (4)$$

$$\text{K utilization efficiency (KUE, \%)} = \frac{(KUG + KUS) - (KUG_0 + KUS_0)}{TKA} \times 100 \quad (5)$$

GY and *SY* refer to the rice grain yield and straw yield (kg·ha^{-1}); *GN* and *SN* refer to N content of rice grain and straw (%); *GP* and *SP* refer to P content of rice grain and straw (%); *GK* and *SK* refer to K content of rice grain and straw (%); *NUG*, *NUS* (*PUG*, *PUS*, *KUG*, *KUS*) refer to N (P, K) uptake by rice grains and straws in fertilized area (kg·ha^{-1}); NUG_0, NUS_0 (PUG_0, PUS_0, KUG_0, KUS_0) refer to N (P, K) uptake by rice grains and straws in non-fertilized area (kg·ha^{-1}); *TNA* (*TPA*, *TKA*) refer to total N (P, K) application (kg·ha^{-1}).

The IBM-SPSS Statistics 23 software package (SPSS, Chicago, IL, USA) was used to perform a one-way analysis of variance (LSD) between processing treatments, and $p < 0.05$ was considered a significant difference. Origin 2023 (Systat Software, Inc., Washington, DC, USA) software was used for mapping.

3. Results

3.1. Analysis of the Nutrient Content of Biogas Slurries in Lujiang County

The nutrient content of biogas slurries sampled from 12 townships of Lujiang County (including Guohe Town, Jinniu Town, Ketan Town, and so on) in 2022 and 2023 was analyzed. The field experiment site, location, and quantity of biogas slurry sampling are shown in Figure 1.

As shown in Figure 2, the N content of the biogas slurry was 0.08–0.18%, and the P and K contents were 0.01–0.05%. The soil organic C (SOC) was 0.01–0.05%, and the pH was 7.08–8.40.

3.2. Rice Yield

The grain and straw yields of rice in 2022 were 6737.39–9527.44 and 6654.98–9019.71 kg·ha^{-1}, respectively (Figure 3). In 2023, rice grain and straw yields were 6508.60–8346.14 and 5857.97–7548.34 kg·ha^{-1}, respectively, and the rice grain yield under all treatments was higher in 2022 than in 2023. Compared with CF treatment, grain and straw yields of OF treatment increased in 2022 and 2023, with grain yield increasing by 1.67% and 7.87%, respectively, and straw yield increasing by 7.20% and 14.58%, respectively. In 2022 and 2023, the rice grain yield under 15% OFB and 30% OFB treatments significantly increased ($p < 0.05$) compared with CF treatment. The rice grain yield was the highest after 30% OFB treatment in two years. In 2022, the grain yield of 50% OFB treatment was significantly lower than that of CF treatment ($p < 0.05$) by 4.52%, while there was no significant difference between 50% OFB treatment and CF treatment in 2023. In comparison to the OF treatment, the yields of rice grain and straw showed an increasing trend in both the 15% OFB and 30% OFB treatments during 2022. Furthermore, in 2023, only the 30% OFB treatment led to an elevation in both rice grain and straw yields. With the increase in the proportion of biogas slurry replacing fertilizer, the rice yield in 2022 and 2023 showed a trend of first increasing and then decreasing (30% OFB > 15% OFB > 50% OFB).

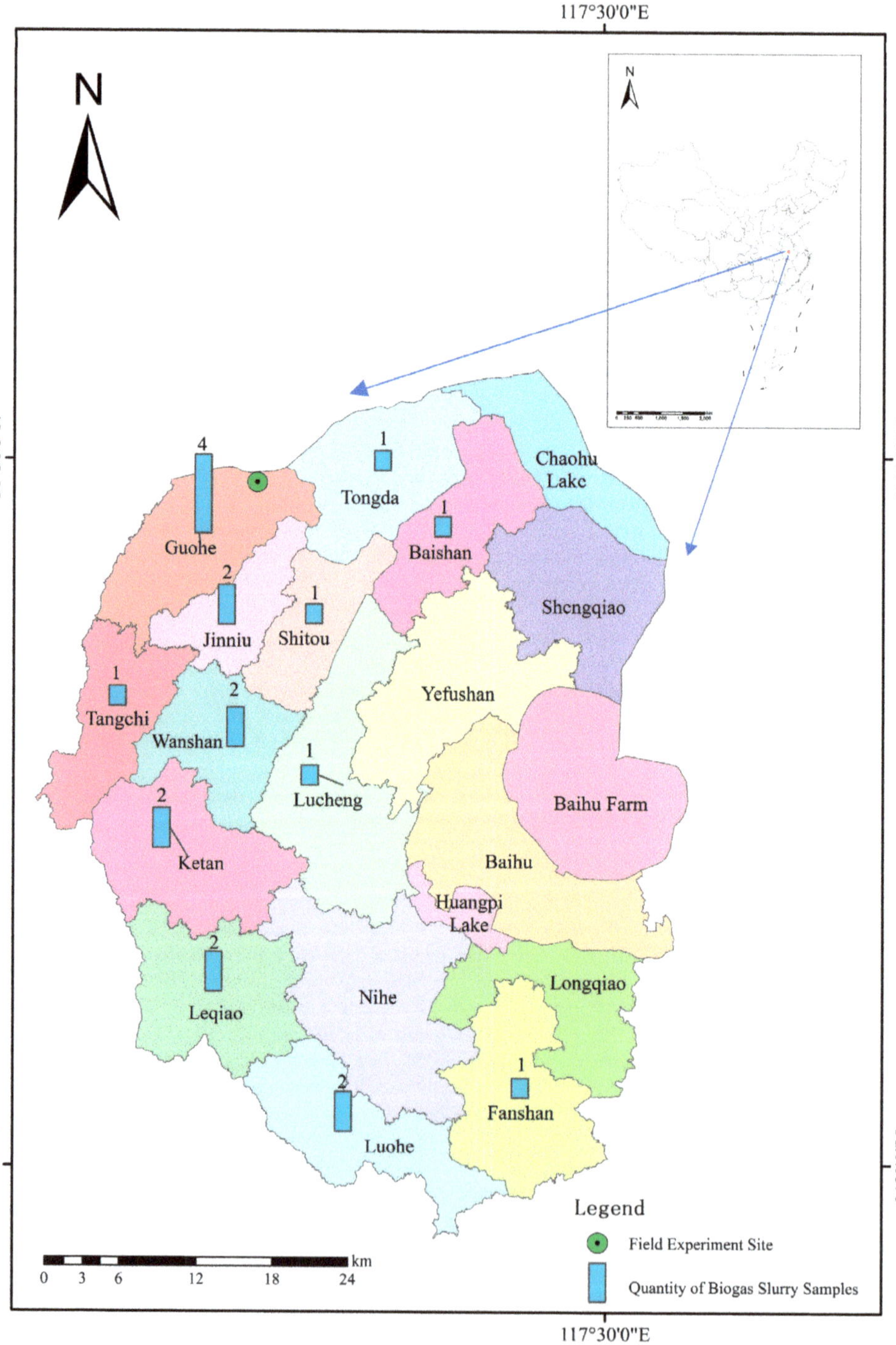

Figure 1. Field experiment site, location, and quantity of biogas slurry samples.

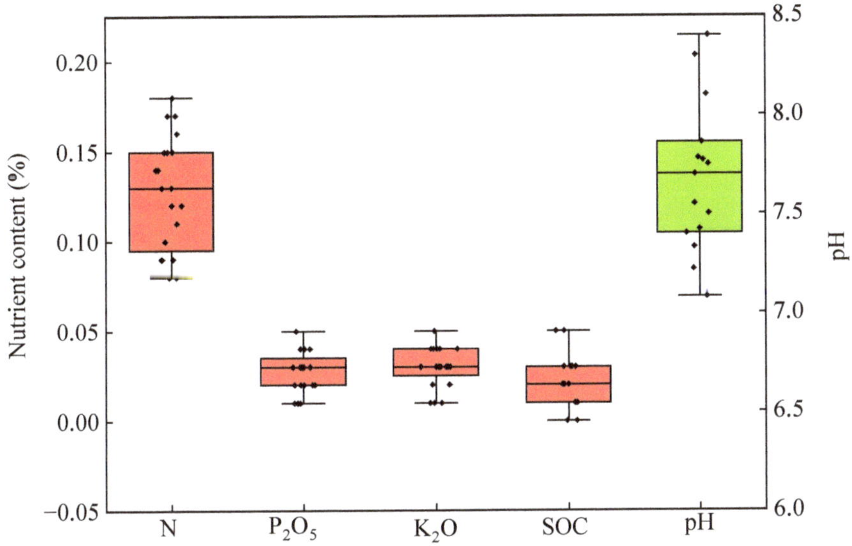

Figure 2. Analysis of nutrient content of biogas slurry in twelve townships of Lujiang County.

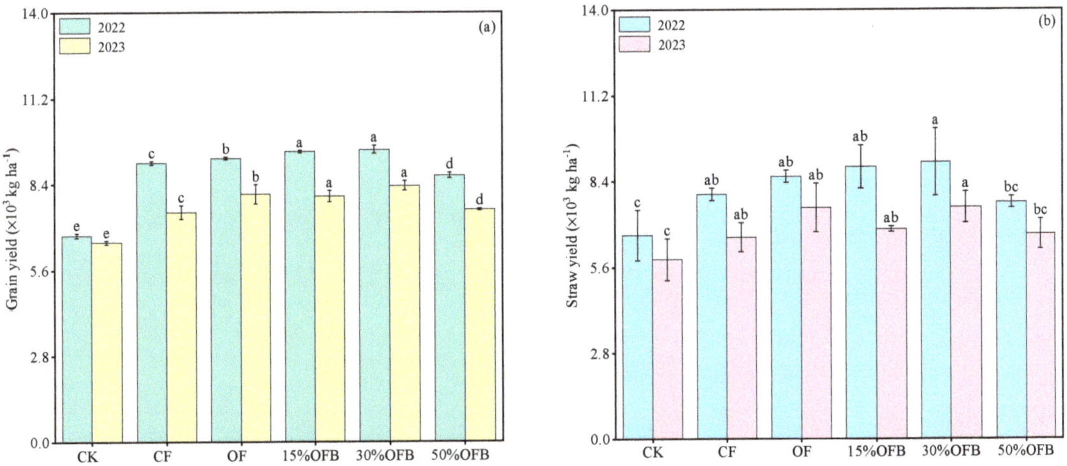

Figure 3. Effects of biogas slurry substitution for fertilizer on grain yield (**a**) and straw yield (**b**) of rice. Bars represent the SD of the means (n = 3), and the different lowercase letters above the column indicate that the differences among treatments are significant at $p < 0.05$.

In 2022 and 2023, the plant height, 1000-grain weight, and kernels per spike of rice under the OF treatment increased compared with CF treatment, but the differences were not statistically significant (Table 4). Compared with CF treatment, the plant height, 1000-grain weight, and kernels per spike of rice treated with 15% OFB, 30% OFB, and 50% OFB increased in 2022 and 2023. The 1000-grain weight and kernels per spike of rice treated with 30% OFB showed significant differences ($p < 0.05$), with the 1000-grain weight increasing by 25.72% and 6.82% and kernels per spike increasing by 61.19% and 36.81%, respectively. Compared with OF treatment, the 1000-grain weight of rice treated with 30% OFB in 2022 and 2023 significantly increased ($p < 0.05$) by 25.72% and 6.50%, respectively. With the increase in the proportion of biogas slurry replacing fertilizer, the composition factors of rice yield in 2022 and 2023 showed a trend of first increasing and then decreasing.

Table 4. Effects of biogas slurry substitution for fertilizer on rice yield components.

Years	Treatment	Plant Height (cm)	Thousand Grain Weight (g)	Kernels per Spike (kernel)
2022	CK	70.53 ± 2.84 b	20.81 ± 0.92 c	124.33 ± 5.13 e
	CF	77.73 ± 2.73 a	22. 51 ± 0.34 b	146.00 ± 3.00 d
	OF	78.40 ± 2.03 a	22. 91 ± 0.44 b	172.67 ± 6.81 c
	15% OFB	79.73 ± 1.30 a	23.05 ± 0.92 b	196.00 ± 5.20 b
	30% OFB	81.80 ± 2.80 a	28.30 ± 0.84 a	235.33 ± 5.69 a
	50% OFB	78.53 ± 1.67 a	22.68 ± 0.42 b	191.00 ± 6.93 b
2023	CK	78.00 ± 4.36 b	23.27 ± 1.32 c	122.00 ± 20.07 b
	CF	90.00 ± 2.00 a	24.53 ± 0.55 bc	144.00 ± 11.53 b
	OF	90.67 ± 1.15 a	24.60 ± 0.66 bc	170.00 ± 34.70 ab
	15% OFB	91.67 ± 2.08 a	25.67 ± 0.32 ab	193.33 ± 8.62 a
	30% OFB	94.00 ± 2.00 a	26.20 ± 0.79 a	197.00 ± 46.60 a
	50% OFB	92.67 ± 1.15 a	24.87 ± 0.15 ab	184.00 ± 40.85 a

Note: Data are presented as mean ± SD (n = 3). Different lowercase letters within each column indicate that the differences among treatments are significant at $p < 0.05$.

3.3. Fertilizer Utilization Efficiency

Compared with CF treatment, the N, P, and K uptake of rice grains under OF treatment increased by 12.83%, 4.77%, and 33.20%, respectively, in 2022 and by 4.46%, 23.96%, and 7.64%, respectively, in 2023 (Table 5). In contrast to CF treatment, the N uptake of rice straws under OF treatment increased by 5.50% and 3.90%, the P uptake increased by 15.76% and 27.89%, and the K uptake increased by 8.78% and 45.65% in 2022 and 2023, respectively. Compared with CF treatment, in 2022, 15% OFB and 30% OFB treatments significantly increased ($p < 0.05$) the N, P, and K uptake of rice grains, and the 30% OFB treatment significantly increased ($p < 0.05$) the N, P, and K uptake in rice straw. In 2023, the N, P, and K uptake of rice grains and straws treated with 15% OFB, 30% OFB, and 50% OFB were significantly increased ($p < 0.05$) compared with those treated with CF. When compared to OF treatment, in 2022 and 2023, 30% OFB treatment significantly increased ($p < 0.05$) the N, P, and K uptake in grains and straws, and the 15% OFB and 50% OFB treatments significantly increased ($p < 0.05$) the K uptake of grain and straw. With the increase in the proportion of biogas slurry replacing fertilizer, the N, P, and K uptake in grain and straw in 2022 and 2023 showed a trend of first increasing and then decreasing (30% OFB > 15% OFB > 50% OFB).

Table 5. Effects of biogas slurry substitution for fertilizer on nutrient uptake of rice.

Years	Treatment	Grain Nutrient Uptake (kg·ha^{-1})			Straw Nutrient Uptake (kg·ha^{-1})		
		N	P	K	N	P	K
2022	CK	51.28 ± 9.34 d	10.05 ± 2.91 d	6.26 ± 1.09 e	28.12 ± 8.76 c	7.43 ± 1.05 d	22.33 ± 2.85 d
	CF	77.54 ± 7.98 c	16.77 ± 1.00 c	15.18 ± 1.46 d	41.67 ± 4.83 b	9.39 ± 2.00 cd	31.42 ± 1.32 c
	OF	87.49 ± 9.66 bc	17.57 ± 0.51 bc	20.22 ± 3.04 c	43.96 ± 1.32 b	10.87 ± 0.23 bc	34.18 ± 1.25 c
	15% OFB	94.05 ± 8.34 b	20.45 ± 1.44 ab	26.86 ± 0.57 b	47.07 ± 4.34 b	12.80 ± 1.34 b	53.94 ± 5.07 b
	30% OFB	111.89 ± 7.59 a	23.28 ± 0.46 a	35.22 ± 2.12 a	63.84 ± 3.38 a	17.27 ± 0.81 a	67.76 ± 5.92 a
	50% OFB	90.25 ± 2.30 bc	18.90 ± 1.95 bc	23.69 ± 1.56 b	44.99 ± 2.39 b	11.65 ± 0.75 b	49.38 ± 1.68 b
2023	CK	75.29 ± 1.60 e	11.89 ± 1.27 d	14.50 ± 1.43 c	18.77 ± 3.91 d	5.45 ± 0.75 e	23.36 ± 5.50 e
	CF	95.68 ± 3.14 d	15.36 ± 0.89 cd	22.77 ± 1.38 b	27.98 ± 4.27 c	6.97 ± 0.46 d	33.10 ± 1.42 d
	OF	99.95 ± 5.62 cd	19.04 ± 2.15 bc	23.40 ± 1.34 b	29.07 ± 1.35 c	8.50 ± 0.91 c	48.21 ± 3.98 c
	15% OFB	106.58 ± 2.34 b	21.58 ± 1.88 b	30.28 ± 2.51 a	36.68 ± 1.94 b	9.81 ± 0.14 b	57.53 ± 2.72 b
	30% OFB	111.82 ± 0.74 a	27.16 ± 3.92 a	33.23 ± 4.77 a	65.97 ± 4.79 a	11.76 ± 1.01 a	80.34 ± 7.10 a
	50% OFB	104.40 ± 0.87 bc	20.63 ± 0.96 b	29.03 ± 0.22 a	36.05 ± 5.21 b	8.82 ± 0.49 bc	56.43 ± 3.55 b

Note: Data are presented as mean ± SD (n = 3). Different lowercase letters within each column indicate that the differences among treatments are significant at $p < 0.05$.

Compared with CF treatment, the N use efficiency (NUE), P use efficiency (PUE), and K use efficiency (KUE) of rice under OF treatment increased by 29.11%, 27.44%, and 43.31% in 2022, and by 15.10%, 67.71%, and 87.34% in 2023, respectively (Figure 4). The NUE, PUE, and KUE of rice treated with 15% OFB, 30% OFB, and 50% OFB were significantly increased ($p < 0.05$) compared with CF treatment. In contrast to OF treatment, the NUE and KUE of rice treated with 15% OFB, 30% OFB, and 50% OFB were significantly increased ($p < 0.05$). With the increase in the proportion of biogas slurry replacing fertilizer, the NUE, PUE, and KUE in rice in 2022 and 2023 showed a trend of first increasing and then decreasing (30% OFB > 15% OFB > 50% OFB).

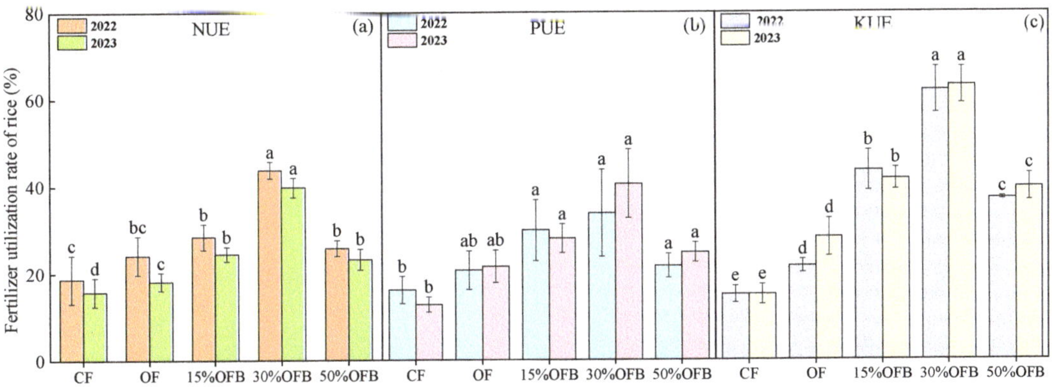

Figure 4. Effects of biogas slurry on N use efficiency (NUE) (**a**), P use efficiency (PUE) (**b**) and K use efficiency (KUE) (**c**) of rice. Bars represent the SD of the means ($n = 3$). The different lowercase letters above the column indicate that the differences among treatments are significant at $p < 0.05$.

3.4. Soil Nutrients

Compared with CF treatment, the contents of soil total N (TN), NH_4^+-N, NO_3^--N, and hydrolysable N (HN) under OF treatment in 2022 increased by 1.87%, 6.25%, 26.85%, and 18.50%, respectively (Table 6). In 2023, the content of NH_4^+-N, NO_3^--N, and HN in soil under OF treatment increased by 16.33%, 14.13%, and 2.29%, respectively. In 2022, the contents of NH_4^+-N and HN in soil under 30% OFB treatment significantly increased ($p < 0.05$) compared with CF treatment, showing increases of 16.03% and 32.86%, respectively. In contrast to CF treatment, the contents of NH_4^+-N, NO_3^--N, and HN increased under 15% OFB, 30% OFB, and 50% OFB treatments in 2023, while the contents of TN, NH_4^+-N, and HN under 30% OFB treatment significantly increased ($p < 0.05$) by 31.00%, 38.90%, and 21.80%, respectively. There were no significant differences in soil NO_3^--N content between CF and other treatments. Over the course of two years, the soil TN content increased under 15% OFB, 30% OFB, and 50% OFB treatments when compared to the OF treatment. Additionally, the soil NH_4^+-N content rose under 15% OFB and 30% OFB treatments as well. Compared with OF treatment, in 2022, the soil NH_4^+-N and HN contents significantly increased ($p < 0.05$) by 9.21% and 12.12%, respectively, under 30% OFB treatment. When compared to OF treatment, in 2023, the contents of TN and HN significantly increased by 35.05% and 19.07%, respectively, under 30% OFB treatment ($p < 0.05$). With the increase in the proportion of biogas slurry replacing chemical fertilizer, the contents of NH_4^+-N and HN in the soil first increased and then decreased (30% OFB > 15% OFB > 50% OFB) in 2022 and 2023.

Table 6. Effects of biogas slurry replacing fertilizer on soil N content.

Years	Treatment	TN ($g{\cdot}kg^{-1}$)	NH_4^+-N ($mg{\cdot}kg^{-1}$)	NO_3^--N ($mg{\cdot}kg^{-1}$)	HN ($mg{\cdot}kg^{-1}$)
2022	CK	0.97 ± 0.11 b	33.19 ± 2.63 c	2.51 ± 0.54 b	90.90 ± 3.10 c
	CF	1.07 ± 0.08 ab	42.10 ± 0.66 b	2.98 ± 0.62 ab	92.83 ± 7.95 c
	OF	1.09 ± 0.07 ab	44.73 ± 3.45 b	3.78 ± 0.49 a	110.00 ± 5.29 b
	15% OFB	1.02 ± 0.02 b	45.15 ± 1.05 b	3.55 ± 0.31 ab	112.67 ± 3.51 ab
	30% OFB	1.23 ± 0.10 a	48.85 ± 0.85 a	3.99 ± 0.57 a	123.33 ± 3.21 a
	50% OFB	1.06 ± 0.12 b	42.67 ± 2.02 b	3.15 ± 0.72 ab	111.13 ± 11.04 b
2023	CK	0.95 ± 0.03 c	23.97 ± 2.97 c	2.60 ± 0.44 b	76.77 ± 4.26 b
	CF	1.00 ± 0.04 c	27.25 ± 6.71 bc	2.83 ± 0.43 ab	77.30 ± 8.69 b
	OF	0.97 ± 0.03 c	31.70 ± 2.20 abc	3.23 ± 0.34 ab	79.07 ± 3.71 b
	15% OFB	0.99 ± 0.01 c	32.95 ± 4.25 ab	3.12 ± 0.52 ab	91.87 ± 4.14 a
	30% OFB	1.31 ± 0.07 a	37.85 ± 2.95 a	3.60 ± 0.44 a	94.15 ± 1.35 a

Note: Data are presented as mean ± SD (n = 3). Different lowercase letters within each column indicate that the differences among treatments are significant at $p < 0.05$. TN: total N; HN: hydrolysable N.

Compared with CF treatment, OF treatment increased the soil bulk density by 7.08% and 6.48%, pH by 0.76% and 4.89%, SOC by 1.28% and 0.20%, and available P by 19.16% and 2.53% in 2022 and 2023, respectively (Table 7). The available K was reduced by 21.56% and 3.07% in 2022 and 2023, respectively. In 2022, the contents of SOC, available K, and P under 30% OFB treatment significantly increased ($p < 0.05$) by 14.77%, 43.10%, and 17.67%, respectively, compared with CF treatment. In 2023, soil available K and SOC contents under 15% OFB and 30% OFB treatments significantly increased ($p < 0.05$). In contrast to OF treatment, the SOC and available K contents increased under 15% OFB, 30% OFB, and 50% OFB treatments in two years, and the available P content also increased under 15% OFB and 30% OFB treatments. The 15% OFB, 30% OFB, and 50% OFB treatments significantly increased ($p < 0.05$) the soil available K content in 2022 and 2023, and the 15% OFB and 30% OFB treatments significantly increased ($p < 0.05$) the SOC content in 2023. With the increase in the proportion of biogas slurry replacing fertilizer, the contents of SOC, available P, and K showed a trend of first increasing and then decreasing (30% OFB > 15% OFB > 50% OFB) in 2022 and 2023.

Table 7. Effects of biogas slurry replacing fertilizer on soil nutrients.

Years	Treatment	BD ($g{\cdot}cm^{-3}$)	pH (H_2O)	SOC ($g{\cdot}kg^{-1}$)	Available P ($mg{\cdot}kg^{-1}$)	Available K ($mg{\cdot}kg^{-1}$)
2022	CK	1.22 ± 0.04 a	5.50 ± 0.03 a	9.40 ± 1.08 c	29.93 ± 3.72 b	79.67 ± 9.50 c
	CF	1.13 ± 0.00 b	5.29 ± 0.05 c	10.17 ± 0.79 bc	30.63 ± 4.90 b	111.33 ± 3.21 b
	OF	1.21 ± 0.03 a	5.33 ± 0.05 c	10.30 ± 0.90 abc	36.50 ± 2.52 ab	87.33 ± 9.71 c
	15% OFB	1.23 ± 0.02 a	5.44 ± 0.05 ab	11.02 ± 0.25 ab	39.07 ± 9.23 ab	127.67 ± 7.77 a
	30% OFB	1.23 ± 0.06 a	5.27 ± 0.07 c	11.75 ± 0.90 a	43.83 ± 2.41 a	131.00 ± 9.85 a
	50% OFB	1.24 ± 0.03 a	5.36 ± 0.07 bc	10.48 ± 0.23 abc	32.33 ± 2.73 b	111.67 ± 2.52 b
2023	CK	1.16 ± 0.08 a	6.57 ± 0.32 a	9.71 ± 0.44 c	21.67 ± 5.33 a	65.00 ± 5.00 d
	CF	1.08 ± 0.10 a	6.34 ± 0.35 a	10.11 ± 0.38 c	22.50 ± 7.37 a	87.00 ± 7.00 c
	OF	1.15 ± 0.03 a	6.65 ± 0.39 a	10.13 ± 0.52 c	23.07 ± 2.64 a	84.33 ± 4.04 c
	15% OFB	1.19 ± 0.07 a	6.53 ± 0.45 a	11.19 ± 0.24 b	27.83 ± 1.59 a	116.67 ± 5.77 a
	30% OFB	1.16 ± 0.12 a	6.31 ± 0.38 a	12.59 ± 0.91 a	29.77 ± 2.83 a	117.67 ± 7.09 a
	50% OFB	1.15 ± 0.03 a	6.27 ± 0.39 a	10.48 ± 0.38 bc	22.63 ± 5.59 a	102.00 ± 2.00 b

Note: Data are presented as mean ± SD (n = 3). Different lowercase letters within each column indicate that the differences among treatments are significant at $p < 0.05$. BD: bulk density, SOC: soil organic C.

3.5. Correlation Analysis of Yield, Soil Fertility and Fertilizer Utilization Efficiency

As shown in Figure 5, grain yield was significantly positively correlated with soil NH_4^+-N, hydrolysable N, and available P contents ($p < 0.01$), significantly positively correlated with soil NO_3^--N, bulk density (BD), and straw P uptake ($p < 0.05$), significantly negatively correlated with rice plant height and soil pH ($p < 0.01$), and significantly negatively correlated with kernels per spike ($p < 0.05$). There was a significant positive correlation between SOC and N, P, and K utilization efficiency ($p < 0.01$). Soil TN and NO_3^--N contents were significantly positively correlated with the NUE, PUE, and KUE ($p < 0.05$).

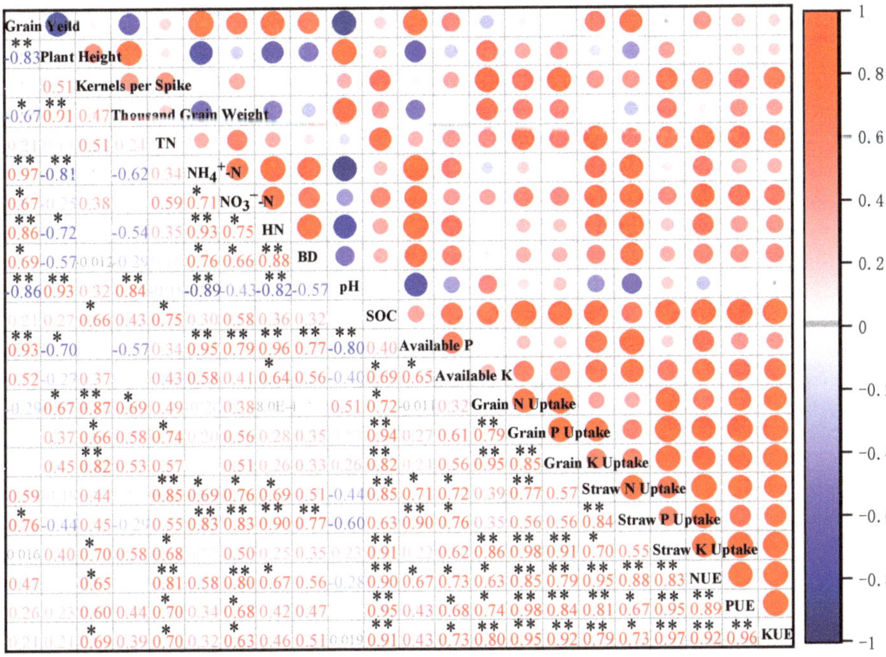

Figure 5. Correlation analysis of yield and yield composition, soil nutrient content and nutrient uptake of rice grain and straw, fertilizer utilization efficiency. Note: * indicates significant correlation ($p < 0.05$), ** indicates significant correlation ($p < 0.01$).

4. Discussion

4.1. Effects of Biogas Slurry Substitution for Fertilizer on Rice Yield

In this study, the rice grain yield was 6737.39–9527.44 and 6508.60–8346.14 kg·ha^{-1} in 2022 and 2023, respectively. The overall rice yield in 2023 was lower than that in 2022. The reason may be that the extreme maximum temperature from June to August in Anhui province was 36–38 °C in 2023. Affected by Typhoon "Du Suri" in late July, there was too much rainfall, the rice seed setting rate decreased, and rice was vulnerable to diseases and pests. Moreover, there was less sunshine during the whole growing period of rice, resulting in a lower rice yield. This resulted in poor environmental conditions, including temperature, rainfall, and sunshine time, in Hefei compared with 2022 [18,19], and they were not suitable for rice growth [20]. In this experiment, compared with CF treatment, OF treatment significantly increased rice yield in 2022 and 2023 ($p < 0.05$), indicating that optimized N and K application can effectively increase rice yield. Studies have applied 10%, 20%, and 40% N fertilizer at the booting stage, resulting in rice yields of 5494.50, 10,400, and 5002.73 kg·ha^{-1}, respectively, which indicated that 20% N fertilizer applied at the booting stage was conducive to increasing rice yield [21–23]. In this study, the proportion of N fertilizer applied at the booting stage increased from 15% in the CF treatment to 20% in the OF treatment, which was one of the reasons why the optimized fertilization treatment resulted in a higher rice yield than the conventional fertilization treatment. The results of previous experiments showed that when the proportion of K fertilizer applied at the booting stage was 40% and 30%, the rice yield was 5002.73 and 3724 kg·ha^{-1}, respectively [23,24]. In addition, K fertilizer applied at the tillering stage can promote N and P uptake, increase lodging resistance and the seed setting rate of rice, and further increase the yield, thus increasing the K fertilizer utilization efficiency [25]. In the present study, K fertilizer was applied as 45% base fertilizer, 15% tillering fertilizer, and 40% booting fertilizer under OF treatment, which was further optimized on the basis of conventional fertilization (50% base fertilizer + 50% booting fertilizer). Therefore, adjusting the proportion of K fertilizer at the booting stage and increasing the proportion of K fertilizer at the tillering stage may be another reason for optimizing fertilization to increase yield and fertilizer utilization efficiency.

Organic fertilizer instead of chemical fertilizer can improve the ability of rice to absorb and transform nutrients, increasing nutrient accumulation and thus achieving high yield [26,27]. When organic fertilizers replaced 15%, 30%, and 50% of fertilizer, the rice yield showed a trend of first increasing and then decreasing, and the yield effect was the best in the 30% replacement treatment [28], which was consistent with the results of this experiment. In this study, the yield of rice treated with 15% and 30% of fertilizer replaced by biogas slurry was significantly higher than that of rice treated with conventional fertilizer, and the yield of rice treated with 30% fertilizer replacement with biogas slurry was the highest. Biogas slurries were rich in N, P, K, and other nutrients required for crop growth, which can meet the needs of soil microorganisms, improve the richness of the soil microbial community, reduce soil water and nutrient losses, promote crop growth, and enhance the root absorption capacity [29]. In this study, the N, P, and K contents of grains and stalks treated with 15% and 30% of fertilizer replacement with biogas slurry increased compared with the optimized fertilization treatment. The contents of SOC, NH_4^+-N, hydrolysable N, available P and K increased compared with the optimized fertilization treatment, which further confirmed that biogas slurry application promoted rice growth and increased the soil nutrient supply [30]. The biogas slurry partially replaced chemical fertilizer treatment, significantly increased the 1000-grain weight and kernels per spike of rice, and increased crop yield by regulating N release in the early stage and supplementing organic nutrition in the later stage [31]. This study reached a consistent conclusion that the plant height, 1000-grain weight, and kernels per spike of rice treated with 15% and 30% of fertilizer replacement with biogas slurry increased compared with the optimal treatment.

Excessive application of biogas slurry cannot significantly improve crop yield because it releases nutrients slowly. Soil microorganisms need time to decompose it, and the nutrients available in soil cannot timely meet the early growth needs of crops, resulting in poor development of vegetative organs, which is not conducive to high and stable crop yield [32]. Compared with the CF treatment, the yield of rice treated with 50% OFB decreased by 4.52% in 2022. Xiao et al. (2023) also showed that wheat yield decreased when 50% of fertilizer was replaced by biogas slurry [14]. This may be because the biogas slurry was applied at the rice tillering stage, and organic fertilizer was not applied at the base fertilizer stage, resulting in insufficient N fertilizer, which affected rice growth. Even if the N content was fully compensated for at the tillering stage, the rice yield was still affected [33]. The nutrient release of biogas slurry organic fertilizer was slow, and an excessive substitution ratio led to an insufficient N supply, affecting rice growth. Therefore, compared with optimal fertilization, rice yield was still reduced [34], and the method by which the biogas slurry was applied may also affect rice yield. Studies have shown that the application of biogas slurry as a base fertilizer can increase yield [30], while application at the tillering stage can reduce yield [35]. In this study, in 2022, when the biogas slurry was applied at the tillering stage, the rice yield under the 50% OFB treatment was significantly lower than that of conventional fertilizer treatment. However, in 2023, when the biogas slurry was applied as a base fertilizer, the rice yield under 50% OFB treatment was not significantly different from that of the conventional fertilizer treatment. This may be because the application of biogas slurry as a base fertilizer can timely provide sufficient nutrients to rice at the tillering stage and promote rice growth and development, while the application of biogas slurry at the tillering stage led to insufficient nutrients for growth in the early growth stage, affecting rice yield [36].

4.2. Effects of Biogas Slurry on the Fertilizer Utilization Efficiency of Rice

The N application level can affect the NUE of rice [37], and the NUE decreased with the increase in the N application rate [38]. In this study, the NUE of rice in 2022 and 2023 were 18.69–43.81% and 15.76–39.82%, respectively, which were higher than 17.79–26.61% observed by Liu et al. (2017) [39] and lower than 25.12–48.16% observed by Zhou et al. (2021) [40]. The reason may be that the N application rate in this experiment (225 kg·ha^{-1}) was lower than that in Liu et al. (2017) (300 kg·ha^{-1}) and higher than that in Zhou et al. (2021) (179.4 kg·ha^{-1}) [39,40].

The substitution of organic fertilizer for chemical fertilizer promoted the uptake of N, P, and K in rice grains and straws, and the utilization efficiency of fertilizer increased [41]. In this study, compared with CF treatment, replacing 15%, 30%, and 50% of fertilizer with biogas slurry increased the N, P, and K uptake of rice grains and straws in 2022 and 2023, further significantly improving the fertilizer utilization efficiency ($p < 0.05$). The SOC content was extremely significantly positively correlated with the fertilizer utilization efficiency, and soil TN and NO_3^--N contents were significantly positively correlated with the NUE, PUE, and KUE (Figure 5). This indicated that applying biogas slurry to the soil promoted the mineralization and decomposition of organic matter and other nutrients by soil microorganisms and promoted the release of a large amount of N, P, and K nutrients, which was conducive to the absorption of N, P, and K by crop grains and stalks, thus improving the fertilizer utilization efficiency [42]. In this study, when biogas slurry replaced 15%, 30%, and 50% of fertilizers, the fertilizer utilization efficiency showed a trend of first increasing and then decreasing (30% OFB > 15% OFB > 50% OFB) in 2022 and 2023. The maximum value was found in the 30% OFB treatment, showing a significant effect ($p < 0.05$). Studies have shown that with the increase in the proportion of organic fertilizer replacing chemical fertilizer, the fertilizer utilization efficiency presented a trend of first increasing and then decreasing [43], and the effect of organic fertilizer replacing 30% chemical fertilizer was the best [40]. This was consistent with the results of this study. This may be because the fertilizer effect of organic fertilizer was slow and lasting. When organic fertilizer replaced chemical fertilizer in a proportion that was too high or too low, the available nutrients cannot meet the needs of rice in time. When organic fertilizer replaced 30% of chemical fertilizer, the nutrients required for rice growth and development could be better supplied, and it was easy to absorb N, P, and K [44].

4.3. Effects of Biogas Slurry Replacement of Fertilizer on Soil Nutrients

There was a significant positive correlation ($p < 0.05$) between NO_3^--N content in soil and N application amount [45]. In this study, the NO_3^--N content in the 0–20 cm soil layer was 2.51–3.99 mg·kg^{-1}, which was higher than 0.48–1.56 mg·kg^{-1} observed by Xiao et al. (2023) in wheat season, possibly because the N application amount was higher in this study (225 kg·ha^{-1}) than that in the study of Xiao et al. (2023) (180 kg·ha^{-1}) [14]. In this experiment, compared with CF treatment, replacing 15%, 30%, and 50% of fertilizer with biogas slurry increased the contents of SOC, NH_4^+-N, and HN. The contents of SOC, NH_4^+-N, and hydrolysable N significantly increased when biogas slurry replaced 30% of the fertilizer ($p < 0.05$). Previous studies have also reached a similar conclusion that, compared with conventional treatment, treatment with organic fertilizer instead of chemical fertilizer can increase the contents of soil SOC, NH_4^+-N, and hydrolysable N, and the difference was significant when 30% of chemical fertilizer was replaced by organic fertilizer [46–48]. The reason may be that organic fertilizer was rich in humic acid and organic acid, which increased the SOC content, provided abundant carbon sources for the soil–crop system [49], and increased the mineralization rate and mineralization potential of N. Furthermore, an increase in SOC content promoted the accumulation of N mineralization amounts [50], increased the content and availability of soil N, and further increased the content of SOC, NH_4^+-N, and hydrolysable N in soil [51].

Compared with the conventional fertilization treatment, the replacement of 15%, 30%, and 50% of fertilizer with biogas slurry increased the available K and P contents in soil, and the replacement of 15% and 30% of fertilizer with biogas slurry significantly increased the available K content in soil ($p < 0.05$). Previous studies have shown that, compared with a single application of chemical fertilizer, biogas slurry, pig manure, commercial organic fertilizer, and cow manure can all increase the soil available K and P contents, and the replacement of 15% and 30% fertilizer with organic fertilizer can significantly increase the soil available K content [46,52–55]. This was consistent with the results of this study. This may be because a reasonable proportion of organic fertilizer replacing chemical fertilizer promoted the after-effect of P, which was conducive to the conversion of accumulated P in the soil to the available state and significantly increased the P content in the soil [56]. Additionally, the K contained in organic fertilizer was highly mobile, and once the organic fertilizer was applied to the soil, it decomposed rapidly, released more organic P, and increased the P and K contents in the soil [57].

5. Conclusions

This study investigated whether optimizing N and K topdressing methods increased rice yield and fertilizer utilization efficiency and improved soil fertility. The substitution of biogas slurry for N fertilizer increased rice yield and yield composition, and replacing 30% N fertilizer with biogas slurry had the highest yield of grain and straw in rice. In addition, the substitution of biogas slurry for N fertilizer promoted the uptake of N, P, and K in rice grains and straws and improved the fertilizer utilization efficiency. Replacing fertilizer with biogas slurry increased the contents of SOC, NH_4^+-N, hydrolysable N, and available K and improved soil fertility, and the replacement of 30% of N fertilizer with biogas slurry had a more obvious effect. The 30% OFB treatment resulted in the highest rice yield and fertilizer utilization efficiency and improved soil fertility, indicating that biogas slurry replacing 30% of fertilizer was the best application mode for rice in this region.

Author Contributions: Conceptualization, Y.L., X.L. and Y.M.; methodology, Y.L., X.L. and Y.M.; formal analysis, Y.L., Q.X., H.Y., T.G., S.W. and T.C.; software, Y.L. and T.C.; writing—original draft preparation, Y.L. and X.L.; writing—review and editing, X.W., X.L. and Y.M.; supervision, X.L. and Y.M. All authors have read and agreed to the published version of the manuscript.

Funding: This work was funded by National Pilot Project of Green Planting and Recycling Agriculture of Ministry of Agriculture and Rural Affairs (202110; LJZY2022-15); Talent Research Funding Project of Anhui Agricultural University (yj2021-24).

Data Availability Statement: The data presented in this study are available on request from the corresponding author.

Acknowledgments: We thank the staff of the experimental field in Lujiang County for their help in carefully managing this experiment.

Conflicts of Interest: The authors declare no conflicts of interest.

References

1. Xin, L.J. Chemical fertilizer rate, use efficiency and reduction of cereal crops in China, 1998–2018. *J. Geogr. Sci.* **2022**, *32*, 65–78. [CrossRef]
2. Liu, Q.P.; Wu, T.Y.; Pu, L.J.; Sun, J.R. Comparison of fertilizer use efficiency in grain production between developing countries and developed countries. *J. Sci. Food Agric.* **2022**, *102*, 2404–2412. [CrossRef] [PubMed]
3. Ladha, J.K.; Tirol-Padre, A.; Reddy, C.K.; Cassman, K.G. Global nitrogen budgets in cereals: A 50-year assessment for maize, rice and wheat production systems. *Sci. Rep.* **2016**, *6*, 19355. [CrossRef] [PubMed]
4. Norse, D.; Ju, X. Environmental costs of China's food security. *Agric. Ecosyst. Environ.* **2015**, *209*, 5–14. [CrossRef]
5. Liu, H.; Li, J.; Li, X.; Zhen, Y.H.; Su, F.F.; Gao, M.J. Mitigating greenhouse gas emissions through replacement of chemical fertilizer with organic manure in a temperate farmland. *Sci. Bull.* **2015**, *60*, 598–606. [CrossRef]

6. Yang, Y.H.; Ma, W.H.; Mohammat, A.; Fang, J.Y. Storage, patterns and controls of soil nitrogen in China. *Pedosphere* **2007**, *17*, 776–785. [CrossRef]
7. Qian, Y.; Song, K.H.; Hu, T.; Ying, T.Y. Environmental status of livestock and poultry sectors in China under current transformation stage. *Sci. Total Environ.* **2018**, *622*, 702–709. [CrossRef]
8. Xu, W.P. Thoughts and suggestions on the preliminary investigation of the project of resource utilization of livestock and poultry waste in Lujiang County. *Mod. Agric. Sci. Technol.* **2020**, *14*, 169–171.
9. Wang, H.X.; Xu, J.L.; Liu, X.J.; Sheng, L.X.; Zhang, D.; Li, L.W.; Wang, A.X. Study on the pollution status and control measures for the livestock and poultry breeding industry in northeastern China. *Environ. Sci. Pollut. Res.* **2018**, *25*, 4435–4445. [CrossRef]
10. Gao, R.Z.; Liu, G.L.; Fan, Y.Z.; Wang, X.Y.; Ren, Z. The effect of perceived value on farmers' livestock manure resource utilization behavior: Evidence from Shandong, China. *Front. Psychol.* **2023**, *14*, 1098587. [CrossRef]
11. Cayuela, M.L.; Mondini, C.; Insam, H.; Sinicco, T.; Franke-Whittle, I. Plant and animal wastes composting: Effects of the N source on process performance. *Bioresour. Technol.* **2009**, *100*, 3097–3106. [CrossRef] [PubMed]
12. Kumar, A.; Verma, L.M.; Sharma, S.; Singh, N. Overview on agricultural potentials of biogas slurry (BGS): Applications, challenges, and solutions. *Biomass Convers. Biorefinery* **2022**, *13*, 13729–13769. [CrossRef] [PubMed]
13. Niyungeko, C.; Liang, X.Q.; Liu, C.L.; Zhou, J.J.; Chen, L.L.; Lu, Y.Y.; Tiimub, B.M.; Li, F.Y. Effect of biogas slurry application on soil nutrients, phosphomonoesterase activities, and phosphorus species distribution. *J. Soils Sediments* **2020**, *20*, 900–910. [CrossRef]
14. Xiao, Q.; Wu, S.; Liu, Y.; Chao, C.; Liao, X.; Ma, Y.H. Effects of different combinations of organic nutrients and chemical fertilizers on wheat yield, nitrogen use efficiency, and soil fertility. *J. Agro-Environ. Sci.* **2023**, *42*, 2291–2300.
15. Wang, X.Q.; Yang, Y.D.; Zhao, J.; Nie, J.W.; Zang, H.; Zeng, Z.D.; Zeng, Z.H.; Olesen, J.E. Yield benefits from replacing chemical fertilizers with manure under water deficient conditions of the winter wheat–summer maize system in the North China Plain. *Eur. J. Agron.* **2020**, *119*, 1261. [CrossRef]
16. Li, X.; Fang, J.; Shagahaleh, H.; Wang, J.F. Impacts of partial substitution of chemical fertilizer with organic fertilizer on soil organic carbon composition, enzyme activity, and grain yield in wheat–maize rotation. *Life* **2023**, *13*, 1929. [CrossRef] [PubMed]
17. Chen, Z.M.; Wang, Q.; Ma, J.W.; Chapman, S.; Zou, P.; Ye, J.; Yu, Q.G.; Yu, W.C.; Lin, H.; Jiang, L.N. Soil microbial activity and community composition as influenced by application of pig biogas slurry in paddy field in southeast China. *Paddy Water Environ.* **2020**, *18*, 15–25. [CrossRef]
18. Li, C.; Li, Z.Z.; Xu, H.; Huang, J.; Zhang, F.M.; Qian, Z.H. Fluctuation characteristics of wheat yield and their relationships with precipitation anomalies in Anhui province, China. *Int. J. Plant Prod.* **2022**, *16*, 483–494. [CrossRef]
19. He, H.; Chen, M.J.; Li, M.; Qu, K.X.; Dang, H.H.; Li, Q.; Hu, Z.H.; Zhang, Q. Impact of climate change on the potential allocation of resources of rice cultivation in Yangtze-Huai Rivers region: A case study of Anhui Province, China. *Theor. Appl. Climatol.* **2024**, *8*, 1–12. [CrossRef]
20. Huang, N.; Song, Y.; Wang, J.L.; Zhang, Z.Y.; Ma, S.Q.; Jiang, K.; Pan, Z.H. Climatic threshold of crop production and climate change adaptation: A case of winter wheat production in China. *Front. Ecol. Evol.* **2022**, *10*, 1019436. [CrossRef]
21. Tang, H.M.; Li, C.; Xiao, X.P.; Tang, W.G.; Cheng, K.K.; Pan, X.C.; Wang, K.; Li, W.Y. Impacts of different manure and chemical fertilizer N input ratios on physiological characteristics of leaves and yield of rice under double-cropping rice field. *J. Agric. Sci. Technol.* **2020**, *22*, 149–160.
22. Wang, J.B.; Zhang, X.M.; Yuan, M.M.; Wu, G.; Sun, Y.X. Effects of partial replacement of nitrogen fertilizer with organic fertilizer on rice growth, nitrogen utilization efficiency and soil properties in the Yangtze river basin. *Life* **2023**, *13*, 624. [CrossRef] [PubMed]
23. Zhao, Y.L.; Yu, H.B.; Lin, W.L.; Yi, X.; Cai, S.Y.; Yang, B.L. Difference of growth, development and processing quality of two main japonica rice cultivars under different nitrogen fertilization. *Barley Cereal Sci.* **2024**, *41*, 49–56+73.
24. Xu, W.T.; Liu, S.R.; Feng, J.C.; Wang, B.; Shao, Z.Z.; Wang, Y.; Hou, W.F.; Gao, Q. Rice canopy light resources allocation, leaf net photosynthetic rate, and yield formation characteristics response to combined application of nitrogen and potassium. *J. Soil Sci. Plant Nutr.* **2023**, *23*, 5257–5269. [CrossRef]
25. Kundu, A.; Raha, P.; Dubey, A.N. Impact of source and method of potassium application on dry matter accumulation and partitioning of potassium in rice (*Oryza sativa* L.). *J. Soil Sci. Plant Nutr.* **2021**, *21*, 2252–2263. [CrossRef]
26. Yang, G.Y.; Ji, H.T.; Liu, H.J.; Zhang, Y.F.; Chen, L.G.; Zheng, J.C.; Guo, Z.; Sheng, J. Assessment of productivity, nutrient uptake and economic benefits of rice under different nitrogen management strategies. *PeerJ* **2020**, *8*, e9596. [CrossRef] [PubMed]
27. Guan, G.; Tu, S.; Yang, J.; Zhang, J.; Li, Y. A field study on effects of nitrogen fertilization modes on nutrient uptake, crop yield and soil biological properties in rice-wheat rotation system. *Agric. Sci. China* **2011**, *10*, 1254–1261. [CrossRef]
28. Gosal, S.K.; Gill, G.K.; Sharma, S.; Walia, S.S. Soil nutrient status and yield of rice as affected by long-term integrated use of organic and inorganic fertilizers. *J. Plant Nutr.* **2018**, *41*, 539–544. [CrossRef]
29. Shi, Z.; Yang, Y.; Fan, Y.; He, Y.; Li, T. Dynamic responses of rhizosphere microorganisms to biogas slurry combined with chemical fertilizer application during the whole life cycle of rice growth. *Microorganisms* **2023**, *11*, 1755. [CrossRef]
30. Xu, M.; Xian, Y.; Wu, J.; Gu, Y.F.; Yang, G.; Zhang, X.H.; Peng, H.; Yu, X.Y.; Xiao, Y.L.; Li, L. Effect of biogas slurry addition on soil properties, yields, and bacterial composition in the rice-rape rotation ecosystem over 3 years. *J. Soils Sediments* **2019**, *19*, 2534–2542. [CrossRef]

31. Wu, T.D.; Li, C.; Xing, X.; Pan, X.P.; Liu, C.H.; Tian, Y.J.; Wang, Z.Y.; Zhao, J.S.; Wang, J.P.; He, B. Straw return and organic fertilizers instead of chemical fertilizers on growth, yield and quality of rice. *Earth Sci. Inform.* **2022**, *15*, 1363–1369. [CrossRef]
32. Iqbal, A.; Tang, X.R.; Ali, I.; Yuan, P.L.; Khan, R.; Khan, Z.; Adnan, M.; Wei, S.Q.; Jiang, L.G. Integrating low levels of organic fertilizer improves soil fertility and rice yields in paddy fields by influencing microbial communities without increasing CH_4 emissions. *Appl. Soil Ecol.* **2023**, *189*, 104951. [CrossRef]
33. Ashrafi Esfahani, A.; Niknejad, Y.; Fallah, H.; Dastan, S. Integrated management of organic manures and chemical fertilizers for enhancing paddy yield and the nutrient content of rice cultivars. *Commun. Soil Sci. Plant Anal.* **2019**, *50*, 570–585. [CrossRef]
34. Iqbal, M.T.; Ortaş, I.; Ahmed, I.A.; Isik, M.; Islam, M.S. Rice straw biochar amended soil improves wheat productivity and accumulated phosphorus in grain. *J. Plant Nutr.* **2019**, *42*, 1605–1623. [CrossRef]
35. Tang, Y.; Wen, G.; Li, P.; Dai, C.; Han, J. Effects of biogas slurry application on crop production and soil properties in a rice–wheat rotation on coastal reclaimed farmland. *Water Air Soil Pollut.* **2019**, *230*, 51. [CrossRef]
36. Li, M.; Liu, Y.X.; Luo, L.P.; Ying, S.S.; Jiang, P.K. Effects of different biogas slurry application patterns on nitrogen and phosphorus losses in a paddy field. *Paddy Water Environ.* **2024**, *9*, 1–13. [CrossRef]
37. Sui, B.; Feng, X.; Tian, G.; Hu, X.; Shen, Q.; Guo, S. Optimizing nitrogen supply increases rice yield and nitrogen use efficiency by regulating yield formation factors. *Field Crops Res.* **2013**, *150*, 99–107. [CrossRef]
38. Xie, J.; Evgenia, B.; Zhang, Y.; Wan, Y.; Hu, Q.J.; Zhang, C.M.; Wang, J.; Zhang, Y.Q.; Shi, X.J. Substituting nitrogen and phosphorus fertilizer with optimal amount of crop straw improves rice grain yield, nutrient use efficiency and soil carbon sequestration. *J. Integr. Agric.* **2022**, *21*, 3345–3355. [CrossRef]
39. Liu, H.J.; Jiang, H.W.; Sun, G.F.; Shen, M.X.; Chen, L.G.; Zhen, J.C. Effects of different proportions of organic-inorganic fertilizer on nitrogen uptake and utilization of rice. *China Soil Fertil.* **2017**, *5*, 61–66.
40. Zhou, Y.L.; Zhang, Y.F.; Chen, S.Q.; Zhang, L.; Wang, W.B. Effect of combined application of organic and inorganic fertilizers on nitrogen uptake and utilization efficiency by paddy rice. *Chin. J. Trop. Agric.* **2021**, *41*, 1–6.
41. Lu, W.D.; Hao, Z.Q.; Ma, X.L.; Gao, J.L.; Fan, X.Q.; Guo, J.F.; Li, L.Q.; Lin, M.; Zhou, Y.Z. Effects of different proportions of organic fertilizer replacing chemical fertilizer on soil nutrients and fertilizer utilization in gray desert soil. *Agronomy* **2024**, *14*, 228. [CrossRef]
42. Ferdous, Z.; Ullah, H.; Datta, A.; Attia, A.; Rakshit, A.; Molla, S.H. Application of biogas slurry in combination with chemical fertilizer enhances grain yield and profitability of maize (*Zea mays* L.). *Commun. Soil Sci. Plant Anal.* **2020**, *51*, 2501–2510. [CrossRef]
43. Sarwar, G.; Hussain, N.; Schmeisky, H.; Suhammad, S.; Ibrahim, M.; Ahmad, S. Efficiency of various organic residues for enhancing rice-wheat production under normal soil conditions. *Pak. J. Bot.* **2008**, *40*, 2107–2113.
44. Gao, P.; Zhang, T.; Lei, X.Y.; Cui, X.W.; Lu, Y.X.; Fan, P.F.; Long, S.P.; Huang, J.; Gao, J.S.; Zhang, Z.H.; et al. Improvement of soil fertility and rice yield after long-term application of cow manure combined with inorganic fertilizers. *J. Integr. Agric.* **2023**, *22*, 2221. [CrossRef]
45. Wang, X.; Shi, Y.; Guo, Z.; Zhang, Y.; Yu, Z. Water use and soil nitrate nitrogen changes under supplemental irrigation with nitrogen application rate in wheat field. *Field Crops Res.* **2015**, *183*, 117–125. [CrossRef]
46. Li, M.J.; Wang, K.T.; Zheng, W.; Maddela, N.R.; Xiao, Y.Y.; Li, Z.L.; Tawfik, A.; Chen, Y.C.; Zhou, Z.B. Metagenomics and network analysis decipher profiles and co-occurrence patterns of bacterial taxa in soils amended with biogas slurry. *Sci. Total Environ.* **2023**, *877*, 162911. [CrossRef]
47. Ju, J.; Gu, Q.; Zhou, H.W.; Zhang, H.; Mao, W.; Yang, H.J.; Mi, W.H.; Zhao, H.T. Effects of organic fertilizer combined with chemical fertilizer on nutrients, enzyme activities, and rice yield in reclaimed soil. *Commun. Soil Sci. Plant Anal.* **2022**, *53*, 3060–3071. [CrossRef]
48. He, H.; Peng, M.W.; Hou, Z.A.; Li, J.H. Unlike chemical fertilizer reduction, organic fertilizer substitution increases soil organic carbon stock and soil fertility in wheat fields. *J. Sci. Food Agric.* **2024**, *104*, 2798–2808. [CrossRef]
49. Tang, H.M.; Li, C.; Xiao, X.P.; Pan, X.C.; Cheng, K.K.; Shi, L.H.; Li, W.Y.; Wen, L.; Wang, K. Effects of long-term fertilizer regime on soil organic carbon and its labile fractions under double cropping rice system of southern China. *Acta Agric. Scand. Sect. B Soil Plant Sci.* **2020**, *70*, 409–418.
50. Ashrae, M.N.; Hu, C.; Wu, L.; Duan, Y.H.; Zhang, W.J.; Aziz, T.; Cai, A.D.; Abrar, M.M.; Xu, M.G. Soil and microbial biomass stoichiometry regulate soil organic carbon and nitrogen mineralization in rice-wheat rotation subjected to long-term fertilization. *J. Soils Sediments* **2020**, *20*, 3103–3113. [CrossRef]
51. Zhao, J.; Ni, T.; Li, J.; Lu, Q.; Fang, Z.Y.; Huang, Q.W.; Zhang, R.F.; Li, R.; Shen, B.; Shen, Q.R. Effects of organic–inorganic compound fertilizer with reduced chemical fertilizer application on crop yields, soil biological activity and bacterial community structure in a rice–wheat cropping system. *Appl. Soil Ecol.* **2016**, *99*, 1–12. [CrossRef]
52. Hanc, A.; Tlustos, P.; Száková, J.; Balík, J. The influence of organic fertilizers application on phosphorus and potassium bioavailability. *Plant Soil Environ.* **2008**, *54*, 247–254. [CrossRef]
53. Liu, J.A.; Shu, A.P.; Song, W.F.; Shi, W.C.; Li, M.C.; Zhang, W.X.; Li, Z.Z.; Liu, G.R.; Yuan, F.S.; Zhang, S.X.; et al. Long-term organic fertilizer substitution increases rice yield by improving soil properties and regulating soil bacteria. *Geoderma* **2020**, *404*, 115287. [CrossRef]

54. Iqbal, A.; He, L.; Ali, I.; Ullah, S.; Khan, A.; Akhtar, K.; Wei, S.Q.; Fahad, S.; Khan, R.; Jiang, L.G. Co-incorporation of manure and inorganic fertilizer improves leaf physiological traits, rice production and soil functionality in a paddy field. *Sci. Rep.* **2021**, *11*, 10048. [CrossRef] [PubMed]
55. Tang, J.; Anthony, J.D.; Wang, W.; Zhang, X.H.; Wu, D.F.; Hu, L.; Yin, J.Z. Effects of Biogas Slurry on Crop Yield, Physicochemical properties and aggregation characteristics of lime concretion soil in wheat–maize rotation in the north China plain. *J. Soil Sci. Plant Nutr.* **2022**, *22*, 2406–2417. [CrossRef]
56. Chen, M.M.; Zhang, S.R.; Wu, L.P.; Fei, C. Organic fertilization improves the availability and adsorptive capacity of phosphorus in saline-alkaline soils. *J. Soil Sci. Plant Nutr.* **2021**, *21*, 487–496.
57. Du, Z.Y.; Zhou, J.M.; Wang, H.Y.; Du, C.W.; Chen, X.Q. Potassium movement and transformation in an acid soil as affected by phosphorus. *Soil Sci. Soc. Am. J.* **2006**, *70*, 2057–2064. [CrossRef]

plants

MDPI

Article

Genetic Evaluation of Water Use Efficiency and Nutrient Use Efficiency in *Populus deltoides* Bartr. ex Marsh. Seedlings in China

Chengcheng Gao [1,2,†], Chenggong Liu [1,2,†], Cun Chen [3], Ning Liu [1,4], Fenfen Liu [1,2], Xiaohua Su [1,2] and Qinjun Huang [1,2,*]

1 State Key Laboratory of Tree Genetics and Breeding, Research Institute of Forestry, Chinese Academy of Forestry, Beijing 100091, China; gaocc@caf.ac.cn (C.G.); liucgwlq@caf.ac.cn (C.L.); ningliu666666@gmail.com (N.L.); lfflff0122@163.com (F.L.); suxh@caf.ac.cn (X.S.)

2 Key Laboratory of Tree Breeding and Cultivation, State Forestry and Grassland Administration, Beijing 100091, China

3 School of Life Sciences, Qilu Normal University, Jinan 250013, China; chencun0610@163.com

4 UGent-Woodlab (Laboratory of Wood Technology), Department of Environment, Ghent University, 9000 Ghent, Belgium

* Correspondence: huangqj@caf.ac.cn; Tel.: +86-010-62889661

† These authors contributed equally to this work.

Abstract: *Populus deltoides* Bartr. ex Marsh. represents a valuable genetic resource for fast-growing plantations in temperate regions. It holds significant cultivation and breeding potential in northern China. To establish an efficient breeding population of poplar, we studied the genetic variation of *P. deltoides* from different provenances. Our focus was on genotypes exhibiting high growth rates and efficient water and nutrient use efficiency (WUE and NUE). We evaluated 256 one-year-old seedlings from six provenances, measuring height, ground diameter, total biomass, and leaf carbon and nitrogen isotope abundance ($\delta^{13}C$ and $\delta^{15}N$). Our analytical methods included variance analysis, multiple comparisons, mixed linear models, correlation analysis, and principal component analysis. The results showed that the coefficient of variation was highest for $\delta^{15}N$ and lowest for $\delta^{13}C$ among all traits. Except for $\delta^{15}N$, the effects of intra- and inter-provenance were highly significant ($p < 0.01$). The rates of variation for all traits ranged from 78.36% to 99.49% for intra-provenance and from 0.51% to 21.64% for inter-provenance. The heritability of all traits in AQ provenance was over 0.65, and all exhibited the highest level except for seedling height. All traits were significantly positively correlated with each other ($p < 0.05$), while ground diameter, total biomass, and WUE were highly significantly negatively correlated with latitude ($p < 0.01$). After a comprehensive evaluation, two provenances and eight genotypes were selected. The genetic gains for seedling height, ground diameter, total biomass, WUE, and NUE were 27.46 cm (178-2-106), 3.85 mm (178-2-141), 16.40 g (178-2-141), 0.852‰ (LA05-N15), and 3.145‰ (174-1-2), respectively. Overall, we revealed that the abundant genetic variation in *P. deltoides* populations mainly comes from intra-provenance differences and evaluated provenances and genotypes. The results of this study will contribute to optimizing and enhancing the breeding process of Chinese poplar and improving the productivity of fast-growing plantations.

Keywords: poplar; genetic variation; $\delta^{13}C$ and $\delta^{15}N$; water and nutrient use efficiency

1. Introduction

Poplar is one of the fastest-growing tree species in temperate regions, playing a significant role in the production of wood raw materials, the supply of bioenergy, and ecological protection [1,2], and is one of the major species cultivated in plantations around the world [3]. China boasts the largest area of poplar plantation forests globally, covering 8.5 million hectares, which accounts for about one-third of the world's poplar plantations [4]. It plays an irreplaceable and important role in the construction of China's national reserve forests, the creation of farmland protection forests, and the maintenance of domestic timber production. *Populus deltoides* Bartr. ex Marsh., as one of the most suitable tree species for the intensive management of industrial timber with short rotation periods in the mid-latitude regions of the world, also has a dominant position in the poplar plantations in China [5]. However, *P. deltoides* is highly sensitive to drought and nutrient stress, which significantly limits its effectiveness in the cultivation of poplars in the arid, infertile, saline, and sandy regions of North China. In addition, as a common parent for hybrid selection, the existing resources of *P. deltoides* in China are few, and all of them are imported from abroad [6]. Therefore, evaluating the physiological characteristics of the existing *P. deltoides*, such as drought resistance and tolerance to nutrient deficiency, as well as selecting germplasm resources with superior traits, has become crucial for the breeding of this poplar species at the present stage.

Throughout the life cycle of plants, environmental stresses can significantly impact their growth and development [7]. In particular, the participation of water and nitrogen is indispensable for various metabolic activities and physiological reactions [8,9]. Optimal levels of water and nitrogen are crucial for enhancing plant growth and photosynthetic capacity [10], and the effective supply of water and nitrogen is the core element for improving plant primary productivity [9,11]. With the effects of global warming and human activities in recent years, the global landmass is generally heading towards increased aridity [12–14], which inevitably leads to greater evaporative loss of soil–water from otherwise arid regions, and also exacerbates the degree of mineralization of soils, which in turn restricts plant survival and growth [15]. Providing adequate water and applying nitrogen fertilizer can significantly enhance the growth and yield of crops such as maize (*Zea mays* L.) [16], wheat (*Triticum aestivum* L.) [17], cotton (*Gossypium hirsutum* L.) [18], and others [19,20]. However, for fast-growing, short-rotation tree species, which tend to have higher water and nutrient requirements, frequent harvesting can limit nutrient cycling and reduce soil fertility [21–23]. Additionally, the excessive use of nitrogen fertilizers can in turn lead to reduced yield, quality, and nutrient effectiveness, as well as water, air, and soil pollution [20,24,25]. Therefore, in a resource-limited environment, the selection and breeding of plant material that uses resources efficiently is an important method of attaining sustainable forest development.

The pattern of plant uptake and utilization of water and nutrients determines, to a certain extent, the outcome of plant responses to changes in environmental water status [26]. Water use efficiency (WUE) and nitrogen use efficiency (NUE) are widely used to assess the water and nutrient utilization status of plants and are key physiological parameters reflecting the relationship between plant productivity and water and nutrient utilization [27,28]. However, traditional research methods for evaluating water and nutrient utilization often face limitations due to their destructive nature or restrictive conditions [29]. With the development of testing methods such as mass spectrometry, the isotope natural abundance method has become a new technical tool to quantitatively study plant WUE and NUE due to its advantages of being unlimited in time and space, non-invasive, and easy to measure [30–33]. In C_3 plants, the abundance of carbon stable isotope ($\delta^{13}C$), which is primarily associated with the ratio of leaf internal to atmospheric CO_2 concentration (C_i/C_a), serves as a reliable indicator of long-term internal WUE [34]. Nitrogen stable isotope abundance ($\delta^{15}N$) can detect and quantify plant N inputs and losses and is used to characterize plant NUE [35,36]. Currently, $\delta^{13}C$ and $\delta^{15}N$ are widely used in crop and forest breeding studies [37–41]. However, studies on forest trees have primarily concentrated on investigating the effects of artificially controlled environmental factors on WUE or NUE, including drought stress [42], salinity stress [43], and nutrient addition [44]. Chinese researchers have studied the WUE and NUE physiological mechanisms of *P. tomentosa* Carrière [45], *P.* × *canadensis* Moench [46], and *P. alba* L. × *P. glandulosa*) [47]. Nevertheless, current poplar research in China has not adequately addressed the genetic variation characteristics of provenances, which limits the selection of breeding parents with high WUE and NUE, thereby hindering the extension of poplar cultivation to more arid and barren regions.

P. deltoides originates from the lower Mississippi River in North America and has a natural distribution from southern Canada to the southeastern United States [5,48], and was introduced to China for the first time in the 1950s [49]. Currently, Chinese researchers are selecting and breeding several genotypes of *P. deltoides* with improved traits, successfully propagating them in China, but the number of bred cultivars is still insufficient, and most of the cultivars are mainly cultivated in the southern region, where precipitation is plentiful and soil fertility is relatively high [50]. In regions characterized by short growing seasons, low precipitation, and low soil fertility, there are limited suitable cultivars of *P. deltoides*, resulting in generally low-yielding stands and a low application rate of cultivars [51]. This situation constrains the yield of Chinese poplar wood, making efficient poplar breeding an urgent and significant task in contemporary research. Building on the core germplasm construction previously established by our research team [52], this study aims to explore the genetic variation patterns of WUE and NUE among provenances of *P. deltoides*. We hope to obtain breeding materials of *P. deltoides* for arid and semi-arid regions in northern China. The results will also address the following key questions: Is there significant genetic variation in growth, WUE, and NUE among provenances? Which provenance exhibits higher heritability? Is there consistency in growth, WUE, and NUE? What are the best provenances and genotypes? How much genetic gain can we achieve?

2. Results

2.1. Genetic Variation in Growth and Biomass

Table 1 shows that at inter-provenance level, seedling height, ground diameter and total biomass ranged from 61.56 cm to 77.67 cm (mean: 68.12 cm), 7.02 mm to 8.20 mm (mean: 7.48 mm), and 19.10 g to 23.28 g (mean: 20.65 g), respectively. Seedling height was maximum in AW, ground diameter, and total biomass in AL. The mean CV for seedling height, ground diameter, and total biomass was 22.48%, 13.57%, and 27.11%, respectively. The CVs across different provenance sources ranged from 17.23 % to 29.91%, 9.54% to 17.89%, and 18.91% to 31.92%, respectively. There were large variations in seedling height, ground diameter, and total biomass at the provenance level, with total biomass > seedling height > ground diameter in the order of variability.

Table 1. Multiple comparisons of growth traits. The data in brackets are the coefficient of variation values. Different lowercase letters indicate significant differences at the 0.05 level. Presented here are Iowa America (AI), Louisiana America (AL), Missouri America (AM), Tennessee America (AT), Quebec Canada (AQ), and Washington America (AW).

Trait	AW	AT	AQ	AM	AL	AI	Mean
Height (cm)	77.67 a (21.27%)	61.56 d (29.91%)	62.49 cd (22.7%)	70.67 b (17.23%)	70.16 b (25.1%)	66.17 c (18.66%)	68.12 (22.48%)
Ground Diameter (mm)	7.38 bc (9.54%)	7.64 b (14.1%)	7.02 d (16.98%)	7.54 b (12.74%)	8.20 a (17.89%)	7.12 cd (10.19%)	7.48 (13.57%)
Total Biomass (g)	20.12 bc (26.49%)	20.67 bc (31.92%)	19.18 c (30.62%)	21.53 b (23.63%)	23.28 a (31.1%)	19.10 c (18.91%)	20.65 (27.11%)

Table 2 shows that there were highly significant differences ($p < 0.01$) in plant height, ground diameter, and total biomass in intra- and inter-provenance, with intra-provenance variation accounting for 90.53%, 83.48%, and 93.75%, respectively, which were higher than that of inter-provenance variation. Furthermore, seedlings from AW exhibited greater height, while AL showed significantly a larger ground diameter and total biomass compared to other provenances.

Table 2. Analysis of variance of growth traits. ** represents extremely significant correlation (p less than 0.01).

Trait	Quadratic Sum		Mean Square		F		Percentage of Variation (%)	
	Intra-Provenances	Inter-Provenances	Intra-Provenances	Inter-Provenances	Intra-Provenances	Inter-Provenances	Intra-Provenances	Inter-Provenances
Height (cm)	196,889.5	14,969.44	772.122	2993.89	6.500 **	25.222 **	90.53	9.47
Ground Diameter (mm)	1142.81	136.64	4.48	27.33	6.29 **	38.34 **	83.48	16.52
Total Biomass (g)	29,099.52	1597.38	114.12	319.48	5.28 **	14.79 **	93.75	6.25

2.2. Genetic Variation in Leaf $\delta^{13}C$ and $\delta^{15}N$

The results show (Table 3) that at the inter-provenance level, the variability of $\delta^{13}C$ ranged from −30.555‰ to −29.804‰ (average: −30.164‰) and $\delta^{15}N$ ranged from −0.967‰ to −0.656‰ (average: −0.831‰), with $\delta^{13}C$ and $\delta^{15}N$ being the largest in AI. The average CVs of $\delta^{13}C$ and $\delta^{15}N$ were 2.11% and 94.9%, respectively, with ranges of 1.81% to 2.48% and 85.54% to 114.89%.

Table 3. WUE and NUE of *P. deltoides*. Different lowercase letters indicate significant differences at the 0.05 level.

Trait	AW	AT	AQ	AM	AL	AI	Mean
$\delta^{13}C$ (‰)	−30.277 b (1.84%)	−29.937 a (2.48%)	−30.514 c (2.35%)	−30.555 c (1.90%)	−29.894 a (2.26%)	−29.804 a (1.81%)	−30.164 (2.11%)
$\delta^{15}N$ (‰)	−0.798 (97.90%)	−0.916 (88.36%)	−0.967 (114.89%)	−0.814 (85.54%)	−0.834 (91.49%)	−0.656 (91.24%)	−0.831 (94.90%)

Table 4 shows that leaf $\delta^{13}C$ was significantly different at intra- and inter-provenance levels ($p < 0.01$), while leaf $\delta^{15}N$ was significantly different at intra-provenance level ($p < 0.01$), but not at inter-provenance level ($p < 0.05$). The percentage of variation in $\delta^{13}C$ and $\delta^{15}N$ at intra-provenance level was 78.36% and 99.49%, respectively, which were all higher than inter-provenance level. Additionally, the leaf $\delta^{13}C$ of AI, AL, and AT provenances were significantly higher than those of the provenances, as shown in Table 3.

Table 4. Analysis of variance of WUE and NUE. ** represents extremely significant correlation (p less than 0.01).

Trait	Quadratic Sum		Mean Square		F		Percentage of Variation (%)	
	Intra-Provenances	Inter-Provenances	Intra-Provenances	Inter-Provenances	Intra-Provenances	Inter-Provenances	Intra-Provenances	Inter-Provenances
$\delta^{13}C$	428.19	69.29	1.68	13.86	9.85 **	81.31 **	78.36	21.64
$\delta^{15}N$	1055.17	6.27	4.14	1.25	5.51 **	1.67	99.49	0.51
Mean							88.93	11.08

2.3. Heritability of Traits

Table 5 shows that the heritability for all traits in AQ was above 0.65 and was at the highest level for all traits in this provenance except seedling height. WUE was almost the highest heritability trait for all the provenances, except for AT, where the heritability of WUE was second only to the heritability of seedling height. The heritability for WUE was the most constant among all the traits. The lowest heritability for seedling height (0.36) was recorded for AM, and the lowest heritability for NUE was recorded for AI, AW, and AT with 0.42, 0.39, and 0.32, respectively. The lowest heritability for total biomass was recorded for AQ and AL, with 0.66 and 0.48, respectively.

Table 5. Heritability of traits.

Heritability (h^2)	Height	Ground Diameter	Total Biomass	WUE	NUE
AM	0.36	0.49	0.48	0.62	0.60
AI	0.71	0.60	0.55	0.73	0.42
AW	0.61	0.47	0.42	0.73	0.39
AQ	0.69	0.67	0.66	0.77	0.71
AL	0.51	0.58	0.48	0.69	0.66
AT	0.68	0.51	0.50	0.61	0.32

2.4. Correlation Analysis of Parameters

Figure 1 shows that seedling height, ground diameter, total biomass, leaf $\delta^{13}C$, and leaf $\delta^{15}N$ were significantly ($p < 0.05$) or extremely significantly ($p < 0.01$) positively correlated with one another. All traits, with the exception of seedling height and leaf $\delta^{15}N$, exhibited a highly significant ($p < 0.01$) negative correlation with LAT. With the exception of total biomass and leaf $\delta^{15}N$, the rest of the traits were extremely significantly ($p < 0.01$) positively correlated with LNG.

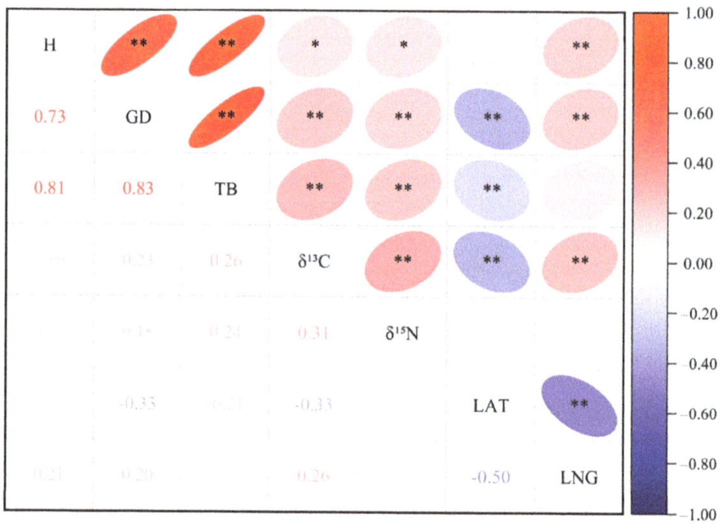

Figure 1. Correlation coefficient among plant traits, longitude, and latitude. The lower left corner is the correlation coefficient. The red font indicates a positive correlation and the blue font indicates a negative correlation. The color of the font indicates the strength of the correlation: the deeper the color, the stronger the correlation. The upper right corner is a highly significant level. The red indicates a positive correlation and blue indicates a negative correlation. The elliptical eccentricity size and the color depth indicate the correlation strength: the greater the elliptical eccentricity, the deeper the color, the stronger the correlation. * indicates p less than 0.05 and ** means p less than 0.01. The right color column represents the correlation coefficient. Presented here are the latitude, the longitude (LNG), the height (H), the ground diameter (GD), the total biomass (TB), the carbon isotope composition ($\delta^{13}C$), and the nitrogen isotope composition ($\delta^{15}N$).

2.5. Comprehensive Evaluation

A total of two factors were extracted from the principal component analysis with a cumulative contribution of 78.01% (Table 6). The first principal component included the total biomass, ground diameter, and seedling height with a contribution of 55.10%, characterizing growth and total biomass. The second principal component included $\delta^{13}C$ and $\delta^{15}N$ with a contribution of 22.91%, characterizing WUE and NUE (Figure 2).

Table 6. Principal component analysis of traits. Presented here are principal component 1 (PC_1), principal component 2 (PC_2), carbon isotope composition ($\delta^{13}C$), and nitrogen isotope composition ($\delta^{15}N$).

Trait	Total Biomass	Ground Diameter	Height	$\delta^{15}N$	$\delta^{13}C$	Eigenvalue	Contribution Rate/%	Cumulative Contribution Rate/%
PC_1	0.94	0.90	0.88	0.37	0.40	2.76	55.10	55.10
PC_2	−0.15	−0.19	−0.27	0.73	0.70	1.14	22.91	78.01
Comprehensive evaluation model: $P_{CE-i} = 0.707 \times y(PC_{i1}) + 0.293 \times y(PC_{i2})$								

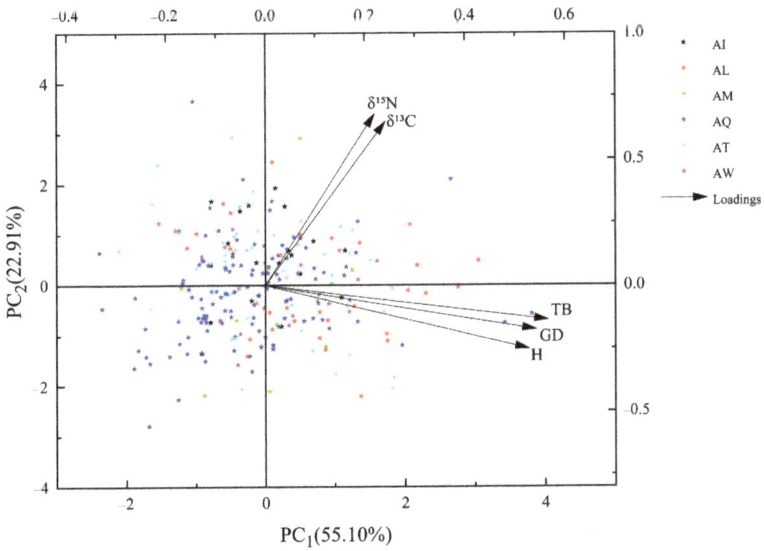

Figure 2. PCA analysis of 256 genotypes in 6 provenances. The x-axis and y-axis represent principal component 1 (PC_1) and principal component 2 (PC_1) with the proportions. Presented here are the following: height (H), ground diameter (GD), total biomass (TB), carbon isotope composition ($\delta^{13}C$), nitrogen isotope composition ($\delta^{15}N$), Iowa America (AI), Louisiana America (AL), Missouri America (AM), Tennessee America (AT), Quebec Canada (AQ), and Washington America (AW).

Combined with the results of the principal component analysis, a comprehensive evaluation model was established using the fuzzy mathematical affiliation function method. Subsequently, the genotypes were screened according to the comprehensive score, and as shown in Table 7, there were eight genotypes with superior WUE and NUE, of which five were from the AL provenance and three from the AQ provenance. Except for genotype LA05-N27, there were two or more single traits with excellent performance (Table S2). Moreover, the genetic gain of traits was calculated for each genotype and all the traits had positive genetic gain. The highest genetic gain for seedling height, ground diameter, total biomass, WUE, and NUE were 27.46 cm, 3.85 mm, 16.40 g, 0.852‰, and 3.145‰, respectively, corresponding to genotypes 178-2-106, 178-2-141, 178-2-141, LA05-N15, and 174-1-2, in that order.

Table 7. Comprehensive evaluation of superior genotypes. Presented here are Louisiana America (AL) and Quebec Canada (AQ).

Genotype Number	Population	Comprehensive Score	Ranking	Genetic Gain (ΔG)				
				Height (cm)	Ground Diameter (mm)	Total Biomass (g)	WUE (‰)	NUE (‰)
178-2-141	AQ	0.81	1	22.17	3.85	16.40	0.221	1.265
174-1-2	AQ	0.80	2	23.09	1.41	8.86	0.721	3.145
LA05-N15	AL	0.77	3	23.53	1.82	7.49	0.852	1.104
178-2-106	AQ	0.76	4	27.46	2.80	15.78	0.008	1.167
LA05-N25	AL	0.71	5	22.34	1.48	7.96	0.532	0.807
LA05-N27	AL	0.70	6	9.93	1.44	4.89	0.774	1.374
LA09-N23	AL	0.67	7	11.46	1.60	6.08	0.110	1.499
LA01-N3	AL	0.65	8	13.50	1.66	6.94	0.447	0.615
Mean				19.19	2.01	9.30	0.458	1.372

3. Discussion

3.1. Genetic Variation of *P. deltoides*

Inheritance and variation are the basis of tree breeding and its genetic improvement, where the coefficient of variation provides an assessment of the genetic variability of a trait or population [53,54]. Generally, the level of genetic variation is categorized into three classes, namely, low (<10%), medium (10% to 20%), and high (>20%) [55]. A higher classification indicates that an individual or population possesses a greater level of genetic diversity and enhanced capacity for environmental adaptation [56,57]. In this study, the coefficients of the variation in phenotypic traits (seedling height, ground diameter, and total biomass) of 256 *P. deltoides* genotypes were all at a medium-high level, suggesting that these individuals possess diversified genetic information and a large selection potential. This result has similarity with previous findings in *Widdringtonia whteir* (Rendle) Silba [58], *Larix olgensis* A. Henry [59] and *Pinus sibirica* (Ledeb.) Turcz. [60]. In addition, the coefficient of variation for intra-provenance NUE ($\delta^{15}N$) was as high as 94.90%, which provided a material basis for selecting genotypes that are suitable for efficient nutrient utilization in soil-poor areas of northern China, thereby supporting the hypothesis posited by Villani et al. [61] that the wide distribution of a species is associated with large genetic variation among populations. However, some traits in plants are genetically conserved during evolution or are influenced by the consistency of environmental conditions and exhibit low levels of genetic variation [62]. In this study, poplars had a low level of variation in WUE ($\delta^{13}C$) (2.11%), which may be due to the fact that carbon is a basic structural substance that constitutes the plant skeleton [63] and is in high abundance as an energy source for physiological activities, such as metabolism, growth, development, and reproduction. Consequently, carbon is highly abundant in plants and exhibits low variability [64], which is consistent with the findings of Müller et al. [65].

Analysis of variance (ANOVA) is also an important method for assessing the magnitude of variation in tree breeding studies [66]. In this study, the differences in the phenotypic traits, WUE, and NUE of seedlings reached extremely significant levels across inter-provenance; except for NUE, the differences in each trait in inter-provenance also reached extremely significant levels, suggesting that long-term natural selection caused poplars to exhibit high intra-provenance variation in growth and physiological traits, which is similar to that in a related study on cedar [67]. As an important indicator of genetic differentiation among provenances, variance components can be used to further reveal the extent of genetic variation within and between provenances [68,69]. In our study, the percentage of intra-provenance variation for each trait in poplars was higher than 78.36%, and the percentage of inter-provenance variation was less than 21.64%, indicating that the variation for each trait was mainly derived from intra-provenance variation, and inter-provenance variation had less influence on the traits of poplars, which is similar to the results of the previous study conducted by our team [70]. Furthermore, low differentiation in inter-provenances was found in studies of genetic variation in *P. tomentosa* Carrière [71],

P. simonii simonii [72], and *P. trichocarpa* Torr. & Gray [73], but it differs from the natural *P. euphratica* Oliv. [74]. This may be related to the fact that *P. deltoides* is a heterozygous wind-borne plant, and its provenance sites are mostly distributed in the vast plains and close to the Mississippi, Columbia, and St. Lawrence Rivers. These conditions enhance seed dispersal and facilitate frequent gene exchange among inter-provenances through pollen and seeds, thereby limiting genetic differentiation among inter-provenances to a certain extent [75].

3.2. Heritability of Traits in *P. deltoides*

In general, heritability, as one of the most crucial genetic parameters, reflects the extent of genetic control over plant growth traits. A higher heritability indicates a greater stability of these traits, enhanced parental ability to transmit them, reduced environmental influence, and improved selection effectiveness [76,77]. The degree of genetic control was broadly categorized based on the magnitude of broad-sense heritability: high (>0.80), medium-high (0.60 to 0.79), medium (0.40 to 0.59), and low (<0.40) [55,78]. We found that the heritability of *P. deltoides* provenance traits ranged from 0.36 to 0.77, which is low to medium-high heritability under strong genetic control, which is in line with previous findings [79,80]. *P. deltoides* WUE (0.61 to 0.77) had medium to high heritability among inter-provenances, which was similar to the magnitude of heritability of *P. nigra* L. [81] and *P. trichocarpa* WUE and NUE, suggesting that genetics is the main factor influencing WUE across provenances and genotypes, further supporting the feasibility of early selection for *P. deltoides* provenance and genotypes. However, NUE heritability was much higher in this study than in *P. trichocarpa* [73], either because of differences in plant type or life type [82,83] or because our experimental soil conditions were consistent and, compared to tracking soil $\delta^{15}N$ in leaves, measuring $\delta^{15}N$ in leaves is less susceptible to phenotypic plasticity [84]. These results indicate that the materials selected in this study exhibit considerable variation among provenances and genotypes, demonstrating strong heritability and significant potential for genetic improvement and the selection of new cultivars.

3.3. Correlation of Traits and Geographic Location in *P. deltoides*

A correlation analysis responds to the associations that exist between traits and plays an important role in understanding the relationships among different traits [85]. We found that $\delta^{13}C$ and $\delta^{15}N$ were significantly and positively correlated with seedling height, and highly significantly and positively correlated with ground diameter and biomass, which indicated that improvements in any of the growth traits would bring positive improvements in WUE and NUE. Notably, higher ground diameter and total biomass resulted in more desirable WUE and NUE genotypes as compared to plant height, which was observed similarly in the study of *P. balsamifera* L. with respect to the covariate trait relationship between growth traits (seedling height) and physiological traits (isotope) [86]. In addition, an extremely significant positive correlation between $\delta^{13}C$ and $\delta^{15}N$ was also found in this study, which is consistent with the findings of Chen et al. [87] and Perid et al. [88], which suggests that there is a strong coupling between $\delta^{13}C$ and $\delta^{15}N$ driven by water dynamics.

Phenotypic characteristics and physiological traits of plants are often related to the pattern of variations in the geographic latitude and longitude of their growing sites [44,89,90]. In addition, climatic factors can influence C_i/C_a by affecting leaf stomatal conductance (C_i) and chloroplastic conductance (C_a), which subsequently drives variability in leaf $\delta^{13}C$ [91]. Due to high levels of summer light, high temperatures at low latitudes and increased water stress, plants close some stomata to reduce water transpiration losses, which decreases stomatal conductance and intercellular CO_2 concentration, decreases C_i/C_a, and increases $\delta^{13}C$ [34]. Lower winter temperatures in high-latitude regions decrease leaf sarcolemmal conductance with a decreasing temperature, which increases the resistance to CO_2 diffusion within the leaf and decreases C_i/C_a, leading to an increase in $\delta^{13}C$ with a decreasing temperature [85,92]. Our findings indicated that the ground diameter, total biomass, and $\delta^{13}C$ of *P. deltoides* were extremely significantly negatively correlated with latitude, and seedling height, ground diameter, and $\delta^{13}C$ were extremely significantly positively correlated with longitude. This suggests that their WUE and growth capacity have similar patterns, which may be due to higher WUE and faster growth in low latitude and high longitude provenance areas, and limited energy recharge in high-latitude and low-longitude provenance areas restricts WUE and growth [93]. This result aligns with the climatic characteristics of *P. deltoides* provenance sites. Specifically, the AL provenance site, located at a low latitude, experiences a humid subtropical monsoon climate characterized by abundant rainfall and high levels of summer sunlight. In contrast, the AQ provenance site, situated in a high longitude region, falls within the cool temperate zone and features a humid continental climate with predominantly cloudy and wet weather. In addition, plants are associated with a trade-off between long-term $\delta^{13}C$ and NUE, with higher $\delta^{13}C$ coming at the cost of lower NUE [94,95]. However, this trade-off relationship between plant WUE and NUE may be broken under different climatic contexts, showing no significant correlation between the two [96,97]. In the present study, it was similarly found that $\delta^{15}N$ was not significantly correlated with $\delta^{13}C$ and latitude and longitude. It is evident that the relationship between WUE and NUE, as well as their respective associations with latitude and longitude, is complex, and influenced by a multitude of factors. Plants may exhibit adaptive changes to adjust water utilization and carbon and nitrogen allocation when faced with environmental stresses. Based on these findings, we conclude that *P. deltoides* from AL and AQ provenances exhibit enhanced growth performance and water acquisition strategies, making them suitable candidates for efficient water utilization.

3.4. A Comprehensive Evaluation of *P. deltoides*

The criteria for selecting suitable materials vary based on different breeding objectives. To achieve the goal of jointly selecting multiple traits, principal component analysis and fuzzy affiliation function methods have been widely employed in the comprehensive evaluation of multiple traits in plants and tree selection studies [98,99]. We utilized a comprehensive index selection focusing on the WUE and NUE of *P. deltoides*, while requiring fast growth and high productivity. The objective was to screen and breed a new generation of *P. deltoides* genotypes suitable for northern China, and to obtain excellent parental materials for Populus hybrid breeding. Based on this method, eight efficient genotypes were obtained in this study, which were from the excellent provenances AQ and AL, among which the two genotypes with the best overall evaluation (178-2-141 and 174-1-2) were from the provenance AQ, and the heritability of each trait from the AQ provenance source was above 0.65. Considering that high heritability does not imply high genetic gain for a specific trait [100]. Genetic gain is a crucial parameter for assessing the effectiveness of breeding, as it reflects the extent to which the breeding population surpasses the existing population, thereby indicating the success of the breeding efforts [101]. Therefore, we evaluated the genetic gain of the selected superior genotypes, and the average genetic gains for seedling height, ground diameter, total biomass, $\delta^{13}C$, and $\delta^{15}N$ were 19.19 cm, 2.01 mm, 9.30 g, 0.458‰, and 1.372‰, respectively. The results align with findings from previous poplar studies [102–104], suggesting significant potential for early selection. All superior genotypes exhibited positive genetic gain, which we recommend as promising candidates for subsequent breeding.

4. Materials and Methods

4.1. Test Materials

From 2005 to 2009, we collected resources through the phenotypic selection of superior trees within the natural distribution area of *P. deltoides*. Six provenances (Figure 3), 31 families, and 61 clones were finally conserved through selection and seedling multiplication. Between 2008 and 2014, we established genebanks in several climate zones in China, specifically at Junshan Forestry in Yueyang, Hunan; Shishou Poplar Research Institute in Hubei; Gaoqiao Forestry in Ningyang, Tai'an, Shandong; and Dalinghe Forestry in Linghai, Liaoning. The test materials were sourced from these genebanks (Tables 8 and S1). In April

2016, one-year-old branches were collected, and cuttings were completed in April 2016 in the greenhouse at the Tongzhou Experimental Nursery Base (39°44′01″ N, 116°45′06″ E), under the auspices of the Research Institute of Forestry, Chinese Academy of Forestry. In early August, nine well-grown plants were selected per genotype and subsequently transferred to the field. The field experiment was conducted in a completely randomized block design consisting of three blocks with three replications in each block. The plants received 1200 mL of water every two days and were manually weeded and treated for pests every two weeks.

Table 8. Provenances information of *P. deltoides*.

Provenance	Longitude (W)	Latitude (N)	Type of Climate	Genotype Number
Iowa, America (AI)	93°05′60″	41°52′48″	Temperate continental climate	19
Louisiana, America (AL)	91°52′48″	31°18′36″	Subtropical humid climate	37
Missouri, America (AM)	89°50′24″	38°03′36″	Subtropical humid climate	11
Tennessee, America (AT)	89°24′00″	36°09′36″	Subtropical humid climate	51
Quebec, Canada (AQ)	72°29′24″	46°20′24″	Temperate continental climate	124
Washington, America (AW)	119°04′48″	46°13′12″	Temperate continental climate	14
			Total	256

4.2. Trait Measurement

4.2.1. Subsubsection Carbon and Nitrogen Isotope Ratios in Leaves

In September 2016, three replicates of three to five mature functional leaves were collected from each genotype. The leaves were dried at 75 °C and ground, and the ratios of ^{13}C to ^{12}C and ^{15}N to ^{14}N in the samples were determined using a DELTA V Advantage isotope ratio mass spectrometer (Thermo Fisher Scientific, Inc., Waltham MA, USA).

4.2.2. Growth Trait

In October 2016, the seedling height, ground diameter, and number of leaves were measured for each genotype. Subsequently, all leaves, roots, and stems were harvested and weighed after being dried at 75 °C until a constant mass was achieved for leaf, stem, and root biomass.

4.3. Data Processing

A nested ANOVA, Duncan's multiple comparisons, correlation analysis, and principal component analysis were performed using SPSS 21.0; a mixed-effects modeling analysis was performed using the R package Asreml 4.0 [105]; and plotting was performed using Origin 2021. The experimental data were recorded using the Excel software application, and the parameters were calculated as follows:

Coefficient of variation formula:

$$CV = \sigma/\overline{x} \times 100\%, \quad (1)$$

where CV is the coefficient of variation of traits, σ is the standard deviation, and $\overline{x}$ is the mean value.

Total biomass formula:

$$TB = LN \times LB + SB + RB, \quad (2)$$

where TB is the total biomass, LN is the amount of leave, LB is the single leaf biomass, SB is the stem biomass, and RB is the root biomass.

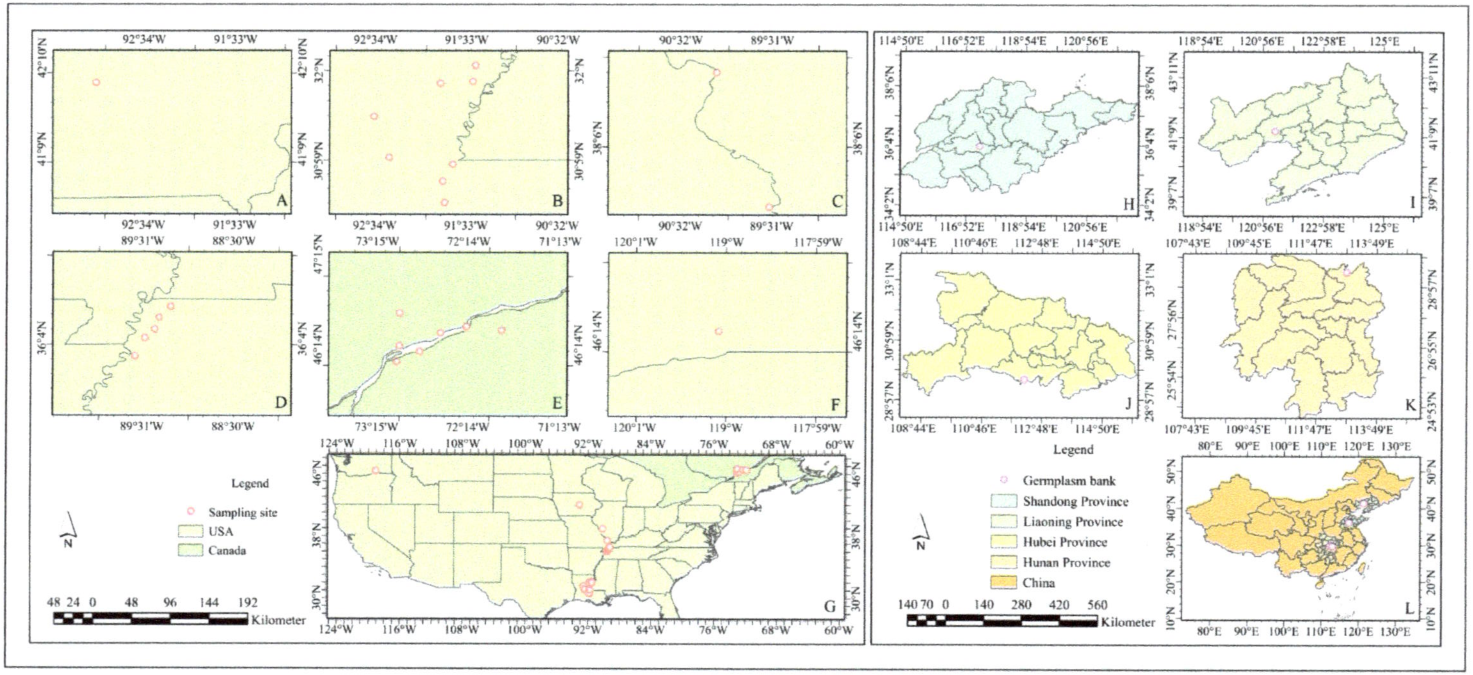

Figure 3. Distribution of provenances and genebanks. The areas marked (**A–F**) are the geographical locations of the sampling points of the six provenances of *P. deltoides*: (**A**) Iowa, America (AI); (**B**) Louisiana, America (AL); (**C**) Missouri, America (AM); (**D**) Tennessee, America (AT); (**E**) Quebec, Canada (AQ); and (**F**) Washington, America (AW). The area marked (**G**) is the distribution of the sampling points in the USA and Canada, and the legend, the compass, and the scale are on its left side. The areas marked (**H–K**) are the geographical locations of the germplasm resources of *P. deltoides*: (**H**) Ningyang Gaoqiao Forest Farm; (**I**) Daling-river Forest Farm; (**J**) Shishou Poplar Research Institute; (**K**) Junshan Forest Farm. The area marked (**L**) is the distribution of the germplasm resources in China, and the legend, the compass, and the scale are on its left side.

Carbon/nitrogen isotope abundance formula:

$$\delta^{13}C/\delta^{15}N = (R_{Sample} - RPDB)/RPDB \times 1000‰, \quad (3)$$

where R_{Sample} represents the samples' $^{13}C/^{12}C$ or $^{15}N/^{14}N$ ratios, and RPDB is the $^{13}C/^{12}C$, $^{15}N/^{14}N$ of the international standard substance PDB (Pee Dee Belemnite); the analytical accuracy of these ratios was ±0.20‰ [34].

Percentage of variation formula:

$$V_{t/s} = \sigma^2{}_{t/s}/(\sigma^2{}_{t/s} + \sigma^2{}_s), V_s = \sigma^2{}_s/(\sigma^2{}_{t/s} + \sigma^2{}_s), \quad (4)$$

where $V_{t/s}$ and V_s are the percentages of trait variation in the inter- and intra-provenance, and $\sigma^2{}_{t/s}$ and $\sigma^2{}_s$ are the variance components of the inter- and intra-provenance [106].

The fuzzy mathematical affiliation function was used to calculate the composite score of the principal component results:

$$y(PC_{ij}) = (PC_{ij} - PC_{jmin})/(PC_{jmax} - PC_{jmin}), \quad (5)$$

$$w_j = r_j/\sum r_j, \quad (6)$$

$$P_{CE-i} = \sum w_j \times y(PC_{ij}), \quad (7)$$

where $y(PC_{ij})$ represents the fuzzy mathematical affiliation function values for the jth principal component of the ith genotype, PC_{ij} is the score of the jth principal component of the ith genotype, PC_{jmin} and PC_{jmax} are the minimum and maximum values of the score of the jth principal component, w_j and r_j are the weight and contribution of the jth principal component, and P_{CE-i} is the composite score of the ith genotype [107].

$$\text{Combined trait superiority genotype}: P_{CE-i} \geq \frac{\sum_1^{256} P_{CE-i}}{256} + 2\sigma_1, \quad (8)$$

where P_{CE-i} is the composite score of the ith genotype, and σ_1 is the corresponding standard deviation.

Mixed-effects model for extracting provenance variance components:

$$y_i = \mu + S_i + R + e_i, \quad (9)$$

where y_i is the observation of the ith provenance, μ is the mean of all observations, S_i is the effect of the provenance, R is the effect of repetition, and e_i is the random error. In the model, R indicates fixed effects, and S_i indicates random effects.

The heritability of a trait formula:

$$h^2 = V_{Si}/[V_{Si} + V_{ei}], \quad (10)$$

where h^2 is the heritability of a trait, V_{Si} is the variance component of the provenance, and V_{ei} is the variance component of the random error [68].

The genetic gain formula:

$$\Delta G = (A_s - A_p) \times h_i{}^2, \quad (11)$$

where ΔG is the trait's genetic gain, A_s and A_p are the mean values of traits in the selected genotypes and the total population of the experiment, and $h_i{}^2$ is the heritability of the selected provenance [105].

5. Conclusions

P. deltoides received its name based on its origin in the Americas. Since its introduction to China, it has been recognized as a widely cultivated species in artificial forests, including timber forests and shelterbelts. However, as its cultivation expands to northern regions—primarily arid and semi-arid areas such as Shandong, Hebei, and Inner Mongolia—its ability to withstand drought and barren conditions gradually diminishes. This decline is evidenced by a significant reduction in survival rates, yield rates, and growth rates. Given the practical challenges and emerging issues, an early screening strategy for provenances and genotypes with optimal resource utilization of *P. deltoides* is particularly important.

Our study revealed significant inter-origin and inter-genotype differences in the growth, water use efficiency, and nutrient use efficiency characteristics of *P. deltoides* germplasm resources, highlighting a high degree of genetic diversity and strong environmental adaptation. Notably, trait variation within provenances emerged as the primary driver of variation, while the low differentiation among provenances suggests that future breeding efforts should prioritize the identification and selection of superior genotypes within these provenances.

The heritability of various traits in *P. deltoides* from the six provenances is generally at a medium to high level, particularly for water and nutrient use efficiency, which are significantly influenced by genetic factors. This suggests a promising potential for cultivating new varieties through genetic improvement. Additionally, the significant positive correlations among the traits of interest indicate that future breeding practices or genetic enhancement efforts could benefit from optimizing single traits, as this may synergistically improve other important traits. Therefore, while focusing on the enhancement of a specific trait in *P. deltoides*, it is essential to consider the comprehensive improvement of other target traits.

Another key result of this study is the successful identification of eight outstanding *P. deltoides* genotypes that demonstrate significant positive genetic gains in growth as well as in the utilization of water and nutrients. Among these, three genotypes originate from Quebec, Canada: 178-2-141, 174-1-2, and 178-2-106. The remaining five genotypes are from Louisiana, USA: LA05-N15, LA05-N25, LA05-N27, LA09-N23, and LA01-N3. These exceptional genotypes represent valuable germplasm resources for the expansion and consolidation of poplar plantations in arid and semi-arid regions globally, including China and their respective provenance areas. Furthermore, they serve as breeding parents for the development of superior new poplar germplasm, thereby establishing a robust material foundation and theoretical support for the ongoing enhancement of poplar genetic improvement.

Supplementary Materials: The following supporting information can be downloaded at: https://www.mdpi.com/article/10.3390/plants13162228/s1, Table S1: Information of provenances and genotypes of *P. deltoides*; Table S2: Clonal character ranking of *P. deltoides*.

Author Contributions: C.G. and C.L. contributed equally to this study. Conceptualization, C.G., C.L., Q.H. and X.S.; methodology, C.G. and C.L.; software, C.G., N.L. and C.L.; validation, C.G., C.C. and N.L.; formal analysis, C.G., C.C., N.L., and F.L.; investigation, C.C. and F.L.; resources, Q.H.; data curation, C.G.; writing—original draft preparation, C.G. and C.L.; writing—review and editing, C.G., C.L., X.S. and Q.H.; visualization, C.G. and C.L.; supervision, X.S.; project administration, Q.H.; funding acquisition, Q.H. All authors have read and agreed to the published version of the manuscript.

Funding: This project was financially supported by the Fourteenth Five-Year National Key Research and Development Program of China, project name: Breeding and Germplasm Innovation of Fast-growing New Varieties of Temperate and North Subtropical Poplar (No. 2022YFD2200301).

Data Availability Statement: The data underlying this article are available in the article and in its Supplementary Materials.

Acknowledgments: We are grateful for the scientific research platform and support provided by the Chinese Academy of Forestry and the State Key Laboratory of Tree Genetics and Breeding.

Conflicts of Interest: The authors declare no conflicts of interest.

References

1. Navarro, A.; Portillo-Estrada, M.; Ceulemans, R. Identifying the best plant water status indicator for bio-energy poplar genotypes. *GCB Bioenergy* **2020**, *12*, 426–444. [CrossRef]
2. Du, K.; Jiang, S.X.; Chen, H.; Xia, Y.F.; Guo, R.H.; Ling, A.Y.; Liao, T.; Wu, W.Q.; Kang, X.Y. Spatiotemporal miRNA and transcriptomic network dynamically regulate the developmental and senescence processes of poplar leaves. *Hortic. Res.* **2023**, *10*, uhad186. [CrossRef]
3. He, D.; Wan, X.; Wang, B.; Wan, X.; Lu, M. *Poplars and Willows, Sustaining Livelihoods in Urban and Periurban Forests in China*; FAO: Rome, Italy, 2018.
4. Tun, T.N.; Guo, J.; Fang, S.Z.; Tian, Y. Planting spacing affects canopy structure, biomass production and stem roundness in poplar plantations. *Scand. J. For. Res.* **2018**, *33*, 464–474. [CrossRef]
5. Fahrenkrog, A.M.; Neves, L.G.; Resende, M.F.R.; Dervinis, C.; Davenport, R.; Barbazuk, W.B.; Kirst, M. Population genomics of the eastern cottonwood (*Populus deltoides*). *Ecol. Evol.* **2017**, *7*, 9426–9440. [CrossRef] [PubMed]
6. Su, X.H.; Ma, C.G.; Ding, C.J. Establish & improve scientific development system of poplar breeding in China. *Prot. For. Sci. Technol.* **2011**, *1*, 11–14. (In Chinese)
7. Zhao, M.Q.; Lei, Y.J.; Wu, L.; Qi, H.R.; Song, Z.H.; Xu, M. The miR159a-*PeMYB33* module regulates poplar adventitious rooting through the abscisic acid signal pathway. *Plant J.* **2024**, *18*, 879–891. [CrossRef] [PubMed]
8. Li, J.P.; Zhang, Z.; Yao, C.S.; Yang, L.; Wang, Z.M.; Fang, B.T.; Zhang, Y.H. Improving winter wheat grain yield and water-/nitrogen-use efficiency by optimizing the micro-sprinkling irrigation amount and nitrogen application rate. *J. Integr. Agric.* **2021**, *20*, 606–621. [CrossRef]
9. Chen, X.; Xing, H.L.; Liu, B.; Wang, Y.S.; Cui, N.B.; Wang, Z.H.; Zhang, Y.X. Changes induced by multi-stage water stress on maize growth, water and nitrogen utilization and hormone signaling under different nitrogen supplies. *Agric. Water Manag.* **2023**, *290*, 108570. [CrossRef]
10. Querejeta, J.I.; Prieto, I.; Armas, C.; Casanoves, F.; Diémé, J.S.; Diouf, M.; Yossi, H.; Kaya, B.; Pugnaire, F.I.; Rusch, G.M. Higher leaf nitrogen content is linked to tighter stomatal regulation of transpiration and more efficient water use across dryland trees. *New Phytol.* **2022**, *235*, 1351–1364. [CrossRef] [PubMed]
11. Xu, Z.W.; Jiang, L.; Ren, H.Y.; Han, X.G. Opposing responses of temporal stability of aboveground and belowground net primary productivity to water and nitrogen enrichment in a temperate grassland. *Glob. Change Biol.* **2024**, *30*, e17071. [CrossRef]
12. Koutroulis, A.G. Dryland changes under different levels of global warming. *Sci. Total Environ.* **2019**, *655*, 482–511. [CrossRef] [PubMed]
13. Lian, X.; Piao, S.L.; Chen, A.P.; Huntingford, C.; Fu, B.J.; Li, L.Z.; Huang, J.P.; Sheffield, J.; Berg, A.M.; Keenan, T.F. Multifaceted characteristics of dryland aridity changes in a warming world. *Nat. Rev. Earth Environ.* **2021**, *2*, 232–250. [CrossRef]
14. Pokhrel, Y.; Felfelani, F.; Satoh, Y.; Boulange, J.; Burek, P.; Gädeke, A.; Gerten, D.; Gosling, S.N.; Grillakis, M.; Gudmundsson, L. Global terrestrial water storage and drought severity under climate change. *Nat. Clim. Change* **2021**, *11*, 226–233. [CrossRef]
15. Meusburger, K.; Trotsiuk, V.; Schmidt-Walter, P.; Baltensweiler, A.; Brun, P.; Bernhard, F.; Gharun, M.; Habel, R.; Hagedorn, F.; Köchli, R. Soil–plant interactions modulated water availability of Swiss forests during the 2015 and 2018 droughts. *Glob. Change Biol.* **2022**, *28*, 5928–5944. [CrossRef] [PubMed]
16. Fan, Z.; Deng, M.Z.; Lin, Y.R.; Liu, P.Z.; Wang, X.L.; Yang, S.F.; Ren, X.L.; Chen, X.L.; Liu, T.N. Effects of the border on yield and water use in wheat/maize intercropping in rain-fed areas with different nitrogen levels. *Field Crops Res.* **2023**, *302*, 109105. [CrossRef]
17. Ye, T.Y.; Ma, J.F.; Zhang, P.; Shan, S.; Liu, L.L.; Tang, L.; Cao, W.X.; Liu, B.; Zhu, Y. Interaction effects of irrigation and nitrogen on the coordination between crop water productivity and nitrogen use efficiency in wheat production on the North China Plain. *Agric. Water Manag.* **2022**, *271*, 107787. [CrossRef]
18. Wang, H.D.; Wu, L.F.; Wang, X.K.; Zhang, S.H.; Cheng, M.H.; Feng, H.; Fan, J.L.; Zhang, F.C.; Xiang, Y.Z. Optimization of water and fertilizer management improves yield, water, nitrogen, phosphorus and potassium uptake and use efficiency of cotton under drip fertigation. *Agric. Water Manag.* **2021**, *245*, 106662. [CrossRef]
19. Zulfiqar, F.; Thapa, G.B. Is 'Better cotton' better than conventional cotton in terms of input use efficiency and financial performance? *Land Use Policy* **2016**, *52*, 136–143. [CrossRef]
20. Shah, A.N.; Javed, T.; Singhal, R.K.; Shabbir, R.; Wang, D.; Hussain, S.; Anuragi, H.; Jinger, D.; Pandey, H.; Abdelsalam, N.R. Nitrogen use efficiency in cotton: Challenges and opportunities against environmental constraints. *Front. Plant Sci.* **2022**, *13*, 970339. [CrossRef]
21. Cannell, M.G. Environmental impacts of forest monocultures: Water use, acidification, wildlife conservation, and carbon storage. *Planted For.* **1999**, *17*, 239–262.

22. Monclus, R.; Dreyer, E.; Villar, M.; Delmotte, F.M.; Delay, D.; Petit, J.M.; Barbaroux, C.; Le Thiec, D.; Bréchet, C.; Brignolas, F. Impact of drought on productivity and water use efficiency in 29 genotypes of *Populus deltoides* × *Populus nigra*. *New Phytol.* **2006**, *169*, 765–777. [CrossRef] [PubMed]
23. Toillon, J.; Dallé, E.; Bodineau, G.; Berthelot, A.; Bastien, J.C.; Brignolas, F.; Marron, N. Plasticity of yield and nitrogen removal in 56 *Populus deltoides*× *P. nigra* genotypes over two rotations of short-rotation coppice. *For. Ecol. Manag.* **2016**, *375*, 55–65. [CrossRef]
24. Galloway, J.N.; Dentener, F.J.; Capone, D.G.; Boyer, E.W.; Howarth, R.W.; Seitzinger, S.P.; Asner, G.P.; Cleveland, C.C.; Green, P.; Holland, E.A. Nitrogen cycles: Past, present, and future. *Biogeochemistry* **2004**, *70*, 153–226. [CrossRef]
25. Khan, A.; Tan, D.K.Y.; Munsif, F.; Afridi, M.Z.; Shah, F.; Wei, F.; Fahad, S.; Zhou, R.Y. Nitrogen nutrition in cotton and control strategies for greenhouse gas emissions: A review. *Environ. Sci. Pollut. Res.* **2017**, *24*, 23471–23487. [CrossRef] [PubMed]
26. Wu, J.E.; Zeng, H.H.; Zhao, F.; Chen, C.F.; Liu, W.J.; Yang, B.; Zhang, W.J. Recognizing the role of plant species composition in the modification of soil nutrients and water in rubber agroforestry systems. *Sci. Total Environ.* **2020**, *723*, 138042. [CrossRef] [PubMed]
27. Xu, G.H.; Fan, X.R.; Miller, A.J. Plant nitrogen assimilation and use efficiency. *Annu. Rev. Plant Biol.* **2012**, *63*, 153–182. [CrossRef] [PubMed]
28. Cooley, S.S.; Fisher, J.B.; Goldsmith, G.R. Convergence in water use efficiency within plant functional types across contrasting climates. *Nat. Plants* **2022**, *8*, 341–345. [CrossRef] [PubMed]
29. Lu, W.W.; Yu, X.X.; Jia, G.D.; Li, H.Z.; Liu, Z.Q. Variation characteristics of long-term water use efficiency based on tree-ring carbon isotope discrimination. *Acta Ecol. Sin.* **2017**, *37*, 2093–2100. (In Chinese)
30. Robinson, D. $\delta^{15}N$ as an integrator of the nitrogen cycle. *Trends Ecol. Evol.* **2001**, *16*, 153–162. [CrossRef] [PubMed]
31. Kahmen, A.; Wanek, W.; Buchmann, N. Foliar $\delta^{15}N$ values characterize soil N cycling and reflect nitrate or ammonium preference of plants along a temperate grassland gradient. *Oecologia* **2008**, *156*, 861–870. [CrossRef]
32. Fang, H.J.; Yu, G.R.; Cheng, S.L.; Zhu, T.H.; Zheng, J.J.; Mo, J.M.; Yan, J.H.; Luo, Y.Q. Nitrogen-15 signals of leaf-litter-soil continuum as a possible indicator of ecosystem nitrogen saturation by forest succession and N loads. *Biogeochemistry* **2011**, *102*, 251–263. [CrossRef]
33. Liu, F.; Liu, P.; Cao, M.; Yang, C.; Chen, T.T.; Zhou, H.K.; Wang, W.Y. Review on application of stable isotope technique to the study of plant water relations. *Ecol. Sci.* **2020**, *39*, 224–232. (In Chinese)
34. Farquhar, G.D.; Ehleringer, J.R.; Hubick, K.T. Carbon isotope discrimination and photosynthesis. *Annu. Rev. Plant Biol.* **1989**, *40*, 503–537. [CrossRef]
35. Schulze, E.D.; Gebauer, G.; Ziegler, H.; Lange, O.L. Estimates of nitrogen fixation by trees on an aridity gradient in Namibia. *Oecologia* **1991**, *88*, 451–455. [CrossRef] [PubMed]
36. Peter, H.; Ian, J.A. Roles of root symbioses in African woodland and forest evidence from ^{15}N abundance and foliar analysis. *J. Ecol.* **1995**, *83*, 217–224.
37. Anyia, A.; Slaski, J.; Nyachiro, J.; Archambault, D.; Juskiw, P. Relationship of carbon isotope discrimination to water use efficiency and productivity of barley under field and greenhouse conditions. *J. Agron. Crop Sci.* **2007**, *193*, 313–323. [CrossRef]
38. Akhter, J.; Monneveux, P.; Sabir, S.; Ashraf, M.; Lateef, Z.; Serraj, R. Selection of drought tolerant and high water use efficient rice cultivars through ^{13}C isotope discrimination technique. *Pak. J. Bot.* **2010**, *42*, 3887–3897.
39. Mininni, A.N.; Tuzio, A.C.; Brugnoli, E.; Dichio, B.; Sofo, A. Carbon isotope discrimination and water use efficiency in interspecific Prunus hybrids subjected to drought stress. *Plant Physiol. Biochem.* **2022**, *175*, 33–43. [CrossRef]
40. Rabarijaona, A.; Ponton, S.; Bert, D.; Ducousso, A.; Richard, B.; Levillain, J.; Brendel, O. Provenance differences in water-use efficiency among sessile oak populations grown in a mesic common garden. *Front. For. Glob. Change* **2022**, *5*, 914199. [CrossRef]
41. Arifuzzaman, M.; Mamidi, S.; Sanz-Saez, A.; Zakeri, H.; Scaboo, A.; Fritschi, F.B. Identification of loci associated with water use efficiency and symbiotic nitrogen fixation in soybean. *Front. Plant Sci.* **2023**, *14*, 1271849. [CrossRef]
42. Jiang, X.Y.; Song, M.Y.; Qiao, Y.; Liu, M.Z.; Ma, L.; Fu, S.L. Long-term water use efficiency and non-structural carbohydrates of dominant tree species in response to nitrogen and water additions in a warm temperate forest. *Front. Plant Sci.* **2022**, *13*, 1025162. [CrossRef] [PubMed]
43. Wang, J.; Wen, X.F. Limiting resource and leaf functional traits jointly determine distribution patterns of leaf intrinsic water use efficiency along aridity gradients. *Front. Plant Sci.* **2022**, *13*, 909603. [CrossRef] [PubMed]
44. Tang, S.B.; Lai, Y.; Tang, X.L.; Phillips, O.L.; Liu, J.F.; Chen, D.X.; Wen, D.Z.; Wang, S.L.; Chen, L.C.; Tian, X.J. Multiple environmental factors regulate the large-scale patterns of plant water use efficiency and nitrogen availability across China's forests. *Environ. Res. Lett.* **2021**, *16*, 034026. [CrossRef]
45. Cao, X.; Jia, J.B.; Zhang, C.; Li, H.; Liu, T.X.; Jiang, X.N.; Polle, A.; Peng, C.H.; Luo, Z.B. Anatomical, physiological and transcriptional responses of two contrasting poplar genotypes to drought and re-watering. *Physiol. Plant.* **2014**, *151*, 480–494. [CrossRef] [PubMed]
46. Luo, J.; Li, H.; Liu, T.X.; Polle, A.; Peng, C.H.; Luo, Z.B. Nitrogen metabolism of two contrasting poplar species during acclimation to limiting nitrogen availability. *J. Exp. Bot.* **2013**, *64*, 4207–4224. [CrossRef] [PubMed]
47. Siegwolf, R.T.W.; Matyssek, R.; Saurer, M.; Maurer, S.; Günthardt-Goerg, M.S.; Schmutz, P.; Bucher, J.B. Stable isotope analysis reveals differential effects of soil nitrogen and nitrogen dioxide on the water use efficiency in hybrid poplar leaves. *New Phytol.* **2008**, *149*, 233–246. [CrossRef] [PubMed]
48. Qu, Y.S.; Zhou, P.Y.; Tong, C.F.; Bi, C.W.; Xu, L.A. Assembly and analysis of the *Populus deltoides* mitochondrial genome: The first report of a multicircular mitochondrial conformation for the genus *Populus*. *J. For. Res.* **2023**, *34*, 717–733. [CrossRef]

49. Yang, S.L.; Shi, J.; Chen, L.H.; Zhang, J.; Zhang, D.J.; Xu, Z.F.; Xiao, J.J.; Zhu, P.; Liu, Y.; Lin, T.T. Physiological and biomass partitioning shifts to water stress under distinct soil types in *Populus deltoides* saplings. *J. Plant Ecol.* **2020**, *13*, 545–553. [CrossRef]
50. Wei, S.Y.; Wu, H.T.; Li, X.P.; Chen, Y.N.; Yang, Y.H.; Dai, M.L.; Yin, T.M. Identification of genes underlying the resistance to *Melampsora larici-populina* in an *R* gene supercluster of the *Populus deltoides* genome. *Plant Dis.* **2020**, *104*, 1133–1143. [CrossRef]
51. Yan, Y.B.; Pan, H.X. Genetic variation and selection of seedling traits in hybrid progeny of *Populus deltoides*. *J. Zhejiang A&F Univ.* **2021**, *38*, 1144–1152. (In Chinese)
52. Chen, C.; Ding, C.J.; Zhang, J.; Li, B.; Chu, Y.G.; Su, X.H.; Huang, Q.J. Population structure analysis and core collection construction of *Populus deltoides*. *Sci. Silvae Sin.* **2020**, *56*, 67–76. (In Chinese)
53. Mwase, W.F.; Savill, P.S.; Hemery, G. Genetic parameter estimates for growth and form traits in common ash (*Fraxinus excelsior*, L.) in a breeding seedling orchard at Little Wittenham in England. *New For.* **2008**, *36*, 225–238. [CrossRef]
54. Metougui, M.L.; Mokhtari, M.; Maughan, P.J.; Jellen, E.N.; Benlhabib, O. Morphological variability, heritability and correlation studies within an argan tree population (*Argania spinosa* (L.) Skeels) preserved in situ. *Int. J. Agric. For.* **2017**, *7*, 42–51.
55. Ewool, M.B.; Akromah, R. Genetic variability, coefficient of variance, heritability and genetic advance of pro-vitamin a maize hybrids. *Int. J. Agric. Inn. Res.* **2017**, *6*, 84–90.
56. Scribner, K.T.; Uhrig, G.; Kanefsky, J.; Sard, N.M.; Holtgren, M.; Jerome, C.; Ogren, S. Pedigree-based decadal estimates of lake sturgeon adult spawning numbers and genetic diversity of stream-side hatchery produced offspring. *J. Great Lakes Res.* **2022**, *48*, 551–564. [CrossRef]
57. Li, A.R.; Ma, M.; Li, H.T.; He, S.F.; Wang, S.G. Genetic diversity and population differentiation of a Chinese endangered plant *Ammopiptanthus nanus* (M. Pop.) Cheng f. *Genes* **2023**, *14*, 1020. [CrossRef] [PubMed]
58. Fred, F.; Mwabumba, L.; Mhango, J.; Missanjo, E.; Kadzuwa, H.; Likoswe, M. Genetic and phenotypic parameters for growth traits of *Widdringtonia whytei*-Rendle translocation provenance Trials in Malawi. *J. Glob. Ecol. Environ.* **2023**, *17*, 32–48. [CrossRef]
59. Zhang, H.; Zhou, X.; Gu, W.; Wang, L.; Li, W.; Gao, Y.; Wu, L.; Guo, X.; Tigabu, M.; Xia, D.; et al. Genetic stability of *Larix olgensis* provenances planted in different sites in northeast China. *For. Ecol. Manag.* **2021**, *485*, 118988. [CrossRef]
60. Zhao, X.Y.; Wang, C.; Li, S.C.; Hou, W.; Zhang, S.Q.; Han, G.J.; Pan, D.; Wang, P.; Cheng, Y.F.; Liu, G.F. Genetic variation and selection of introduced provenances of Siberian Pine (*Pinus sibirica*) in frigid regions of the Greater Xing'an Range, Northeast China. *J. For. Res.* **2014**, *25*, 549–556. [CrossRef]
61. Villani, F.; Sansotta, A.; Cherubini, M.; Cesaroni, D.; Sbordoni, V. Genetic structure of natural populations of *Castanea sativa* in Turkey: Evidence of a hybrid zone. *J. Evol. Biol.* **1999**, *12*, 233–244. [CrossRef]
62. Zemke, N.R.; Armand, E.J.; Wang, W.; Lee, S.; Zhou, J.; Li, Y.E.; Liu, H.; Tian, W.; Nery, J.R.; Castanon, R.G.; et al. Conserved and divergent gene regulatory programs of the mammalian neocortex. *Nature* **2023**, *624*, 390–402. [CrossRef] [PubMed]
63. Wang, M.; Kong, D.; Mo, X.; Wang, Y.; Yang, Q.; Kardol, P.; Barrantes, O.J.V.; Simpson, M.J.; Zeng, H.; Reich, P.B.; et al. Molecular-level carbon traits underlie the multidimensional fine root economics space. *Nat. Plants* **2024**, *10*, 901–909. [CrossRef] [PubMed]
64. Sterner, R.W.; Elser, J.J.; Vitousek, P. *Ecological Stoichiometry: The Biology of Elements from Molecules to the Biosphere*; Princeton University Press: Princeton, NJ, USA, 2017; p. 464.
65. Müller, A.; Horna, V.; Kleemann, F.; Vornam, B.; Leuschner, C. Physiological vs. morphological traits controlling the productivity of six aspen full-sib families. *Biomass Bioenergy* **2013**, *56*, 274–283. [CrossRef]
66. Safavi, S.A.; Pourdad, S.S.; Taeb, M.; Khosroshahli, M. Assessment of genetic variation among safflower (*Carthamus tinctorius* L.) accessions using agro-morphological traits and molecular markers. *J. Food Agric. Envir. Sci. Pollut. Res.* **2010**, *8*, 616–625.
67. Xu, R.; Cheng, S.; Zhou, J.; Tigabu, M.; Ma, X.; Li, M. Intraspecific variations in leaf functional traits of *Cunninghamia lanceolata* provenances. *BMC Plant Biol.* **2023**, *23*, 92. [CrossRef] [PubMed]
68. White, T.L.; Adams, W.T.; Neale, D.B. *Forest Genetics*; CABI: London, UK, 2007.
69. Gomaa, N.H.; Picó, F.X. Depicting the phenotypic space of the annual plant *Diplotaxis acris* in hyperarid deserts. *Ecol. Evol.* **2021**, *11*, 15708–15719. [CrossRef] [PubMed]
70. Chen, C.; Chu, Y.G.; Ding, C.G.; Su, X.H.; Huang, Q.J. Genetic diversity and population structure of black cottonwood (*Populus deltoides*) revealed using simple sequence repeat markers. *BMC Genet.* **2020**, *21*, 2. [CrossRef] [PubMed]
71. Han, Z.; Han, Q.; Xia, Y.; Geng, X.; Du, K.; Yang, J.; Kang, X. Construction of a breeding parent population of *Populus tomentosa* based on SSR genetic distance analysis. *Sci. Rep.* **2020**, *10*, 18573. [CrossRef] [PubMed]
72. Wei, Z.Z.; Zhao, X.; Pan, W.; Zhang, J.F.; Li, B.L.; Zhang, D.Q. Phenotypic variation among five provenances of *Populus simonii* in northern China. *For. Stud. China* **2011**, *13*, 97–103. [CrossRef]
73. McKown, A.D.; Guy, R.D.; Klápště, J.; Geraldes, A.; Friedmann, M.; Cronk, Q.C.; El-Kassaby, Y.A.; Mansfield, S.D.; Douglas, C. Geographical and environmental gradients shape phenotypic trait variation and genetic structure in *Populus trichocarpa*. *New Phytol.* **2014**, *201*, 1263–1276. [CrossRef] [PubMed]
74. Dong, X.; Chen, X.; Gai, Z.; Zhai, J.; Guo, X.; Han, X.; Zhang, S.; Wu, Z.; Li, Z. Phenotypic diversity and variation in natural *Populus euphratica* populations shaped by environmental factors. *Contemp. Probl. Ecol.* **2023**, *16*, 230–252. [CrossRef]
75. Difazio, S.P.; Slavov, G.T.; Joshi, C.P. *Genetics Genomics and Breeding of Poplar*; Science Publishers: New York, NY, USA; CRC Press: New York, NY, USA, 2011.
76. Baltunis, B.S.; Gapare, W.J.; Wu, H.X. Genetic parameters and genotype by environment interaction in radiata pine for growth and wood quality traits in Australia. *Silvae Genet.* **2009**, *59*, 113–124. [CrossRef]

77. Galeano, E.; Thomas, B.R. Unraveling genetic variation among white spruce families generated through different breeding strategies: Heritability, growth, physiology, hormones and gene expression. *Front. Plant Sci.* **2023**, *14*, 1052425. [CrossRef] [PubMed]
78. Singh, B.D. *Plant Breeding: Principles and Methods*, 6th ed.; Kalyani Publishers: New Delhi, India, 2001.
79. Kanaga, M.K.; Ryel, R.J.; Mock, K.E.; Pfrender, M.E. Quantitative-genetic variation in morphological and physiological traits within a quaking aspen (*Populus tremuloides*) population. *Can. J. For. Res.* **2008**, *38*, 1690–1694. [CrossRef]
80. Lojewski, N.R.; Fischer, D.G.; Bailey, J.K.; Schweitzer, J.A.; Whitham, T.G.; Hart, S.C. Genetic basis of aboveground productivity in two native *Populus* species and their hybrids. *Tree Physiol.* **2009**, *29*, 1133–1142. [CrossRef] [PubMed]
81. Guet, J.; Fabbrini, F.; Fichot, R.; Sabatti, M.; Bastien, C.; Brignolas, F. Genetic variation for leaf morphology, leaf structure and leaf carbon isotope discrimination in European populations of black poplar (*Populus nigra* L.). *Tree Physiol.* **2015**, *35*, 850–863. [CrossRef] [PubMed]
82. Su, B.Q.; Shangguan, Z.P. Response of water use efficiency and plant-soil C:N:P stoichiometry to stand quality in *Robinia pseudoacacia* on the Loess Plateau of China. *Catena* **2021**, *206*, 105571. [CrossRef]
83. Chen, J.S.; Chen, Y.P.; Wang, K.B.; Wang, G.L.; Wu, J.H.; Zhang, Y.Y. Differences in soil water storage, consumption, and use efficiency of typical vegetation types and their responses to precipitation in the Loess Plateau, China. *Sci. Total Environ.* **2023**, *869*, 161710. [CrossRef]
84. Houlton, B.Z.; Sigman, D.M.; Schuur, E.A.; Hedin, L.O. A climate-driven switch in plant nitrogen acquisition within tropical forest communities. *Proc. Natl. Acad. Sci. USA* **2007**, *104*, 8902–8906. [CrossRef]
85. Flexas, J.; Carriquí, M.; Coopman, R.E.; Gago, J.; Galmés, J.; Martorell, S.; Morales, F.; Diaz-Espejo, A. Stomatal and mesophyll conductances to CO_2 in different plant groups: Underrated factors for predicting leaf photosynthesis responses to climate change? *Plant Sci.* **2014**, *226*, 41–48. [CrossRef]
86. Soolanayakanahally, R.Y.; Guy, R.D.; Silim, S.N.; Drewes, E.C.; Schroeder, W.R. Enhanced assimilation rate and water use efficiency with latitude through increased photosynthetic capacity and internal conductance in balsam poplar (*Populus balsamifera* L.). *Plant Cell Environ.* **2009**, *32*, 1821–1832. [CrossRef] [PubMed]
87. Chen, M.; Liu, S.; Xu, G.X.; Chen, J.; Xing, H.S.; Li, F.F.; Zhang, M.M.; Cao, X.W.; Shi, Z.M. Differences and drivers of leaf stable carbon and nitrogen isotope in herbs under different vegetation types on the eastern Qinghai-Tibetan Plateau. *Chin. J. Appl. Ecol.* **2024**, *35*, 877. (In Chinese)
88. Peri, P.L.; Ladd, B.; Pepper, D.A.; Bonser, S.P.; Laffan, S.W.; Amelung, W. Carbon ($\delta^{13}C$) and nitrogen ($\delta^{15}N$) stable isotope composition in plant and soil in Southern Patagonia's native forests. *Global Change Biol.* **2012**, *18*, 311–321. [CrossRef]
89. Li, J.M.; Du, L.S.; Guan, W.B.; Yu, F.H.; Mark, V.K. Latitudinal and longitudinal clines of phenotypic plasticity in the invasive herb *Solidago canadensis* in China. *Oecologia* **2016**, *182*, 755–764. [CrossRef] [PubMed]
90. Tang, S.; Dawson, H.R.; Silva, L.C.R.; Peñuelas, J.; Sardans, J.; Lambers, H.; Zeng, F.; Lai, Y.; Jia, Y.; Zhou, G.; et al. Atmospheric factors outweigh species traits and soil properties in explaining spatiotemporal variation in water-use efficiency of tropical and subtropical forest species. *Agric. For. Meteorol.* **2022**, *323*, 109056. [CrossRef]
91. Chen, M.; Shi, Z.M.; Liu, S.; Xu, G.X.; Cao, X.W.; Chen, J.; Zhang, M.M.; Feng, Q.H.; Centritto, M.; Cao, J. Leaf functional traits have more contributions than climate to the variations of leaf stable carbon isotope of different plant functional types on the eastern Qinghai–Tibetan Plateau. *Sci. Total Environ.* **2023**, *871*, 162036. [CrossRef]
92. Evans, J.R.; Von Caemmerer, S. Temperature response of carbon isotope discrimination and mesophyll conductance in tobacco. *Plant Cell Environ. Sci. Pollut. Res.* **2013**, *36*, 745–756. [CrossRef] [PubMed]
93. Ji, Y.D.; Li, Y.; Yao, N.; Biswas, A.; Zou, Y.F.; Meng, Q.T.; Liu, F.G. The lagged effect and impact of soil moisture drought on terrestrial ecosystem water use efficiency. *Ecol. Indicators* **2021**, *133*, 108349. [CrossRef]
94. Field, C.; Merino, J.; Mooney, H.A. Compromises between water-use efficiency and nitrogen-use efficiency in five species of California evergreens. *Oecologia* **1983**, *60*, 384–389. [CrossRef]
95. Kloeppel, B.; Gower, S.; Vogel, J.; Reich, P. Leaf-level resource use for evergreen and deciduous conifers along a resource availability gradient. *Funct. Ecol.* **2000**, *14*, 281–292. [CrossRef]
96. Du, B.; Zheng, J.; Ji, H.; Zhu, Y.; Yuan, J.; Wen, J.; Kang, H.; Liu, C. Stable carbon isotope used to estimate water use efficiency can effectively indicate seasonal variation in leaf stoichiometry. *Ecol. Indicators* **2021**, *121*, 107250. [CrossRef]
97. Shao, X.; Chen, S.; Chen, Y.; Yang, Z.; Liu, Q.; Zhang, G.; Chen, D.; Sui, M.; Zang, L. A Review of the relationship between plant water use efficiency and plant functional community structure. *World For. Res.* **2024**, *37*, 37–44. (In Chinese)
98. Cao, Q.J.; Lu, B.R.; Xia, H.; Rong, J.; Sala, F.; Spada, A.; Grassi, F. Genetic diversity and origin of weedy rice (*Oryza sativa* f. *spontanea*) populations found in North-eastern China revealed by simple sequence repeat (SSR) markers. *Ann. Bot.* **2006**, *98*, 1241–1252. [PubMed]
99. Singh, A.; Kumar, A.; Kumar, R.; Prakash, J.; Kumar, N.; Verma Arvind, K. Evaluation of salt tolerance in jamun (*Syzygium cumini* L. Skeels) using morpho-physiological traits and membership function analysis. *Sci. Hortic.* **2024**, *326*, 112742. [CrossRef]
100. Ramesh, K.J. A study of variability, associations, and path analysis in poplar (*Populus deltoides* Bartr. ex Marsh). *J. Sustain. For.* **2012**, *31*, 185–204.
101. Xu, Y.; Li, P.; Zou, C.; Lu, Y.; Xie, C.; Zhang, X.; Prasanna, B.M.; Olsen, M.S. Enhancing genetic gain in the era of molecular breeding. *J. Exp. Bot.* **2017**, *68*, 2641–2666. [CrossRef] [PubMed]

102. Gylander, T.; Hamann, A.; Brouard, J.S.; Thomas, B.R. The potential of aspen clonal forestry in Alberta: Breeding regions and estimates of genetic gain from selection. *PLoS ONE* **2012**, *7*, e44303. [CrossRef] [PubMed]
103. Niemczyk, M.; Thomas, B.R. Growth parameters and resistance to *Sphaerulina musiva*-induced canker are more important than wood density for increasing genetic gain from selection of *Populus* spp. hybrids for northern climates. *Ann. For. Sci.* **2020**, *77*, 26. [CrossRef]
104. Jiang, L.P.; Pei, X.N.; Hu, Y.B.; Chiang, V.L.; Zhao, X.Y. Effects of environment and genotype on growth traits in poplar clones in Northeast China. *Euphytica* **2021**, *217*, 169. [CrossRef]
105. Liu, N.; Ding, C.J.; Li, B.; Ding, M.; Su, X.H.; Huang, Q.J. Analysis of the genotype interaction of four-year-old *Populus euramericana* using the BLUP-GGE technique. *Forests* **2021**, *12*, 1759. [CrossRef]
106. Nagamitsu, T.; Nagasaka, K.; Yoshimaru, H.; Tsumura, Y. Provenance tests for survival and growth of 50-year-old Japanese larch (*Larix kaempferi*) trees related to climatic conditions in central Japan. *Tree Genet. Genom.* **2013**, *10*, 87–99. [CrossRef]
107. Porth, I.; Klápště, J.; McKown, A.D.; La Mantia, J.; Guy, R.D.; Ingvarsson, P.K.; Hamelin, R.; Mansfield, S.D.; Ehlting, J.; Douglas, C.J.; et al. Evolutionary quantitative genomics of *Populus trichocarpa*. *PLoS ONE* **2015**, *10*, e0142864. [CrossRef] [PubMed]

plants

MDPI

Article

Organic Fertilization and Biostimulant Application to Improve Yield and Quality of Eggplant While Reducing the Environmental Impact

Luigi Giuseppe Duri [1], Roberta Paradiso [1,*], Ida Di Mola [1], Eugenio Cozzolino [2], Lucia Ottaiano [1], Roberta Marra [1] and Mauro Mori [1]

[1] Department of Agricultural Sciences, University of Naples Federico, 80055 Naples, Italy; luigigiuseppe.duri@unina.it (L.G.D.); ida.dimola@unina.it (I.D.M.); lucia.ottaiano@unina.it (L.O.); robmarra@unina.it (R.M.); mauro.mori@unina.it (M.M.)

[2] Council for Agricultural Research and Economics (CREA), Research Center for Cereal and Industrial Crops, 81100 Caserta, Italy; eugenio.cozzolino@crea.gov.it

* Correspondence: roberta.paradiso@unina.it

Abstract: Environmental sustainability is a crucial issue in modern agriculture and special attention needs to be paid to soil health preservation. Eggplant (*Solanum melongena* L.) cultivation implies the supply of relevant quantities of chemical fertilizers, since the crop has high nutrient requirements. This study investigated the combined effects of two common organic amendments—compost and digestate—and two types of biostimulant—a plant-based product and a microbe-based product—on fruit production and quality of eggplant, to highlight the potential synergistic effects of fertilization and biostimulation. The experiment was carried out in a Mediterranean greenhouse in the winter/spring period, assessing early and total marketable yield and fruit qualitative traits (firmness, color, nitrogen, ascorbic acid, carotenoid and phenol content, and antioxidant activity). Results showed that the fertilization strategy significantly influenced plant productivity, with digestate promoting the early fruitification and mineral fertilizers resulting in a higher total yield. Biostimulants, particularly the microbial type, improved the fruit quality in terms of carotenoid content and antioxidant activity. These findings highlight the potential benefits of combining organic amendments with biostimulants in eggplant cultivation, enhancing the economic value of the product through the increase in the early production and fruit nutraceutical value while realizing sustainable practices.

Keywords: *Solanum melongena* L.; compost; digestate; plant-based biostimulant; microbial biostimulant

1. Introduction

Among the sustainable development goals of the Agenda 2030, endorsed by governments of the 193 member countries of the United Nations and approved by the General Assembly, the planet's health and the sustainability of production and consumption processes require wide attention.

Accordingly, environmental sustainability has gained importance in many aspects of daily life during the last years. In agriculture, this concept is mainly associated with the rational use of pesticides and chemicals directly applied on plants. However, it is often neglected that sustainability should start from the soil health. Indeed, in the Mediterranean basin, intensive farming has resulted in several negative effects, including the reduction in soil organic matter (hence, in soil fertility) with consequent soil degradation [1].

The enhancement of soil health can be achieved through proper management strategies. For instance, as stated by Amsili et al. [2], a positive impact can be obtained through those practices, which positively influence carbon and nutrient cycling [3].

According to Clapp et al. [4], a notable increase in the use of organic amendments occurred in response to the decline in organic matter in agricultural soils, with compost and digestate among the most common products. Composting is an effective and ecologically friendly method for recovering and valuing organic biomasses, including bio-wastes, within the agro-food chain [5]. Digestate is a derivative of biodegradable residues and by-products of livestock and agro-industrial sectors, and it results in a mixture of partially degraded organic matter, inorganic compounds, and microbial biomass [6].

Organic fertilizers offer the advantage of bonding nutrients to the organic matrix, limiting their leaching but also reducing their water solubility and availability for plants [7]. This imposes a long-term application, allowing environmental processes and management factors (such as precipitation, tillage, and irrigation) to modify the natural cycle of organic matter, driving the aggregate disruption and the nutrient redistribution [8]. To overcome this constraint, nutrients can be made available earlier using specific microorganisms or plant-derived substances that also exhibit growth-promoting activity. The application of biostimulants to plant leaves and seeds or to soil can stimulate root growth and help nutrient uptake and use efficiency [9], while increasing the beneficial microbial population [10,11].

The combined application of organic amendments and biostimulants can enhance soil health and productivity while reducing the supply of synthetic fertilizers and chemicals, hence promoting environmental sustainability and the circular economy [12], reducing harmful leaks and environmental pollution [13].

Eggplant (*Solanum melongena* L.) is an Asian native crop currently cultivated worldwide [14]. In Europe, Italy is the main producer of eggplant fruits, largely used in the Mediterranean diet, as they are low in calories and have high nutrient potential and are rich in chlorogenic acid and anthocyanin pigments, which are antioxidant compounds [15,16]. Deep and well-drained soils with a high organic matter content are essential for the root development of eggplant, and high levels of nitrogen (N) and phosphorus are required for the proper growth of the aerial part [17].

Based on the above stated principles, this paper delves into the agronomic implications of two organic amendments commonly used in Mediterranean agriculture—compost and digestate—combined with two different types of biostimulants, one derived from a tropical plant extract and one containing the beneficial microorganism *Trichoderma afroharzianum*, on eggplant fruit production and nutraceutical quality. To our knowledge, the synergistic effects of organic fertilization and biostimulation and its possible application to reduce the use of synthetic fertilizers have never been evaluated in eggplant cultivation.

2. Results

2.1. Plant Growth and Fruit Production

The fertilization strategy affected the plant height, with digestate determining taller plants than the other treatments at all the measurement dates (Figure 1). Plant height showed a similar trend in the two biostimulant treatments, with higher values than the control (CTR) with both the microbial (MIC) and the plant-based (BIO) products, in the first two measurements (65 and 107 days after transplanting (DAT), corresponding to the vegetative growth and production stage, respectively), while no difference was observed in completely developed plants (139 DAT, senescence phase).

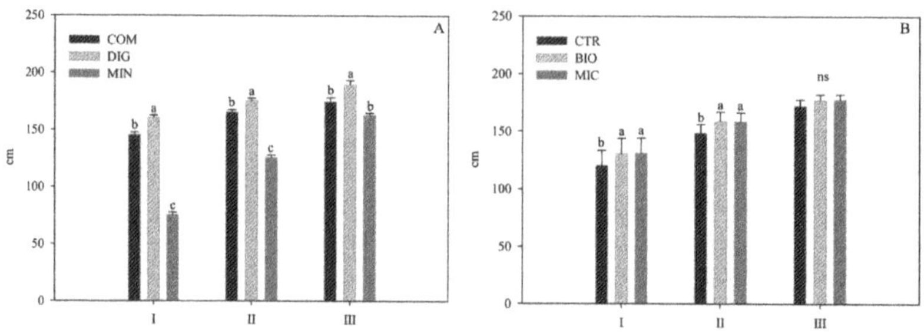

Figure 1. Plant height of eggplant plants in 3 different times of cycle (65, 107, and 139 DAT, corresponding to vegetative, productive, and senescence phases, namely I, II, and III, respectively) as affected by fertilization strategies (digestate—DIG; compost—COM; mineral—MIN) (**A**) and biostimulants application (plant-derived biostimulant—BIO; microbial biostimulant—MIC; not treated control—CTR) (**B**). Different letters indicate significant differences per $p \leq 0.05$ according to Tukey's test. ns—not significant. Vertical bars represent the standard error (n = 3).

The leaf greenness, expressed as the SPAD index, showed a clear decrease over time (Figure 2). The fertilization strategy had a significant effect only at the first sampling (65 DAT, vegetative growth), with the highest value in digestate (Figure 2A) and the lowest in mineral treatment.

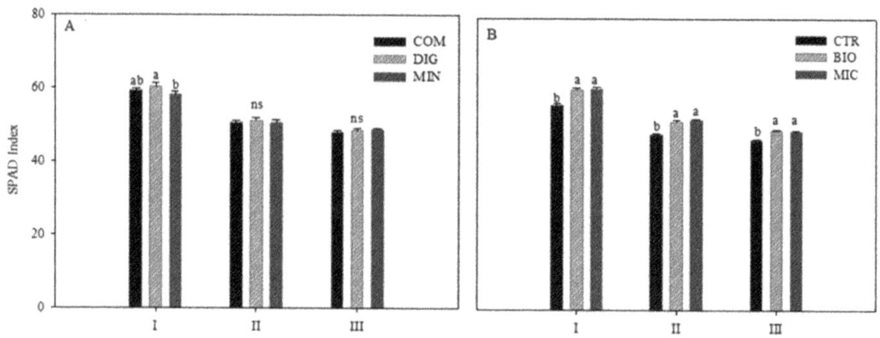

Figure 2. SPAD index in leaves of eggplant plants in 3 different times of cycle (65, 107, and 139 DAT, corresponding to vegetative, productive, and senescence phases, namely I, II, and III, respectively) as affected by fertilization strategies (digestate—DIG; compost—COM; mineral—MIN) (**A**) and biostimulants application (plant-derived biostimulant—BIO; microbial biostimulant—MIC; not treated control—CTR) (**B**). Different letters indicate significant differences per $p \leq 0.05$ according to Tukey's test. ns: not significant. Vertical bars represent the standard error (n = 3).

Biostimulants affected the SPAD index in all three measurement times, always determining higher values compared with CTR (Figure 2B).

The two experimental factors, fertilization strategy (F) and biostimulant application (B), did not show a significant interaction on the main parameters of fruit production; hence, Table 1 reports the average effects of each factor.

Fertilization affected all the main parameters of fruit yield and quality, except firmness (Table 1). The early marketable production (sum of the first five harvests) reached the highest value in plants fertilized with digestate (9.5 fruits m^{-2}), followed by those treated with compost and minerals. The fruit weight was higher in digestate and mineral fertilization (+7% compared with compost).

Table 1. Number and average fresh weight of eggplant fruits of early and total marketable production, dry matter (DM) percentage, firmness, and N-Kjeldahl content, as affected by fertilization strategy (compost (COM), digestate (DIG), mineral (MIN)), and biostimulant application (untreated control (CTR), plant-based biostimulant (BIO), microbial biostimulant (MIC)).

Treatments	Early Marketable Yield		Total Marketable Fruits		DM	Firmness	N-Kjeldahl
	n° m^{-2}	g fruit^{-1}	n° m^{-2}	g fruit^{-1}	%	kg cm^{-2}	%
Fertilization							
COM	7.8 ± 0.7 b	199.5 ± 3.9 b	25.5 ± 1.3 b	186.2 ± 4.7 b	6.53 ± 0.1 b	1.89 ± 0.02	1.92 ± 0.04 b
DIG	9.5 ± 0.7 a	213.5 ± 3.8 a	25.9 ± 1.2 b	193.4 ± 2.3 ab	6.87 ± 0.1 ab	1.89 ± 0.02	2.08 ± 0.02 a
MIN	5.3 ± 0.3 c	213.5 ± 6.8 a	27.4 ± 1.1 a	198.7 ± 2.6 a	7.14 ± 0.1 a	1.90 ± 0.02	1.98 ± 0.07 ab
Biostimulant							
CTR	6.0 ± 0.6 b	200.3 ± 4.8 b	22.2 ± 0.7 b	189.3 ± 1.9	6.73 ± 0.15	1.84 ± 0.01 b	1.95 ± 0.05
BIO	8.1 ± 0.6 a	208.9 ± 5.9 ab	27.7 ± 0.6 a	193.3 ± 2.6	6.91 ± 0.12	1.90 ± 0.01 ab	2.01 ± 0.04
MIC	8.5 ± 0.9 a	217.3 ± 4.1 a	28.8 ± 0.8 a	195.7 ± 2.5	6.91 ± 0.11	1.94 ± 0.02 a	2.02 ± 0.05
Significance							
Fertilization (F)	**	*	**	*	**	ns	*
Biostimulant (B)	**	*	**	ns	ns	**	ns
F × B	ns	ns	ns	ns	ns	ns	ns

Within each column, different letters indicate significant differences according to Tukey's test at $p \leq 0.05$; ns, *, and **: not significant and significant at $p < 0.05$ and $p < 0.01$, respectively. Values are mean ± standard error (n = 3).

In terms of total production, mineral fertilization gave the highest number of fruits (27.4 fruits m^{-2}), while no difference occurred between the two organic treatments. The average fruit weight and dry matter (DM) percentage showed a significant difference only between mineral and compost, with the first giving heavier fruits (+6.7%). Differently, the nitrogen content was highest in DIG and MIN plants.

The biostimulant application increased the total number of fruits per plant compared with the control (Table 1), and determined better productive results, in terms of both early and total yield, with no difference between the two types of biostimulant (Figure 3B). Fruit firmness increased significantly only with the microbial biostimulant (+2.1% than CTR) (Table 1).

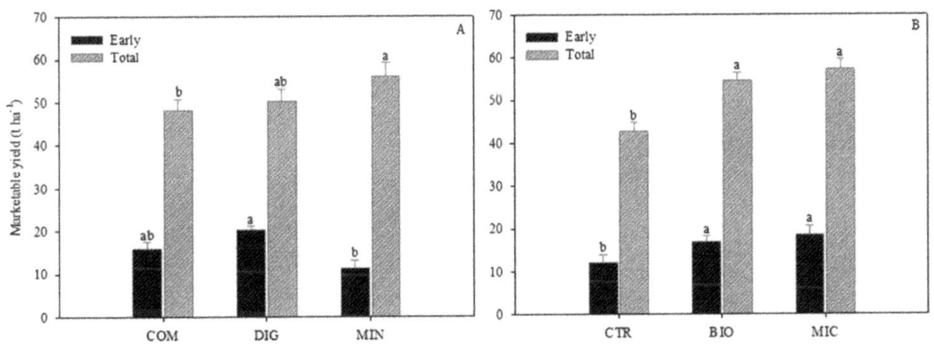

Figure 3. Early (sum of the first 5 harvests) and total (sum of all harvests) marketable yield of eggplant as affected by fertilization strategy (digestate—DIG; compost—COM; mineral—MIN) (**A**) and biostimulant application (plant-derived biostimulant—BIO; microbial biostimulant—MIC; not treated with biostimulant—CTR) (**B**). Different letters indicate significant different per $p \leq 0.05$ according to Tukey's test. Vertical bars indicate standard error (n = 3).

Fertilization strategies affected the fruit yield (Figure 3A), showing a better response of plants fertilized with digestate (about 20 t ha^{-1}) in the early production, but better results in the mineral treatment (about 55 t ha^{-1}) at the end of the cycle.

2.2. Eggplant Fruit Quality

2.2.1. Color Parameters

Table 2 shows that no significant interaction between the two experimental factors was found in the fruit color, expressed as CIELAB color space parameters (L*, a*, and b*). The brightness parameter L* was unaffected by either the fertilization strategy or the biostimulant application. The colorimetric parameter a* (green to red component) was negatively influenced by the BIO biostimulant application compared with CTR. On the other hand, the colorimetric parameter b* (blue to yellow component) was influenced by both fertilization and the biostimulant, with the highest values in plants under DIG fertilization and untreated by the biostimulant, respectively.

Table 2. Color parameters (L*: brightness, ranging from 0 = black to 100 = white; a*: chroma component from green (−60) to red (+60); b*: chroma component from blue (−60) to yellow (+60)) in eggplant fruits as affected by fertilization strategy (compost (COM), digestate (DIG), mineral (MIN)) and biostimulant application (untreated control (CTR), plant based biostimulant (BIO), microbial biostimulant (MIC)).

Treatments	L*	a*	b*
Fertilization			
COM	25.7 ± 0.31	7.21 ± 0.58	0.96 ± 0.17 ab
DIG	26.3 ± 0.44	7.38 ± 0.40	1.07 ± 0.32 a
MIN	25.7 ± 0.18	7.22 ± 0.45	0.60 ± 0.09 b
Biostimulant			
CTR	26.5 ± 0.46	8.26 ± 0.36 a	1.40 ± 0.28 a
BIO	25.5 ± 0.22	6.31 ± 0.23 b	0.51 ± 0.10 b
MIC	25.7 ± 0.17	7.24 ± 0.53 ab	0.71 ± 0.10 b
Significance			
Fertilization (F)	ns	ns	*
Biostimulant (B)	ns	*	**
F × B	ns	ns	ns

Within each column, different letters indicate significant differences according to Tukey's test; ns, *, and ** not significant or significant at $p < 0.05$ and $p < 0.01$, respectively.

2.2.2. Antioxidant Activity and Main Nutraceutical Compounds

The interaction between the fertilization strategy and the biostimulant application also did not produce significant effects for qualitative traits.

The carotenoid content was not affected by the fertilization strategy, only by biostimulants, with a significant increase in the MIC treatment compared with the control (Table 3). The influence of fertilization was relevant only in the ascorbic acid (AsA) content and the hydrophilic antioxidant activity (HAA), with higher values under the organic fertilizers (Table 3). The biostimulant affected the phenol content and ABTS antioxidant activity, which were higher in CNTR and MIC treatments (Table 3).

Table 3. Main bioactive compounds and antioxidant activity in eggplant fruits affected by fertilization strategy (compost (COM), digestate (DIG), mineral (MIN)) and biostimulant application (untreated control (CTR), plant-based biostimulant (BIO), microbial biostimulant (MIC)).

Treatments	Carotenoids µg g^{-1} fw	AsA mg g^{-1} fw	Phenols mg Gallic Acid g^{-1} dw	HAA mmol AA 100 g^{-1} dw	ABTS mmol Trolox 100 g^{-1} dw
Fertilization					
COM	0.017 ± 0.001	45.42 ± 5.45 a	2.80 ± 0.09	5.26 ± 0.35 ab	12.87 ± 0.51
DIG	0.019 ± 0.001	32.82 ± 5.06 ab	2.44 ± 0.16	5.71 ± 0.28 a	13.11 ± 0.65
MIN	0.020 ± 0.002	29.15 ± 3.61 b	2.73 ± 0.17	4.88 ± 0.14 b	12.47 ± 0.80

Table 3. *Cont.*

Treatments	Carotenoids µg g^{-1} fw	AsA mg g^{-1} fw	Phenols mg Gallic Acid g^{-1} dw	HAA mmol AA 100 g^{-1} dw	ABTS mmol Trolox 100 g^{-1} dw
Biostimulant					
CTR	0.016 ± 0.001 b	36.87 ± 4.77	2.62 ± 0.08 ab	5.38 ± 0.32	12.81 ± 0.69 ab
BIO	0.018 ± 0.001 ab	37.44 ± 5.46	2.46 ± 0.17 b	5.25 ± 0.26	12.11 ± 0.58 b
MIC	0.022 ± 0.002 a	33.08 ± 5.72	2.89 ± 0.16 a	5.22 ± 0.30	13.53 ± 0.65 a
Significance					
Fertilization (F)	ns	*	ns	*	ns
Biostimulant (B)	*	ns	*	ns	*
F × B	ns	ns	ns	ns	ns

Within each column, different letters indicate significant differences according to Tukey's test; ns, and * not significant or significant at $p < 0.05$, respectively. AsA—ascorbic acid; HAA—hydrophilic antioxidant activity; ABTS—antioxidant activity; dw—dry weight; fw—fresh weight. Analyses were performed on samples from the 8th harvest. Values are mean (n = 3) ± standard error.

3. Discussion

Intensive farming can reduce the content of organic matter and overall soil fertility, resulting in soil degradation [1]. Our study evaluated the agronomic outputs achieved using practices that increase the organic matter content in a loamy/sandy soil, in eggplant, a nutrient demanding crop, grown in a Mediterranean greenhouse. We studied the impact of two largely used amendments, a compost from urban solid residues and a solid digestate from livestock waste, in combination with two different types of biostimulants, one from tropical plant fermentation and one containing spores of the fungal biocontrol agent *Trichoderma afroharzianum* T-22.

As is known, eggplant thrives better in sandy loam and clay loam soils, with regular irrigation, ensuring constant soil moisture. Proper water management positively influences fruit yield and quality, and the microbial community linked to soil fertility [18]. Also, the plant genotype can influence the crop response to cultural practices such as fertilization. The ecotype "Napoletana", tested in our experiment, has been used in two previous experiments [19,20], but none of them in combination with biostimulants to evaluate the environmental benefit related to the reduction in chemicals. Morra et al. [19] investigated plant response to different fertilization strategies, comparing two doses of compost from urban solid waste with inorganic fertilization. On the other hand, Pane et al. [20] applied eggplant extracts as a biocontrol agent against *Sclerotinia*.

Our results showed that the fertilization strategy influenced eggplant productivity. Both the organic fertilizers elicited higher early production, through a higher number of fruits per plant. However, mineral fertilization determined the highest total yield. This result could be due to the constant supply of nitrogen during the whole cycle, guaranteed by the application of minerals in fertigation. However, the average fruit weight did not differ statistically between mineral and digestate fertilization, presumably because of several reasons. For instance, the higher C/N ratio, which can prevent leaching and increase the fixation of N and the presence of functional elements (such as magnesium) in digestate, could have guaranteed sufficient N availability for the proper growth of the lower number of fruits [21–23].

Plant productive performance was enhanced by the application of biostimulants, as the plant-based product stimulates the plant primary metabolism [24], being rich in amino acids (the main carriers of organic nitrogen, used for protein synthesis) [25], and the microbial product increases the nutrient availability [7].

In our experiment, dry matter and N accumulation were influenced by fertilizers, with lower values in compost and no differences between digestate and mineral fertilizer. Dordas et al. [26] observed in various organs of maize a comparable dry matter production in plants grown with inorganic fertilizers and liquid manure. In contrast, in barley, Arduini et al. [27] obtained a higher N content and dry matter accumulation in plants fertilized with digestate, suggesting that these effects could be attributed to a supply of micronutrients and improved soil biological activity. On this topic, Dion et al. [28] reported that the nutritional composition of organic fertilizers affected microbial communities and their nitrogen mineralization rate.

Regarding fruit firmness, in their review, Rodrigues et al. [29] reported the effects of biostimulants in various examples on fruit plants (including vegetables), showing a varying response due to both the crop and the biostimulant product. Our results align with Consentino et al. [30], who observed a significant increase in fruit firmness due to the application of *T. afroharzianum* in grafted eggplant.

In the present study, the pigment content was influenced by the biostimulant application. Leaf greenness, expressed as the SPAD index, was higher in plants treated with both biostimulants, and a significant increase in carotenoid content was observed in the MIC treatment compared with the untreated control. *Trichoderma* is a fungal genus, which includes several species, widely employed in agriculture, alone and in consortium, as a biofertilizer, biostimulant, and bio-control agent [31]. The SPAD index is a non-destructive method for determining the health status of the plant through the indirect measurement of the chlorophyll and nitrogen content in leaves [32,33]. Our data show that it was positively affected by the biostimulant application but, to determine whether this is due to an effect on the chlorophyll content or nitrogen, further studies should be done. The scientific literature reports that the application of *Trichoderma* in consortium increased the carotenoid content in different plant species [34–36].

In our study, the type of fertilization influenced the amount of ascorbic acid and the hydrophilic antioxidant activity of eggplant fruits, which increased in plants grown with organic fertilization. Regarding the ascorbic acid in eggplant fruits, Michalojc and Buczkowska [37] reported that no differences occurred due to nitrogen sources, while Hassan et al. [38] showed a higher content under organic compared with mineral fertilizers, also hypothesizing that the composition of the organic fertilizer influences the plant metabolism and its interaction with soil. As mentioned, ascorbic acid has a hydrophilic antioxidant nature, so we can hypothesize that the HAA trend is dependent on ascorbic acid due to their high correlation [39]. It is known that compounds with antioxidant capacity, such as phenols, vitamins, and carotenoids, are an efficient countermeasure in the prevention of degenerative diseases [40,41], protecting biological systems against the harmful effects of oxidation [42]. In our experiment, the biosynthesis of phenols as well as the ABTS antioxidant activity were positively influenced by the biostimulant application. The positive effect of plant-based and microbial biostimulants on phenolic compounds was also observed in lettuce [43]. Concerning ABTS, we may hypothesize that its trend is triggered by the greater biosynthesis of carotenoids, as these compounds have a lipophilic behavior [16]; however, the specific mechanisms involved remain to be elucidated.

4. Materials and Methods

4.1. Plant Material, Growth Condition, and Experimental Treatments

The experiment was carried out in a plastic tunnel at the Department of Agricultural Sciences of the University of Naples Federico II (Portici, Italy; 40°49′ N, 14°15′ E, 72 m a.s.l.).

Seedlings of eggplant (*Solanum melongena* L.) ecotype "Napoletana" were transplanted on 22 February 2022, at the density of 2.1 plants m^{-2}, in a loamy/sandy soil, with 1.69% organic matter, 0.11% total nitrogen (N), 97.3 ppm P_2O_5, and 1584.8 ppm K_2O.

This study followed a 2-factor experimental design with 3 replicates, testing the combined effects of a fertilization strategy and biostimulant application. Three fertilization strategies were compared: compost (COM), the solid fraction of a digestate (DIG), and a traditional mineral fertilizer (MIN). The compost originated from the organic fraction of urban solid residues (Progeva s.r.l., Laterza, Taranto, Italy), containing 2% N (94.5% organic N) and 28.3% organic carbon, whereas the digestate, obtained by the anaerobic digestion of livestock waste, had 0.66% N (88% organic N) and 24.3% organic carbon (Power Rinasce S.p.A., Santa Maria la Fossa, Caserta, Italy). Fertilizer rates were determined based on the Fertilization Plain of Campania Region, to obtain the nitrogen dose for eggplant (280 kg ha^{-1}). Accordingly, the compost was applied at 17 t ha^{-1} (fresh weight), while the digestate at 90 t ha^{-1} (fresh weight), both incorporated into the soil before transplanting. To avoid leaching, the mineral fertilizer was applied by a 10 fertigations using calcium nitrate (15.5% N). Based on the soil analyses, the P_2O_5 and K_2O content were adequate to sustain the eggplant growth cycle; hence, no additional fertilizers were supplied.

The biostimulant treatments consisted of the plant-based biostimulant (BIO) and the microbial biostimulant (MIC). BIO (Auxym®, Italpollina S.p.A., Rivoli Veronese, Italy) was obtained by fermenting tropical plants such as hibiscus (*Hibiscus* spp. L., 1753), and applied 4 times as a foliar spray, starting on February 28. MIC (Trianum-P®, Koppert Biological Systems, Berkel en Rodenrijs, The Netherlands) contained the microorganism *Trichoderma afroharzianum* (ex *T. harzianum*) strain Rifai KRL-AG2 (T-22) at a minimum concentration of 1 × 10^9 CFU g^{-1}. MIC was applied 5 times: the 1st one as root immersion (at the dose of 2.5 kg ha^{-1}) on the day of transplant, followed by 4 soil applications (at the dose of 1.0 kg ha^{-1}). No biostimulant was applied in the control (CNT). The experiment consisted of 27 plots (3 fertilization strategies × 3 biostimulant treatments × 3 replicates).

The air temperature in the tunnel was measured by dataloggers (Vantage Pro2, Davis Instruments, Hayward, CA, USA) distributed randomly and located 30 cm above the canopy. The air temperature (minimum and maximum) trend is shown in Figure 4.

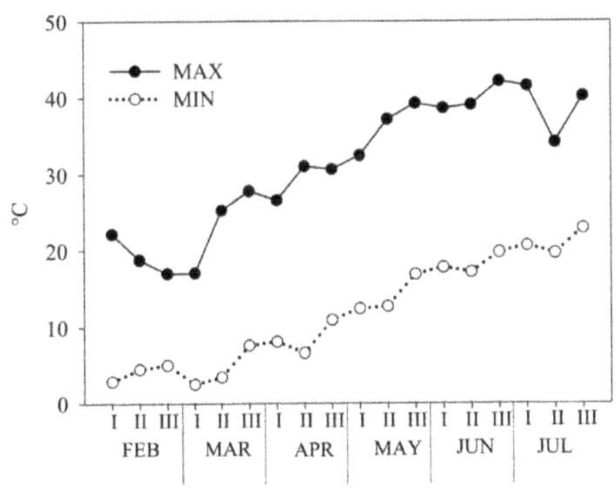

Figure 4. Trend of minimum and maximum temperatures recorded in the greenhouse during the experiment. I, II, and III indicate the 10-day intervals of each month.

4.2. Biometric Measurements, Leaf Greenness, and Fruit Yield

Three times during the growth cycle, the height (cm) of eggplant plants was measured, specifically at 65, 107, and 139 days after transplanting (DAT, and corresponding to vegetative, full production, and senescence phases, respectively), starting from the bottom until the shoot tip.

On the same days as height measurement, soil plant analysis development (SPAD) was also performed using a portable Konica Minolta chlorophyll meter (SPAD-502, Tokyo, Japan). Measurements were carried out on the fifth fully expanded leaf starting from the apex, using three plants per repetition.

The harvests were carried out 14 times, starting on May 19 until July 21 (the last one), on the fruits that reached marketable size. We considered the fruits harvested during the first five harvests as early, while the sampled fruits from the eighth harvest were frozen for the following qualitative analysis.

At each harvest, the number and the average weight of fruits were measured and used to calculate the yield, while a subsample was collected, weighted, and dried in an air oven at 72 °C until constant weight to determine the dry matter (DM) percentage.

4.3. *Qualitative Traits*

4.3.1. Color Parameters and Firmness

The color parameters, expressed as stated by Commission International de l'eclairage (CIELAB) L* (brightness), a* (redness component), and b* (yellowness component), were measured using a CR-300 Chroma Meter (Minolta Camera Co., Ltd., Osaka, Japan) in the middle portion of eggplant fruits at each harvest, in 4 points around the fruit circumference.

On the same fruits, the firmness was measured, expressed in kg cm^{-2}, using a digital penetrometer (T.R. Turoni s.r.l., Forlì, Italy) equipped with an 8 mm diameter probe.

4.3.2. Nitrogen and Carotenoid Content

The frozen eggplant fruits of the eighth harvest were used for various determinations, such as the nitrogen content (expressed in percentage of dry weight) by the Kjeldahl method [44] after sample drying.

Moreover, a portion (1 g) of frozen fruits was extracted in pure acetone and centrifugated at 3000 G (for 5 min). Subsequently, the absorbance of the supernatant was measured with a Hach DR 2000 spectrophotometer (Hach Co., Loveland, CO, USA) at 470 nm to determine the carotenoid content [45].

4.3.3. Ascorbic Acid and Total Phenol Content, and Antioxidant Activity

The frozen fruits were lyophilized and used to determine the following analyses. The ascorbic acid (AsA) assay was performed according to the protocol proposed by Kampfenkel et al. [46], using a spectrophotometer. It was expressed as mg g^{-1} fw. Two hundred mg of frozen dried samples were extracted by two different procedures to assess the eggplant fruit's antioxidant capacity. The first occurred with distilled water to determine the hydrophilic fraction and was measured using the method proposed by Fogliano et al. [47], involving N, N-dimethyl-p-phenylenediamine (DMPD), while the lipophilic fraction was extracted by methanol and the extract was measured by the 2,20-azinobis 3-ethylbenzothiazoline-6-sulfonic acid ABTS method [48]. Both antioxidant activities were determined by UV-Vis spectrophotometry at the absorbance of 505 and 734 nm, respectively. Hydrophilic and lipophilic fractions were expressed as mmol ascorbic acid and Trolox (6-hydroxy-2,5,7,8-tetramethylchro man-2-carboxylic acid) per 100 g^{-1} dw, respectively.

Total phenolics were quantified in methanol extracts using the Folin–Ciocalteu procedure [49], and the absorption was measured at 765 nm using the same spectrophotometer. The result was expressed as mg gallic acid per g dw.

4.4. *Statistical Analysis*

SPSS 2022 software v. 22.0 (IBM SPSS Inc., Chicago, IL, USA) was used for statistical data analysis. A two-way ANOVA was conducted to evaluate the effects of fertilization (F) and biostimulant (B) factors in a two-factorial experimental design (with three replicates per treatment). Additionally, a one-way ANOVA was used to compare the mean effects of individual factors. Significant statistical differences were determined by Tukey's HSD test at the level of $p \leq 0.05$.

5. Conclusions

Our results highlight that, in eggplants, fertilization with digestate allowed to reach a total yield consistent with that obtained with mineral fertilizers. Interestingly, both the digestate and compost elicited early production compared with mineral fertilization, with consequent economic advantage for farmers, since early fruits are usually worth more than the ordinary ones. Also, organic fertilizers improved some qualitative traits of eggplant fruits.

Biostimulant treatments always determined better productive performance compared with non-treated plants, along with a superior product quality, particularly when the microbial-based inoculant was used.

In conclusion, the substitution of mineral fertilizers with digestate and the use of both plant-based and microbial biostimulants are sustainable practices in eggplant cultivation, contributing to maintaining a proper soil nutrient balance without detrimental effects on yield, and possible improvement on fruit quality and useful economical outputs due to the higher early production.

Author Contributions: Conceptualization, M.M., R.M. and R.P.; methodology, E.C. and I.D.M.; formal analysis, L.O. and I.D.M.; investigation, L.O. and E.C.; resources, M.M.; data curation, L.O. and E.C.; writing—original draft preparation, L.G.D.; writing—review and editing, L.G.D. and R.P.; funding acquisition, M.M. All authors have read and agreed to the published version of the manuscript.

Funding: This research received no external funding.

Data Availability Statement: The original contributions presented in this study are included in the article. Further inquiries can be directed to the corresponding author.

Acknowledgments: The authors thank Sabrina Nocerino and Daniele Todisco for their support in laboratory work.

Conflicts of Interest: The authors declare no conflicts of interest.

References

1. Montemurro, F.; Ciaccia, C.; Leogrande, R.; Ceglie, F.; Diacono, M. Suitability of Different Organic Amendments from Agro-Industrial Wastes in Organic Lettuce Crops. *Nutr. Cycl. Agroecosystems* **2015**, *102*, 243–252. [CrossRef]
2. Amsili, J.P.; van Es, H.M.; Schindelbeck, R.R. Cropping System and Soil Texture Shape Soil Health Outcomes and Scoring Functions. *Soil Secur.* **2021**, *4*, 100012. [CrossRef]
3. Rubio, V.; Núñez, A.; Berger, A.; van Es, H. Biomass Inputs Drive Agronomic Management Impacts on Soil Health. *Agric. Ecosyst. Environ.* **2025**, *378*, 109316. [CrossRef]
4. Clapp, C.E.; Hayes, M.H.B.; Ciavatta, C. Organic Wastes in Soils: Biogeochemical and Environmental Aspects. *Soil Biol. Biochem.* **2007**, *39*, 1239–1243. [CrossRef]
5. Verrillo, M.; Khan, M.R.; Volpe, S.; Spaccini, R.; Torrieri, E. Valorization of Organic Biomass through the Production of Active Biopolymer Film Based on Sodium Caseinate, Guar Gum, and Beeswax. *Food Biosci.* **2023**, *53*, 102757. [CrossRef]
6. Tambone, F.; Scaglia, B.; D'Imporzano, G.; Schievano, A.; Orzi, V.; Salati, S.; Adani, F. Assessing Amendment and Fertilizing Properties of Digestates from Anaerobic Digestion through a Comparative Study with Digested Sludge and Compost. *Chemosphere* **2010**, *81*, 577–583. [CrossRef]
7. Halpern, M.; Bar-Tal, A.; Ofek, M.; Minz, D.; Muller, T.; Yermiyahu, U. The Use of Biostimulants for Enhancing Nutrient Uptake. In *Advances in Agronomy*; Academic Press: Cambridge, MA, USA, 2015; pp. 141–174. [CrossRef]
8. Harrison-Kirk, T.; Beare, M.H.; Meenken, E.D.; Condron, L.M. Soil Organic Matter and Texture Affect Responses to Dry/Wet Cycles: Effects on Carbon Dioxide and Nitrous Oxide Emissions. *Soil Biol. Biochem.* **2013**, *57*, 43–55. [CrossRef]
9. Di Mola, I.; Cozzolino, E.; Ottaiano, L.; Nocerino, S.; Rouphael, Y.; Colla, G.; El-Nakhel, C.; Mori, M. Nitrogen Use and Uptake Efficiency and Crop Performance of Baby Spinach (*Spinacia oleracea* L.) and Lamb's Lettuce (*Valerianella locusta* L.) Grown under Variable Sub-Optimal N Regimes Combined with Plant-Based Biostimulant Application. *Agronomy* **2020**, *10*, 278. [CrossRef]
10. Calvo, P.; Nelson, L.; Kloepper, J.W. Agricultural Uses of Plant Biostimulants. *Plant Soil* **2014**, *383*, 3–41. [CrossRef]

11. Cirillo, V.; Romano, I.; Woo, S.L.; Di Stasio, E.; Lombardi, N.; Comite, E.; Pepe, O.; Ventorino, V.; Maggio, A. Inoculation with a Microbial Consortium Increases Soil Microbial Diversity and Improves Agronomic Traits of Tomato under Water and Nitrogen Deficiency. *Front. Plant Sci.* **2023**, *14*, 1304627. [CrossRef]
12. Telo da Gama, J. The Role of Soils in Sustainability, Climate Change, and Ecosystem Services: Challenges and Opportunities. *Ecologies* **2023**, *4*, 552–567. [CrossRef]
13. Geissdoerfer, M.; Savaget, P.; Bocken, N.M.P.; Hultink, E.J. The Circular Economy—A New Sustainability Paradigm? *J. Clean. Prod.* **2017**, *143*, 757–768. [CrossRef]
14. Taher, D.; Solberg, S.Ø.; Prohens, J.; Chou, Y.; Rakha, M.; Wu, T. World Vegetable Center Eggplant Collection: Origin, Composition, Seed Dissemination and Utilization in Breeding. *Front. Plant Sci.* **2017**, *8*, 1484. [CrossRef] [PubMed]
15. Adamczewska-Sowińska, K.; Krygier, M. Yield Quantity and Quality of Field Cultivated Eggplant in Relation to Its Cultivar and the Degree of Fruit Maturity. *Acta Sci. Pol. Hortorum Cultus* **2013**, *12*, 13–23.
16. Consentino, B.B.; Vultaggio, L.; Allevato, E.; Sabatino, L.; Ntatsi, G.; Ciriello, M.; Rouphael, Y.; Di Miceli, G. Plant Protein Hydrolysate and Arbuscular Mycorrhizal Fungi Synergistically Orchestrate Eggplant Tolerance to Iodine Supply: A Two-Year Study. *Sci. Hortic.* **2024**, *336*, 113437. [CrossRef]
17. Caruso, G.; Pokluda, R.; Sękara, A.; Kalisz, A.; Jezdinský, A.; Kopta, T.; Grabowska, A. Agricultural Practices, Biology and Quality of Eggplant Cultivated in Central Europe. A Review. *Hortic. Sci.* **2017**, *44*, 201–212. [CrossRef]
18. Ji, T.; Guo, X.; Wu, F.; Wei, M.; Li, J.; Ji, P.; Wang, N.; Yang, F. Proper Irrigation Amount for Eggplant Cultivation in a Solar Greenhouse Improved Plant Growth, Fruit Quality and Yield by Influencing the Soil Microbial Community and Rhizosphere Environment. *Front. Microbiol.* **2022**, *13*, 981288. [CrossRef]
19. Morra, L.; Bilotto, M.; Raimo, F.; Pizzolongo, G.; Zaccardelli, M.; Pentangelo, A. Yield Response of an Eggplant Ecotype Fertilized with Compost in Two Localities of Campania (Italy). *Italus Hortus* **2010**, *17*, 99–101.
20. Pane, C.; Francese, G.; Raimo, F.; Mennella, G.; Zaccardelli, M. Activity of Foliar Extracts of Cultivated Eggplants against Sclerotinia Lettuce Drop Disease and Their Phytochemical Profiles. *Eur. J. Plant Pathol.* **2017**, *148*, 687–697. [CrossRef]
21. Yan, L.; Liu, C.; Zhang, Y.; Liu, S.; Zhang, Y. Effects of C/N Ratio Variation in Swine Biogas Slurry on Soil Dissolved Organic Matter: Content and Fluorescence Characteristics. Ecotoxicol. *Environ. Saf.* **2021**, *209*, 111804. [CrossRef]
22. Jin, K.; Ran, Y.; Alengebawy, A.; Yang, G.; Jia, S.; Ai, P. Agro-Environmental Sustainability of Using Digestate Fertilizer for Solanaceous and Leafy Vegetables Cultivation: Insights on Fertilizer Efficiency and Risk Assessment. *J. Environ. Manag.* **2022**, *320*, 115895. [CrossRef] [PubMed]
23. Castro Marín, I.; Loef, I.; Bartetzko, L.; Searle, I.; Coupland, G.; Stitt, M.; Osuna, D. Nitrate Regulates Floral Induction in Arabidopsis, Acting Independently of Light, Gibberellin and Autonomous Pathways. *Planta* **2011**, *233*, 539–552. [CrossRef] [PubMed]
24. Ertani, A.; Pizzeghello, D.; Francioso, O.; Sambo, P.; Sanchez-Cortes, S.; Nardi, S. *Capsicum chinensis* L. Growth and Nutraceutical Properties Are Enhanced by Biostimulants in a Long-Term Period: Chemical and Metabolomic Approaches. *Front. Plant Sci.* **2014**, *5*, 375. [CrossRef] [PubMed]
25. Colla, G.; Nardi, S.; Cardarelli, M.; Ertani, A.; Lucini, L.; Canaguier, R.; Rouphael, Y. Protein Hydrolysates as Biostimulants in Horticulture. *Sci. Hortic.* **2015**, *196*, 28–38. [CrossRef]
26. Dordas, C.A.; Lithourgidis, A.S.; Matsi, T.; Barbayiannis, N. Application of liquid cattle manure and inorganic fertilizers affect dry matter, nitrogen accumulation, and partitioning in Maize. *Nutr. Cycl. Agroecosystems* **2008**, *80*, 283–296. [CrossRef]
27. Arduini, I.; Cardelli, R.; Pampana, S. Biosolids Affect the Growth, Nitrogen Accumulation and Nitrogen Leaching of Barley. *Plant Soil Environ.* **2018**, *64*, 95–101. [CrossRef]
28. Dion, P.-P.; Jeanne, T.; Thériault, M.; Hogue, R.; Pepin, S.; Dorais, M. Nitrogen Release from Five Organic Fertilizers Commonly Used in Greenhouse Organic Horticulture with Contrasting Effects on Bacterial Communities. *Can. J. Soil Sci.* **2020**, *100*, 120–135. [CrossRef]
29. Rodrigues, M.; Baptistella, J.L.C.; Horz, D.C.; Bortolato, L.M.; Mazzafera, P. Organic Plant Biostimulants and Fruit Quality—A Review. *Agronomy* **2020**, *10*, 988. [CrossRef]
30. Consentino, B.B.; Sabatino, L.; Vultaggio, L.; Rotino, G.L.; La Placa, G.G.; D'Anna, F.; Leto, C.; Iacuzzi, N.; De Pasquale, C. Grafting Eggplant Onto Underutilized Solanum Species and Biostimulatory Action of *Azospirillum Brasilense* Modulate Growth, Yield, NUE and Nutritional and Functional Traits. *Horticulturae* **2022**, *8*, 722. [CrossRef]
31. Woo, S.L.; Hermosa, R.; Lorito, M.; Monte, E. *Trichoderma*: A multipurpose, plant-beneficial microorganism for eco-sustainable agriculture. *Nat. Rev. Microbiol.* **2023**, *21*, 312–326. [CrossRef]
32. Hu, H.; Liu, H.; Zhang, H.; Zhu, J.; Yao, X.; Zhang, X.; Zheng, K. Assessment of Chlorophyll Content Based on Image Color Analysis, Comparison with SPAD-502. In Proceedings of the 2010 2nd International Conference on Information Engineering and Computer Science, IEEE, Hangzhou, China, 4–6 December 2010; pp. 1–3. [CrossRef]
33. Jiang, C.; Johkan, M.; Hohjo, M.; Tsukagoshi, S.; Maruo, T. A Correlation Analysis on Chlorophyll Content and SPAD Value in Tomato Leaves. *HortResearch* **2017**, *71*, 37–42. [CrossRef]

34. Jain, A.; Singh, A.; Singh, S.; Singh, H.B. Biological Management of *Sclerotinia sclerotiorum* in Pea Using Plant Growth Promoting Microbial Consortium. *J. Basic Microbiol.* **2015**, *55*, 961–972. [CrossRef] [PubMed]
35. Yadav, S.K.; Singh, S.; Singh, H.B.; Sarma, B.K. Compatible Rhizosphere-Competent Microbial Consortium Adds Value to the Nutritional Quality in Edible Parts of Chickpea. J. Agric. *Food Chem.* **2017**, *65*, 6122–6130. [CrossRef] [PubMed]
36. Kaur, T.; Devi, R.; Negi, R.; Kumar, S.; Singh, S.; Rustagi, S.; Shreaz, S.; Rai, A.K.; Kour, D.; Yadav, A.N. Microbial Consortium with Multifunctional Attributes for the Plant Growth of Eggplant (*Solanum melongena* L.). *Folia Microbiol.* **2024**, *69*, 1255–1266. [CrossRef]
37. Michalojc, Z.; Buczkowska, H. Yield and Eggplant Fruit Quality (*Solanum melongena* L.) Dependent on Plant Training and Nitrogen Fertlization. *Ecol. Chem. Eng. A* **2011**, *18*, 73–81.
38. Hassan, S.A.; Mijin, S.; Yusoff, U.K.; Ding, P.; Wahab, P.E.M. Nitrate, Ascorbic Acid, Mineral and Antioxidant Activities of Cosmos Caudatus in Response to Organic and Mineral-Based Fertilizer Rates. *Molecules* **2012**, *17*, 7843–7853. [CrossRef]
39. Arnao, M.B.; Cano, A.; Acosta, M. The Hydrophilic and Lipophilic Contribution to Total Antioxidant Activity. *Food Chem.* **2001**, *73*, 239–244. [CrossRef]
40. Aboul-Enein, H.Y.; Berczynski, P.; Kruk, I. Phenolic Compounds: The Role of Redox Regulation in Neurodegenerative Disease and Cancer. *Mini Rev. Med. Chem.* **2013**, *13*, 385–398. [CrossRef]
41. Luna-Guevara, M.L.; Luna-Guevara, J.J.; Hernández-Carranza, P.; Ruíz-Espinosa, H.; Ochoa-Velasco, C.E. Phenolic Compounds: A Good Choice Against Chronic Degenerative Diseases. *Stud. Nat. Prod. Chem.* **2018**, *59*, 79–108. [CrossRef]
42. Krinsky, N.I. Antioxidant Functions of Carotenoids. *Free Radic. Biol. Med.* **1989**, *7*, 617–635. [CrossRef]
43. Avio, L.; Sbrana, C.; Giovannetti, M.; Frassinetti, S. Arbuscular Mycorrhizal Fungi Affect Total Phenolics Content and Antioxidant Activity in Leaves of Oak Leaf Lettuce Varieties. *Sci. Hortic.* **2017**, *224*, 265–271. [CrossRef]
44. Bremner, J.M. Inorganic Forms of Nitrogen. In *Methods of Soil Analysis, Agronomy Monograph*; Black, C.A., Evans, D.D., White, I.L., Ensminger, L.E., Clark, F.E., Eds.; American Society of Agronomy: Madison, WI, USA, 1965; pp. 1179–1237. [CrossRef]
45. Lichtenthaler, H.K.; Wellburn, A.R. Determinations of Total Carotenoids and Chlorophylls a and b of Leaf Extracts in Different Solvents. Biochem. *Soc. Trans.* **1983**, *11*, 591–592. [CrossRef]
46. Kampfenkel, K.; Vanmontagu, M.; Inze, D. Extraction and Determination of Ascorbate and Dehydroascorbate from Plant Tissue. *Anal. Biochem.* **1995**, *225*, 165–167. [CrossRef]
47. Fogliano, V.; Verde, V.; Randazzo, G.; Ritieni, A. Method for Measuring Antioxidant Activity and Its Application to Monitoring the Antioxidant Capacity of Wines. *J. Agric. Food Chem.* **1999**, *47*, 1035–1040. [CrossRef]
48. Pellegrini, N.; Re, R.; Yang, M.; Rice-Evans, C. Screening of Dietary Carotenoids and Carotenoid-Rich Fruit Extracts for Antioxidant Activities Applying 2,2′-Azinobis (3-Ethylenebenzothiazoline-6-Sulfonic Acid Radical Cation Decolorization Assay). In *Methods in Enzymology*; Academic Press: Cambridge, MA, USA, 1999; pp. 379–389. [CrossRef]
49. Singleton, V.L.; Orthofer, R.; Lamuela-Raventós, R.M. Analysis of Total Phenols and Other Oxidation Substrates and Antioxidants by Means of Folin-Ciocalteu Reagent. In *Methods in Enzymology*; Academic Press: Cambridge, MA, USA, 1999; pp. 152–178. [CrossRef]

plants

MDPI

Article

Fertilization Strategies in Huanglongbing-Infected *Citrus latifolia* and Their Physiological and Hormonal Effects

Luis A. Pérez-Zarate [1], Aída Martínez-Hernández [2], Francisco Osorio-Acosta [1], Eliseo García-Pérez [1], Fredy Morales-Trejo [1] and Juan A. Villanueva-Jiménez [1,*]

[1] Colegio de Postgraduados, Campus Veracruz, Km. 88.5 Carretera Fed. Xalapa-Veracruz, Manlio F. Altamirano, Veracruz 91690, Mexico; alfredo.perez@colpos.mx (L.A.P.-Z.); fosorioa@colpos.mx (F.O.-A.); geliseo@colpos.mx (E.G.-P.); fredymt@hotmail.com (F.M.-T.)

[2] Colegio de Postgraduados, Campus Campeche, Carretera Haltún-Edzná Km 17.5, Sihochac, Champotón, Campeche 24450, Mexico; aida.martinez@colpos.mx

* Correspondence: javj@colpos.mx; Tel.: +52-2292-010770 (ext. 3037)

Abstract: Huanglongbing disease (HLB), caused by *Candidatus* Liberibacter asiaticus (CLas), affects all commercial citrus species. Persian lime (*Citrus latifolia* Tanaka), a crop of global economic importance, has shown tolerance to this disease. Efforts are focused on extending the productive life of diseased trees through effective agronomic management. This study aimed to evaluate how different fertilization strategies influence the physiological and hormonal responses of *Citrus latifolia* on both healthy and HLB-affected plants. It compared the effects of low (Ma-1), medium (Ma-2), and high (Ma-3) doses of macronutrients, with and without the addition of micronutrients (Mi-1), using either soil (Mi-2) or foliar (Mi-3) applications. Treatments were applied every 18 days for one year. *C. latifolia* showed tolerance; however, HLB infection negatively affected growth parameters, photosynthetic activity, vascular bundle anatomy, reflectance at 550 and 790 nm, carbohydrate metabolism, and the concentration of salicylic acid and its biosynthetic precursors. The hormonal response showed higher levels of benzoic acid and lower levels of salicylic acid than those reported in susceptible citrus. Plants treated with low doses of macronutrients along with soil-applied micronutrients (Ma-1 + Mi-2) showed a 17.9% increase in growth, a 31.3% larger canopy volume, and an 83.3% reduction in starch accumulation compared to the treatment with high doses of macronutrients and both soil and foliar applied micronutrients (Ma-3 + Mi-3). These findings indicate that split soil fertilization with low-dose macronutrients and micronutrients might influence plant physiological responses, potentially improving disease management and decreasing fertilizer inputs.

Keywords: *Candidatus* Liberibacter asiaticus; Persian lime; phytohormones; macronutrients; micronutrients

1. Introduction

Huanglongbing (HLB) is one of the most devastating diseases in citrus production, causing severe yield losses and substantial economic damage [1]. This disease is caused by a Gram-negative α-proteobacteria that is restricted to the phloem [2]. *Candidatus* Liberibacter asiaticus (CLas) is the species associated with HLB in Mexico and is transmitted by *Diaphorina citri* Kuwayama, the Asian citrus psyllid [3]. Citrus production profitability has declined due to the high costs of managing the disease in affected areas [4]. HLB is now present in nearly all citrus-producing countries; however, in most regions, the disease remains confined to specific areas [5]. It has resulted in substantial economic losses for countries like China, Brazil, India, and Mexico [6]. In the latter, HLB is present in all 24 citrus-producing states [6].

HLB-infected plants exhibit physiological impairments that lead to foliar symptoms, including green islands, blotchy mottle, lightening and thickening of veins, death of branches and secondary roots, deformation, uneven ripening, and premature fruit drop [1,7]. These symptoms, caused by phloem blockage and restricted photoassimilate flow, disrupt carbohydrate metabolism. Excessive starch accumulation in leaves leads to chloroplast degradation and loss of chlorophyll [8]. Photoassimilate flow to the roots decreases, affecting their development with limited water and nutrient absorption. This limitation may arise from the thickening of xylem cell walls, narrowing the vessel's lumen, limiting the flow of water and nutrients, and ultimately causing water deficits [9].

HLB symptoms are different between HLB-tolerant and -susceptible citrus. Reports indicate that tolerant varieties could have a specific recognition system for CLas and a more efficient detoxification system. This system reduces most damage caused by reactive oxygen species (ROS) and promotes the induction of the immune system, leading to a lower expression of foliar symptoms [10]. Furthermore, tolerant varieties could suppress the direct defense response to CLas infection, especially salicylic acid-mediated signaling [11]. However, no CLas effectors directly related to foliar symptoms have been identified, which has led to the hypothesis that HLB may act as an autoimmune disease [12].

In tolerant varieties, such as Persian lime, symptom development is slower, and diseased plants maintain their development and production despite suffering from the disease [13]. Furthermore, fewer effects of HLB on Persian lime are present at the subcellular level than in susceptible varieties such as Mexican lime [14]. Despite the above, Persian lime plants infected with HLB reduce fruit yield by approximately 2.4 t ha^{-1} due to a 17.3% reduction in fruit weight and 18.6% in juice volume, a reduction that has been similar in the states of Yucatan and Veracruz [5,15]. In Mexico, proposed HLB management included (a) monitoring and controlling the insect vector, (b) using plants from a nursery certified to mass produce disease-free plants, and (c) promptly removing diseased plants in the early stages of infection. To promote coexistence with HLB, cultural practices were implemented, such as improved fertilization strategies, all within the framework of the regulatory campaign against regulated citrus pests. This was further supported by the establishment of Phytosanitary Epidemiological Management Areas (AMEFIs, its acronym in Spanish) [6,16].

Balanced nutrition is crucial for maintaining plant health and vigor; healthy plants are known to be less susceptible to pathogen attacks [17]. Mineral nutrients are part of the plant's first line of defense against pathogens since they are required to properly activate defense mechanisms such as producing toxins, metabolites, and lignin. As an example, calcium serves as a key structural component of the cell wall, critical for its integrity and function, while zinc, iron, and manganese act as cofactors in the synthesis of secondary metabolites [18]. Therefore, balanced fertilization has been suggested as a strategy to extend the production of trees affected by HLB. Mineral nutrition can be supplied through granular fertilizers added to the soil, as well as via the foliar application of macro- and micronutrients [19]. It is relevant to note that balanced nutrition is a complementary tool in the HLB management strategy. It works alongside other approaches, such as certified plant material use and biological and chemical vector control [3].

HLB-infected plants show a reduced availability of micronutrients in symptomatic leaves, including Ca^{2+}, Mg^{2+}, Fe^{2+}, Zn^{2+}, and Cu^{2+} [20]. Applying nitrogen, phosphorus, potassium, calcium, magnesium, copper, iron, and zinc to HLB-infected trees might mitigate the severity of the disease and enhance fruit production. This improvement might result from fewer disruptions in carbohydrate metabolism and a reduced impact of the disease on plant growth and yield [21–24]. The foliar and soil application of calcium, magnesium, zinc, and boron also significantly increased root lifespan [25]. High doses of Mn (4X) were associated with reduced CLas titers [26]. Thus, a 45.0% increment in fruit production in diseased plants was achieved with micronutrient foliar applications. However, applying doses that exceed recommended levels might reduce fruit yield by up to 25.0% [27].

Several reports indicate that HLB-infected trees require higher nutrient levels compared to those recommended for healthy plants [26,28,29]. However, other studies suggest a greater nutrient uptake efficiency in HLB-infected plants than in healthy ones [30]. Notably, not many comparisons have been made considering citrus plants with different levels of HLB tolerance. Ghimire et al. [29] assessed the nutrient uptake potential in HLB-tolerant and -susceptible rootstocks. Tolerant rootstock A + Volk × O-19 had the highest nutrient uptake efficiency, while susceptible US-896 exhibited lower nutrient uptake efficiency and a reduced expression of nutrient transporter genes.

Additional studies highlighted differences in the effectiveness of micronutrient application methods, such as granular versus liquid forms, where micronutrient application rates did not correlate with leaf micronutrient concentrations [30]. Additional research is necessary to better understand the nutrient uptake capacity of Persian lime across different rootstocks under CLas infection. This information will assist in designing improved fertilization programs [29].

Applying the correct fertilizer at the right time, in the proper amount and location, and from a suitable source will enhance the effectiveness of fertilization programs [31,32]. Recently, a consistent and gradual application of fertilizers has been adopted for managing HLB in Florida, USA [33,34]. This approach reaches the crop's annual fertilizer requirement through smaller doses administered more frequently rather than all the fertilizer at once. The plant obtains a continuous supply of nutrients, avoiding leaching losses and preventing nutrient deficiencies associated with HLB [32]. Fertigation allows for up to thirty applications each year, whereas growers who depend on rainfed conditions typically have four to six fertilizations each year [33,34].

Identifying the tolerance characteristics of Persian lime is crucial for determining molecular targets that will help develop effective disease management strategies. This information will assist in implementing a balanced fertilization program specifically designed for Persian lime affected by HLB. Balanced fertilization strategies are necessary for practical field applications under irrigation or rainfed conditions. Recent progress has been made in understanding the role of macro- and micronutrients, as well as their interactions and effects on citrus trees affected by HLB, particularly in *C. sinensis* [26,33,35–37]. However, in Mexico, the impact of balanced fertilization has not been thoroughly studied in either tolerant or susceptible citrus species. This highlights the need for developing an HLB management strategy for Persian lime based on local research, to design short-term, applicable management strategies. This study focuses on nitrogen (N), phosphorous (P), and potassium (K) as key macronutrients, complemented by calcium (Ca), magnesium (Mg), and sulfur (S) as secondary macronutrients, as well as zinc (Zn), boron (B), manganese (Mn), and iron (Fe) as micronutrients. Strategies with different levels of macro- and micronutrients were designed based on recommended regional doses and commercial fertilizers used locally. We hypothesize that applying macronutrients and micronutrients in split doses through soil and/or foliar methods can enhance nutrient efficiency. This approach may reduce the amounts needed while improving the physiological and hormonal responses of *C. latifolia* plants infected with CLas. To test this hypothesis, we compared the effects of low, medium, and high doses of individual macronutrients, as well as their combination with micronutrients, using soil and/or foliar application methods. We assessed their impact on foliar symptoms, CLas populations in midribs, and various physiological parameters, including growth and development, photosynthetic rate, leaf spectral reflectance, vascular bundle damage, carbohydrate metabolism, and the accumulation of salicylic acid (SA) and its biosynthetic precursors. This approach will help to develop new tools for Huanglongbing disease management in Persian lime.

2. Results

2.1. Relative Quantification of CLas 16S rDNA Gene After Fertilization Treatments

The amplification of the 16S rDNA gene at the end of summer assessed the change in the relative accumulation of CLas. This variable evaluated the effect of medium and high macro- and micronutrient treatments compared to the Ma-1 + Mi-1 treatment. No increase in the number of bacteria (with a value equal to or less than 1) was observed for the treatments with low and medium doses of macronutrients (Ma-1 and Ma-2) across all micronutrient levels (Mi-1, Mi-2, and Mi-3). In contrast, an increasing trend was observed with high doses of macronutrients at all three micronutrient levels (Figure 1), especially with the absence of micronutrients (Ma-3 + Mi-1).

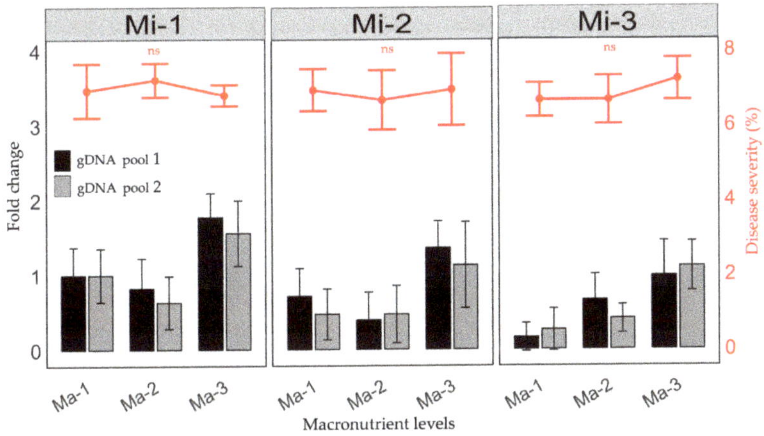

Figure 1. Change in the number of bacteria for high and medium fertilization treatments compared to the Ma-1 + Mi-1 treatment and its relationship with the severity of symptoms on leaves. Ma = macronutrients, Mi = micronutrients. Mean ± propagated error, ns = not significant.

2.2. Fertilization and the Percentage of Severity of HLB Symptoms in Leaves

At the end of the experiment, no significant differences were observed in the percentage of severity among all fertilization treatments ($p = 0.4728$) (Figure 1). The percentage of severity did not exceed 8.0% for any treatment; this rate was lower than results reported for susceptible varieties and those observed in field conditions for Persian lime. Thus, balanced fertilization was effective in delaying the progression of symptoms.

2.3. Fertilization on the Efficiency of Photosystem II (Fv/Fm) in Healthy and Diseased Plants

Before treatment applications, variation in Fv/Fm values was noted among the plant groups. During the summer, means of Fv/Fm were lower in diseased plants (0.791 ± 0.008) than in healthy plants (0.80 ± 0.006) ($p_s \leq 0.0001$). However, after fertilization treatments were applied, Fv/Fm values did not differ significantly between healthy and HLB-infected plants in the winter and autumn ($p_s = 0.1260$ and 0.3689, respectively) (Figure 2).

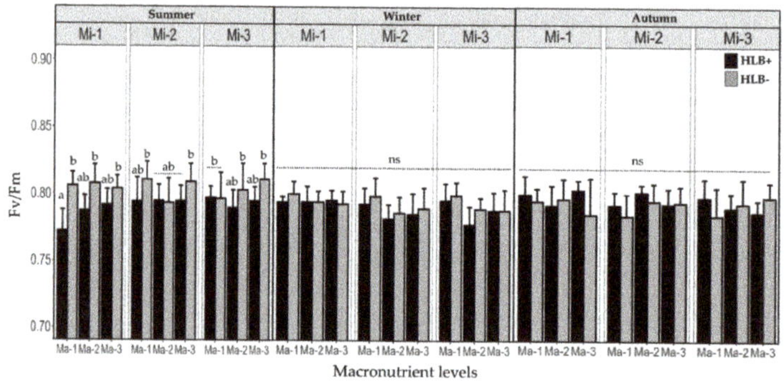

Figure 2. Effect of *C*Las infection and macronutrient and micronutrient levels on the efficiency of photosystem II. ns = not significant. Different letters indicate significant differences (Tukey, $p < 0.05$).

2.4. Reflectance Spectrum on Leaves of Healthy and HLB-Diseased Plants After Fertilization Treatments

Mean reflectance data collected during the experimental period revealed notable differences between healthy and diseased plant groups in both the visible light spectrum (VIS) and near-infrared (NIR) regions (Figure 3A). Wavelengths of 550 nm (visible light) and 790 nm (near-infrared) showed the most significant variability between groups. The first one may have been related to the loss of photosynthetic pigments and the last one to the accumulation of carbohydrates, due to *C*Las infection (Figure 3B,C). Diseased plants had a higher percentage of reflectance at 550 nm compared to healthy plants. Conversely, at 790 nm, the response was the opposite. Unlike the trend in photosynthetic efficiency, at the beginning of the experiment (summer), no differences were observed between treatments in the VIS range ($p_{Ma \times Mi}$ = 0.7194). These results contrast with those obtained during the winter ($p_{Ma \times Mi}$ = 0.0021) and autumn, where only macronutrient levels had a marginally significant effect (p_{Ma} = 0.0589, $p_{Ma \times Mi}$ = 0.5316) (Figure S1A). In the NIR range, no significant differences were found in the summer, winter, or autumn ($p_{Ma \times Mi}$ = 0.8634, 0.3083, and 0.2585, respectively) (Figure S1B). A significant variation in the percentage of reflectance was observed between healthy and diseased leaves. Additionally, no clear correlation was found between the percentage of reflectance and fertilization treatments.

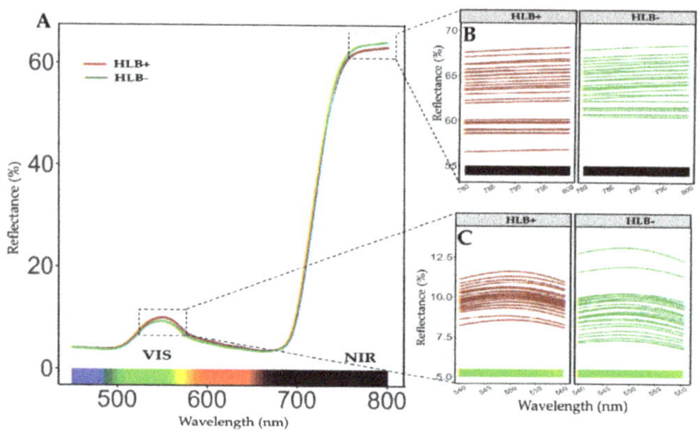

Figure 3. (**A**) Spectrograms with means of reflectance of leaves from healthy (green line) and diseased (red line) plants. (**B**,**C**) Wavelengths of 550 nm in the visible region (VIS) and 790 nm in the near-infrared (NIR) regions were chosen to analyze reflectance on plants treated with different fertilization strategies.

2.5. Fertilization Impact on the Content of Chlorophyll, Sucrose, and Starch in Healthy and HLB-Diseased Leaves

At the end of the experiment, diseased plants were affected in all variables of carbohydrate and chlorophyll metabolism. Healthy plants showed 24.2% more chlorophyll (1.57 mg g^{-1} FW $\pm$ 0.20 SD) compared to diseased plants (1.19 mg g^{-1} FW$\pm$ 0.11 SD). Also, healthy plants presented 75.4% less sucrose concentration (1.68 mg g^{-1} FW $\pm$ 2.18 SD) and 80.0% less starch (0.08 mg g^{-1} FW $\pm$ 0.03 SD) compared to diseased plants (sucrose: 6.87 mg g^{-1} FW $\pm$ 2.39 SD; starch: 0.41 mg g^{-1} FW $\pm$ 0.20 SD) (Figure S2A–C). Also, starch accumulation and several anatomical changes in the central vein of the leaves were observed in diseased plants compared to healthy ones (Figure S3). Results suggest that HLB causes damage to the phloem and alters carbohydrate metabolism. Low doses of macronutrients (Ma-1) were associated with the highest chlorophyll content and the lowest sucrose and starch contents in leaves of HLB-diseased Persian lime plants. In contrast, higher doses of macronutrients showed reduced chlorophyll content in healthy plants and a more significant starch accumulation in diseased plants (Figure 4A,C,E). The micronutrient levels applied did not have a significant effect on chlorophyll content in healthy and HLB-infected plants (p = 0.1271 and 0.1032, respectively). Micronutrients influenced sucrose content in healthy plants (p = 0.0283) but not in infected plants (p = 0.5784). The highest starch accumulation in diseased plants occurred with soil and foliar application (Mi-3) (p = 0.0181) and with soil applications in healthy plants (Mi-2) (p = 0.0001) (Figure 4B,D,F).

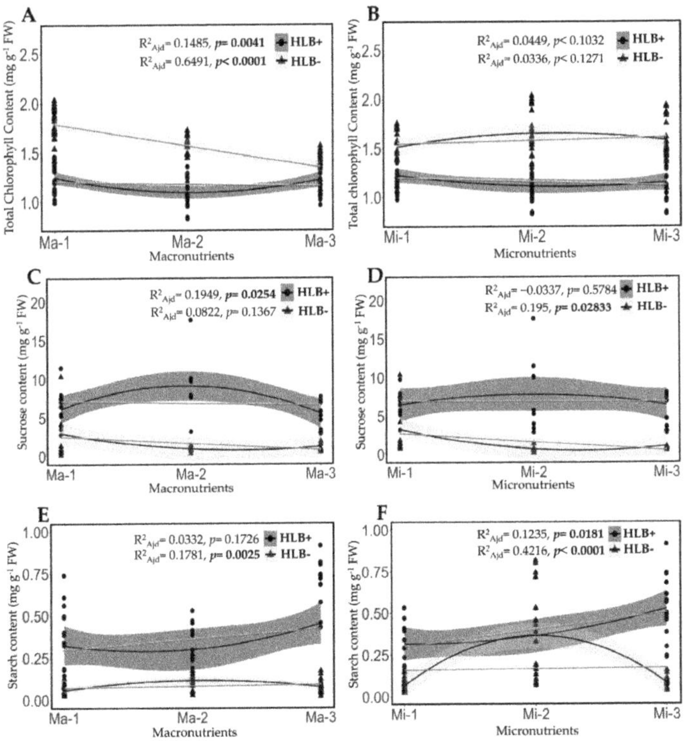

Figure 4. Polynomial regression analysis for macronutrient (**A**,**C**,**E**) and micronutrient (**B**,**D**,**F**) treatment levels and their effect on total chlorophyll, sucrose, and starch content in healthy and HLB-diseased plants. The blue line represents the data linear fit. Values represent four replicates, mean $\pm$ standard deviation. Ma = macronutrients, Mi = micronutrients. Significant p-values are shown in bold.

2.6. Fertilization Impact on Cumulative Growth in Height, Canopy Volume, and Trunk Diameter in Healthy and HLB-Diseased Plants

Infection with CLas significantly affected cumulative plant growth and canopy volume and fostered a larger trunk diameter in healthy plants ($p_s \leq 0.0001$ for all three variables) (Figure S4A–C).

On average, healthy plants showed a 17.0% additional cumulative height (132.58 cm ± 14.86 SD) compared to that in diseased plants (109.03 cm ± 12.04 SD). They also showed 33.7% more canopy volume (2.40 m^3 ± 0.33 SD) than diseased plants (1.59 m^3 ± 0.34 SD), while having a 19.5% smaller trunk diameter (9.44 mm ± 2.11 SD vs. 11.72 mm ± 2 SD in diseased plants). This difference might be attributed to an increase in the phloem area of the stem, similar to that observed in the midrib of leaves (Figure S4A–C).

After fertilizing every 18 days over the span of one year, the lowest doses of macronutrients were associated with the highest cumulative growth, greater canopy volume, and trunk diameter in healthy and diseased plants (Figure 5A,C,E). In contrast, micronutrient levels showed no significant effect in healthy or diseased plants (Figure 5B,D,F). Increased doses of macronutrients plus the application of micronutrients, both to the soil and foliar, do not seem to be favorable for promoting the growth and development of HLB-infected Persian lime plants.

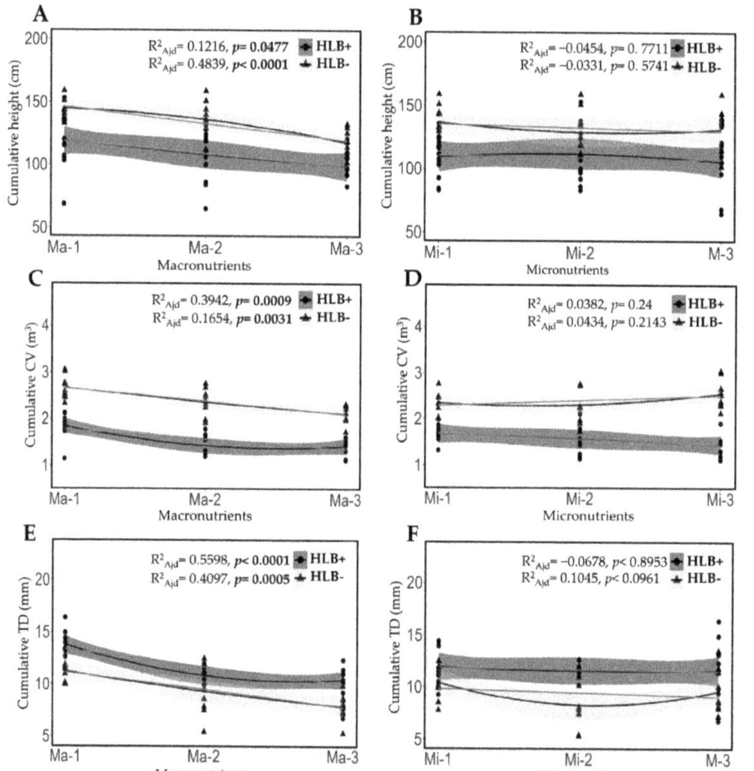

Figure 5. CLas infection and fertilization treatments had an impact on the growth and development of Persian lime plants. Polynomial regression analysis was conducted to evaluate the effects of macronutrient (**A**,**C**,**E**) and micronutrient treatment levels (**B**,**D**,**F**) on cumulative height, canopy volume, and trunk diameter in both healthy and HLB-diseased plants. The blue line indicates the linear fit of the data. Values are expressed as mean ± standard deviation. Ma = macronutrients, Mi = micronutrients, CV = canopy volume, TD = trunk diameter. Significant *p*-values are shown in bold.

2.7. Fertilization Impact on Endogenous Content of Salicylic Acid and Its Biosynthetic Precursors in Healthy and HLB-Diseased Plants

CLas infection increased the endogenous content of *t*-CA, BA, and SA in diseased plants compared to healthy plants. Specifically, healthy plants showed 30.9% lower *t*-CA levels (830.55 ng g^{-1} FW ± 99.84 SD) than diseased plants (1202.15 ng g^{-1} FW ± 149.80 SD). In addition, healthy plants had 29.5% lower BA levels (45,218.49 ng g^{-1} FW ± 15,608.23 SD) than diseased plants (64,104.37 ng g^{-1} FW ± 7315.19 SD). Furthermore, salicylic acid (SA) levels were 41.5% lower in healthy plants (447.64 ng g^{-1} FW ± 122.69 SD) compared to diseased plants (765.72 ng g^{-1} FW ± 103.52 SD) (Figure 6A,D,G).

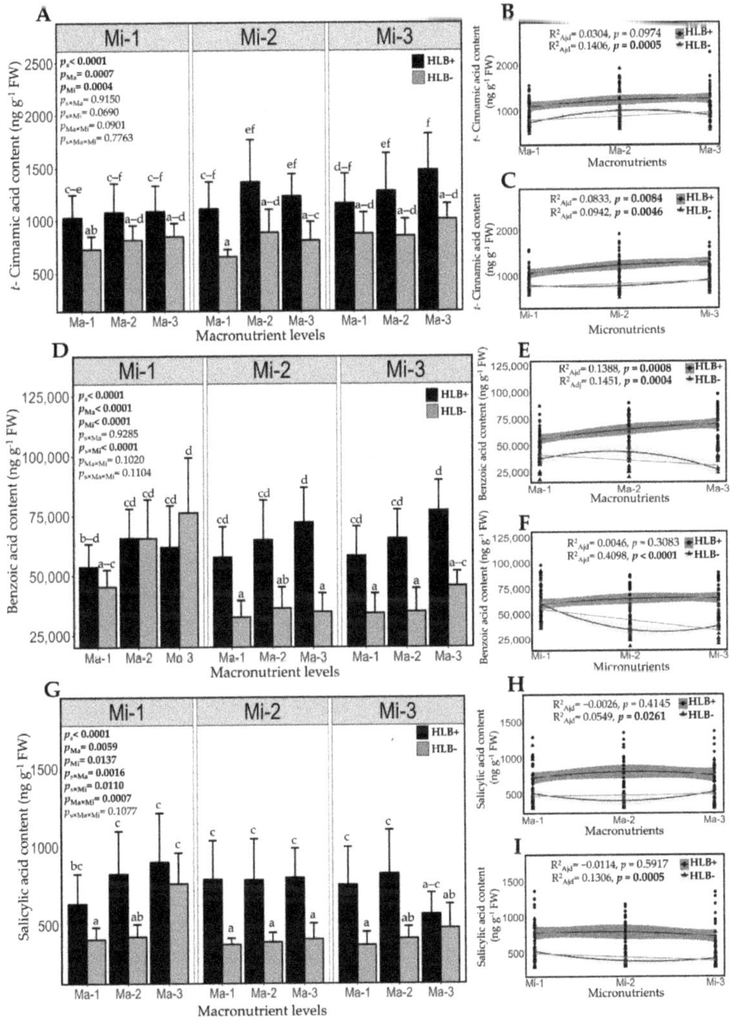

Figure 6. Impact of CLas infection and fertilization treatment on the content of endogenous salicylic acid and its biosynthetic precursor content. (**A**,**D**,**G**) Concentration of *trans*-cinnamic acid, benzoic acid, and salicylic acid in leaves (**B**,**C**,**E**,**F**,**H**,**I**). Polynomial regression analysis for macro- and micronutrient levels in healthy and diseased plants. The blue line represents the linear fit of data. According to the general linear model (ANOVA), the main effects are shown within graphs. Different letters indicate significant differences (Tukey, $p < 0.05$). Values represent mean ± standard deviation. Ma = macronutrients, Mi = micronutrients. Significant *p*-values are shown in bold.

Consistent with the results above, the lowest doses of macronutrients were associated with the lowest concentrations of *t*-CA and BA in diseased plants, with a trend to increase at higher doses. A similar trend regarding *t*-CA was observed in healthy plants; however, higher doses were associated with the lowest concentration of BA. SA concentration was similar at all macronutrient levels in diseased plants, while in healthy plants, it was lower at the medium macronutrient dose (Figure 6B,E,H). Micronutrient deficiency was associated with lower *t*-CA concentrations in healthy and diseased plants. Furthermore, micronutrient application to the soil resulted in the lowest concentration of BA and SA in healthy plants. On the other hand, the concentration of these analytes was similar in diseased plants, regardless of micronutrient levels (Figure 6C,F,I). Healthy and HLB-infected plants exhibit different hormonal responses, indicating that nutrition may affect hormone regulation and the Persian lime's ability to respond to HLB disease progression.

2.8. Variables Associated with Healthy and HLB-Diseased Plant Groups Obtained by Principal Component Analysis (PCA)

The contributions of PC1, PC2, and PC3 were 39.73%, 12.26%, and 9.90%, respectively, accounting for 61.89% of the total variance in the data. The principal components plot showed a clear separation between healthy and diseased plants, indicating differences in the linear properties between the study variables. Healthy plants were associated with higher values for height, canopy volume, photosynthetic efficiency, chlorophyll, and reflectance at 790 nm. In contrast, diseased plants showed elevated values for trunk diameter, glucose, sucrose, starch, *trans*-cinnamic acid, benzoic acid, salicylic acid, and reflectance at 550 nm (Figure S5). These findings suggest that HLB significantly affected most of the variables studied, particularly those linked to growth and development, carbohydrate metabolism, reflectance, and hormone levels, despite the fact that Persian lime is generally considered tolerant to this disease.

3. Discussion

As a species tolerant to HLB, Persian lime shows symptoms more gradually, which enables it to sustain development, cellular homeostasis, and productivity even when infected [13,38]. The tolerance observed in specific citrus cultivars is associated with polyploidy in both rootstocks and commercial varieties, contributing to more efficient detoxification and enhanced phloem regeneration upon CLas infection [14,39]. In this case, the severity of HLB symptoms in leaves remained below 8.0% in all treatments almost two years after the initial inoculation. However, a severity percentage greater than 40.0% has been reported in Persian lime trees in the field [40], suggesting that fertilization may have contributed to the slow development of leaf symptoms.

CLas infection affects chlorophyll fluorescence parameters, including photosystem II efficiency (Fv/Fm) [41]. While diseased plants exhibited a lower total chlorophyll content than healthy plants, their photosynthetic efficiency remained similar across both groups. This finding aligns with the low severity percentage observed in the leaves. Reduced photosynthetic efficiency in diseased plants is linked to increased damage to the photosynthetic apparatus caused by HLB infection [41]. However, fertilization can enhance photosynthetic efficiency by reducing photoinhibition caused by stress [42]. Low doses of macronutrients seem to support stability in chlorophyll content and photosynthetic efficiency. This aligns with previous studies demonstrating how proper fertilization can enhance the chlorophyll index in leaves and promote greater photosynthetic capacity [43]. The decline in chlorophyll content of the leaves in diseased plants was less pronounced than what was reported in symptomatic trees in the field [44]. The loss of chlorophyll may occur due to the buildup of carbohydrates, such as sucrose and glucose, which can suppress certain genes involved in photosynthesis [45]. Furthermore, the buildup of starch in mesophyll chloroplasts can lead the rupture of thylakoids and the degradation of chlorophyll [46].

The blockage of photoassimilate flow in the phloem due to HLB results in an accumulation of carbohydrates in the leaves, including glucose, fructose, sucrose, and starch [47]. This accumulation can affect photosynthesis and reflectance parameters by altering leaf pigments, their biochemical composition, and cellular structure [48]. Although Persian lime shows some tolerance to HLB, research indicates that diseased plants accumulate more glucose, sucrose, and starch in their leaves than healthy plants. This finding confirms that the vascular system of Persian lime is affected by HLB, but to a lesser extent than in more susceptible citrus varieties and species. This buildup of carbohydrates occurs due to the reduced flow of photoassimilates caused by the obstruction of phloem sieve plate pores. This phenomenon was previously observed in diploid varieties, such as the Mexican lime, and triploid varieties, such as the Persian lime. However, reports indicate that Persian lime presents less collapse of phloem cells, as well as an increase in the phloem area of the petioles [14]. This observation aligns with our findings in the leaf veins (Figure S3).

Leaf reflectance is affected by several characteristics, such as the surface, internal structure, and biochemical composition of the leaf [49,50]. Healthy and HLB-diseased Persian lime plants showed distinctive reflectance patterns in the VIS and NIR, consistent with previous reports on reflectance assessments in citrus [48,51]. However, the mild severity of symptoms caused significant variation among plants and sampling times. Still, PCA showed that reflectance at 550 nm is a variable associated with diseased plants, even with disease management actions, which could further mask symptom expression in Persian lime. Although the equipment is capable of distinguishing between healthy and diseased plants, the variation between asymptomatic leaves and those with mild HLB symptoms poses a challenge for early detection. This issue is particularly important for devices that operate solely within the visible light spectrum (400–800 nm) and lack trained software to improve detection accuracy. More advanced reflectance methods could improve efficiency in detecting HLB. Hu et al. [52] combined multicolor fluorescence images with multispectral reflectance and a pre-trained convolutional neural network model, which allowed them to detect HLB in the field with an efficiency greater than 96.0%.

Diseased Persian lime plants showed lower cumulative growth in height and canopy volume, as well as a greater trunk diameter compared to healthy plants. The adverse impact of the disease on the height and canopy volume of affected plants has been well documented, alongside the beneficial effects of foliar micronutrients [35,53]. Our results are aligned with these findings, where the mildest impact on growth variables in diseased plants occurred when low doses of macronutrients were combined with micronutrients in the soil, in contrast to higher doses. Although the increase in trunk diameter in plants infected with HLB has not been previously reported, some studies indicate anatomical damage to the phloem due to the effect of HLB [54,55]. In addition, an increase in phloem thickness has been reported in Persian lime, attributable to more significant regeneration in response to HLB damage, which appears to be one of the tolerance characteristics provided by polyploidy [14].

Phytohormones are crucial in regulating plant growth and development as well as in the signaling networks that coordinate most physiological functions in plants, including the response to pathogens [56]. This work evaluated the endogenous expression of salicylic acid and its biosynthetic precursors in the phenylalanine ammonia-lyase (PAL) pathway, *trans*-cinnamic acid, and benzoic acid. The results showed a higher concentration of *t*-CA, BA, and SA in diseased plants compared to healthy ones, consistent with previous reports in healthy [57] and diseased [58] orange plants. However, there are no reports of the hormonal response of Persian lime to CLas infection. Additionally, the quantification of these analytes in Persian lime leaves revealed lower concentrations of *t*-CA and BA at the lowest doses of macronutrients, and these levels have an upward trend at higher doses. This finding suggests a more significant stress condition at the highest doses, which aligns with alterations in other growth variables, the carbohydrate metabolism, chlorophyll content, and bacterial presence observed by the end of the experiment.

Previous research has shown differences in the phytohormones pathways, such as SA, between tolerant and susceptible citrus varieties [11,46,59,60]. Higher concentrations of SA and *Me*-SA were reported in tolerant grapefruit (*C. paradisi* Macf.) compared to susceptible *C. sinensis*. However, after inoculation with CLas, SA levels were reduced to less than half in grapefruit, while SA concentration in susceptible oranges increased [59]. Similarly, Suh et al. [11] found that metabolites related to the SA signaling pathway decreased in tolerant Sugar Belle® mandarin compared to susceptible Murcott mandarin. In *Citrus reticulata* Blanco, a moderately tolerant species, CLas infection repressed the SA and ethylene signaling pathway [60]. On the other hand, Curtolo et al. [46] did not report changes in the transcription of genes related to SA biosynthesis in *P. trifoliata*. The results suggest that the SA-mediated defense response could be suppressed in tolerant varieties upon CLas infection.

Although in this work, SA levels increased upon CLas infection, the concentration found in HLB-tolerant Persian lime leaves was lower than that reported in susceptible varieties [13]. The increase in BA levels could be associated with the increase in SA levels [61], which differs from our results and leads us to hypothesize a lower conversion of BA to SA. SA synthesis occurs via two chorismate-dependent pathways in the shikimic acid pathway. The first converts chorismate to isochorismate via isochorismate synthase (ICS) and then to SA via isochorismate pyruvate lyase (IPL) [62]. The second pathway is through PAL, which converts L-phenylalanine to cinnamic acid, which is converted into benzoic acid by aldehyde oxidase (AO) and finally into SA via benzoic acid-2-hydroxylase (BA2H) [61,62]. No homologous genes for IPL or BA2H have been found in citrus, suggesting that, in this case, the PAL-dependent pathway could be responsible for SA biosynthesis [58,63]. This statement accounts for the increased concentration of BA observed in Persian lime; however, it does not explain the reduced concentration of SA. All these ideas suggest a lower conversion rate of BA to SA, both in healthy and diseased plants, which may indicate that Persian lime exhibits some tolerance to HLB. Furthermore, we found a significant positive correlation between BA and SA content in healthy plants ($R^2 = 0.54$) but not in diseased plants ($R^2 = 0.07$). A possible characteristic of tolerant varieties is the coexistence with the disease, either restricting or delaying the defense response, thereby reducing damages caused by the disease and supporting the hypothesis that HLB is an autoimmune disease [12]. The hypothesis of SA biosynthesis via the PAL pathway has been raised based on the results in varieties susceptible to HLB; however, citrus species such as *C. aurantifolia* and *C. latifolia* have shown lower PAL activity than *C. sinensis* [64], which may indicate a reduced PAL-dependent pathway activity in HLB-tolerant varieties.

Overall, the analysis of the combination of macronutrient and micronutrient levels suggests that the low doses of macronutrients used in the present study, combined with soil application of micronutrients (Ma-1 + Mi-2), allow diseased plants to show greater height, canopy volume, and chlorophyll concentration, as well as less starch accumulation in leaves. The results with lower-than-recommended doses agree with previous reports, where a potential reduction in N dose of up to 25.0% does not affect vegetative growth, juice yield, and juice quality in HLB-diseased citrus [65]. They also support the relieving effect of micronutrient application to HLB-diseased plants [24,35].

CLas infection induces phosphorus deficiencies in *C. sinensis* due to the overexpression of microRNAs, such as miR399, which encodes a ubiquitin-conjugating enzyme (PHO2) related to the degradation of phosphorus-transporting proteins [66]. A subsequent study confirmed that infection with CLas has a significant effect on the efficiency of phosphorus resorption; however, in tolerant varieties such as *C. limon* and *C. maxima*, infection with CLas did not affect phosphorus levels in leaves, unlike what was observed in HLB-susceptible *C. reticulata* [67]. On the other hand, reports indicate that diseased plants are more efficient in nutrient uptake, despite the reduction in root volume due to the effect of HLB. This is attributed to the overexpression of genes related to the transport of cations such as NH_4^+, K^+, Zn^{2+}, Fe^{2+}, and Cu^{2+} [43].

The results indicate that the increase in macronutrient doses, along with the simultaneous soil and foliar application of micronutrients in short periods, reduces the efficiency of the applications, possibly due to the excessive presence of these micronutrients. This effect was reflected in growth, carbohydrate metabolism, and an increase in the number of bacteria. In addition, greater fertilizer efficiency was observed by applying lower doses, split into more applications per year, compared to the higher doses evaluated, also with a split application. Constant fertilization improves the availability and uptake of nutrients by citrus plants that have lost secondary roots due to the disease; it also reduces leaching losses [23]. Split fertilizer application was evaluated in citrus trees with HLB, and the results have shown an improvement in canopy volume and a reduction in nitrogen requirement with 20 applications per year, in addition to improving the nutritional content at the foliar level and the nitrogen content in the soil in the form of nitrates and ammonium [25]. These results require field validation, particularly with bearing trees, to determine the appropriate dose and number of annual applications in Mexico's citrus-growing areas, complemented with cost–benefit analysis. Some studies have evaluated the economic feasibility of enhanced nutritional programs, with contradictory results. While some authors report benefits in yield and profitability, others have found no significant advantages [68,69]. Ozores-Hampton et al. [70] documented that an enhanced foliar nutrition program applied over three years increased yield by 7.0 to 9.0% in Valencia oranges affected by HLB. In contrast, Bassanezi et al. [71] evaluated various fertilization programs in Valencia oranges with HLB. Surprisingly, the standard nutritional program used by growers provided the best cost–benefit ratio, requiring fewer foliar applications than other treatments. This result aligns with our findings regarding the continuous application of both soil and foliar fertilizers, which does not appear to be beneficial for both healthy and HLB-affected Persian lime trees. Additionally, the economic feasibility of large-scale split applications in tolerant varieties such as Persian lime are needed. Implementing this strategy in rainfed orchards might increase labor costs. However, reducing the total amount of fertilizers could help offset the increased costs.

4. Materials and Methods

4.1. Study Site Description and Plant Material Used

The study was conducted in a greenhouse (290 m^2) in Central Veracruz, Mexico (Lat. 19.194167, Long −96.343611). Persian lime (*Citrus latifolia* Tan.) plants, grafted onto 2-year-old macrophylla (*Citrus macrophylla* Macf.) rootstocks, were purchased from a certified nursery. Plants were transferred to a 40 × 40 cm container with a 2:1 mixture ratio of soil (pH: 5.75 ± 0.07, texture: sandy loam, organic matter: 16.7%, electrical conductivity: 1.9025 dS/m, phosphorus content: 475 mg L^{-1}, potassium content: 700 mg L^{-1}, nitrite content: 2 mg L^{-1}, nitrates content: 500 mg L^{-1}) and Peat moss®. A group of plants were grafted at the end of summer with buds from a tree confirmed positive for Clas by PCR, originating from plots of the National Institute of Forestry, Agricultural, and Livestock

Research (INIFAP), Ixtacuaco Experimental Station. Both healthy and infected groups were grafted in November 2020. A healthy group of plants was grafted with buds produced in the certified nursery. The seasonal variability in macro- and micronutrient demand was associated with Persian lime growth cycles. Therefore, the experiment was conducted over a full year, starting in the summer of 2021 and ending in autumn in 2022.

4.2. Fertilization Treatments

This study evaluated three levels of macronutrients, low (Ma-1), medium (Ma-2), and high (Ma-3), and three conditions for micronutrient application: no additional application (Mi-1), soil-applied (Mi-2), and soil- and foliar-applied (Mi-3) (Table S1). Macronutrient treatments included N, P, and K as essential elements, with Ca and Mg added in medium (Ma-2) and high (Ma-3) doses. Soil-applied micronutrients (S, Zn, B, Mn, and Fe) were selected for their essential roles and potential benefits in enhancing the health of HLB-affected citrus trees [23,37,72]. The medium macronutrient dose was formulated based on regional recommendations designed by INIFAP to meet the nutritional requirements of two-year-old plants [73]. The low macronutrient dose (Ma-1) was defined as half of the medium dose (Ma-2), while the high dose (Ma-3) was determined based on the nutrient requirements of three-year-old plants. The amount of soil-applied micronutrients at medium and high doses of macronutrients was calculated based on the recommendations of Morgan and Kadyampakeni [34] for non-bearing trees, taking into account the amount of nitrogen needed. The treatment with foliar-applied micronutrients associated with low doses of NPK (Ma-1 + Mi-3) consisted of 1.1% B, 1.3% Zn, 6% Fe, 2.4% Mn, 0.25% Cu, and 0.25% Mo. Foliar-applied micronutrient treatments at medium and high macronutrient doses (specifically Ma-2 + Mi-3 and Ma-3 + Mi-3) contained 4.8% N, 4.9% Mg, 4.9% B, and 9.9% Zn. Micronutrient concentrations used in foliar treatments followed recommendations of commercially available products. Commercial granulated fertilizers were used for all soil treatments, while both soil and foliar applications were carried out every 18 days, according to the recommendations of Atta et al. [74].

The experiment used a $2 \times 3 \times 3$ factorial design with two health levels (healthy and diseased plants), three macronutrient levels, and three micronutrient levels. A randomized block design was used for analysis, with nine fertilization treatments applied to both infected and non-infected plants, with four replicates per treatment. Plants were placed in the greenhouse in late July 2021, with an initial spacing of approximately 1 m. Distance increased as required by canopy volume.

4.3. Severity of Foliar Symptoms

The damage scale proposed by Gottwald et al. [75] was used to evaluate foliar symptoms. Each tree was divided into eight sections, individually assessed for severity on a scale of 0 to 5, based on the proportion of leaves showing HLB symptoms. The sum of the results of each section resulted in a range of severity per tree with values from 0 to 40. To standardize assessment criteria, photographs displaying the characteristic symptoms of HLB were printed. The percentage of severity for each tree was calculated using the Shokrollah et al. [76] formula:

$$\text{Disease severity} = \frac{\sum(a \times b)}{N \times Z} \times 100 \tag{1}$$

where $\sum(a \times b)$ = sum of the symptomatic plant and their corresponding rating, N = total number of sampled plants, and Z = highest rating.

4.4. DNA Extraction and CLas Detection

DNA extraction was performed using the CTAB protocol in March–April 2021 [77]. Midribs of four leaves per plant were collected from the middle section of the canopy. Confirmation of healthy and diseased plants at the start of the experiment was determined by PCR, amplifying the 16S gene (rDNA) of CLas with the enzyme MyTaqTM DNA Polymerase (Bioline®, London, UK) and oligos OI/OI2c. PCR protocol started at 95 °C for 5 min, followed by 35 cycles of 30 s at 94 °C, 30 s at 62 °C and 1 min at 72 °C, with a final extension of 10 min at 72 °C according to Jagoueix et al. [78].

At the end of the experiment, the gDNA from the four plants in each treatment was equimolarly mixed, and qPCR amplifications were conducted according to the recommendations of Li et al. [79], with some modifications. Specific primers, CLas-4G [80], HLBr, and the HLBp probe were used to amplify the 16S gene of CLas. Additionally, primers COXF, COXR, and the COXp probe were used to quantify the endogenous COX gene [79]. Two replicates were run for each gDNA pool, and every reaction was performed in triplicate using the Taq PCR Master Mix Kit (Qiagen®, Hilden, Germany). The Ct obtained in qPCR reactions was used for the relative quantification of the CLas genes, following the $2^{-\Delta\Delta Ct}$ method [81]. The relative quantification of the 16S rDNA gene of CLas was assessed at the end of the experiment (autumn 2022). Changes in the amount of CLas under the fertilization treatments were compared to those of the Ma-1 + Mi-1 treatment. The propagated error of the ΔCT values for the 16S and COX genes was calculated [81].

4.5. Growth Variables

The trunk diameter was measured using a digital vernier caliper at a height of 15 cm from the base of the stem. Canopy volume (CV) was determined by assessing the canopy diameter in both the east–west and north–south directions, considering plant height. Then, the prolate spheroid formula VD (m^3) = $(4/3) \times (\pi) \times (\text{height}/2) \times (\text{average canopy radius})^2$ was used [82]. Measurements were taken monthly over the course of one year. Net growth (in cm) per month was calculated by subtracting the previous month's value from the value obtained in the current month for the three parameters evaluated. Monthly values were added, with the final month's value considered as the accumulated growth in height, canopy volume, and trunk diameter.

4.6. Photosystem II (Fv/Fm) Efficiency

Maximum photosystem II efficiency (Fv/Fm) was measured in healthy and diseased Persian lime leaves using a portable OS30p fluorometer (Opti-Sciences) during summer 2021, winter 2022, and autumn 2022. The leaves from healthy and diseased plants were adapted to dark conditions before measurements. Data collection occurred under daylight conditions in the greenhouse at a consistent morning time.

4.7. Chlorophyll Quantification in Leaves

The methodology outlined by Das et al. [83] was applied for chlorophyll extraction, incorporating modifications from Flores-de la Rosa et al. [44]. The absorbance of the solution was measured at wavelengths of 645 and 633 nm using a spectrophotometer, with acetone serving as a comparative blank. Total chlorophyll content was calculated using the following formula [84]:

$$\text{mg of total chlorophyll/g of tissue} = [20.2\,(A645) + 8.02\,(A633)] \times V/1000 \times FW \quad (2)$$

where A = absorbance at a specific wavelength (633 and 645 nm), V = final volume of chlorophyll extracted in 80% acetone, and FW = sample fresh weight.

4.8. Quantification of Sucrose and Starch in Leaves

Sucrose extraction was carried out at the end of the experiment (autumn 2022) using the methodology described by Geigenberger and Stitt [85]. For quantification, commercial kit SCA20 (Sigma-Aldrich®) was used. A calibration curve was constructed with pure rice starch to evaluate starch ($y = 0.0011x + 0.0029$, $R^2 = 0.9997$). The extraction was conducted following the methodology outlined by Zheng et al. [86], with modifications made by Flores-de la Rosa et al. [44]. Sucrose was quantified by measuring the absorbance at 594 nm.

4.9. Reflectance Measurement Using Spectrometry

A spectrometer with a spectral range of 339 to 822 nm and a resolution of 0.45 nm was used to assess changes in leaf reflectance, resulting from the loss of photosynthetic pigments. Based on preliminary results, analyses were focused on reflectance at 550 nm (VIS) and 790 nm (NIR) (Figure 3). The QP600-025-VIS optical fiber is connected to a 74-UV series collimating lens, positioned 5 cm from the module's base. Oceanview® software (version 2.0.8) helped to capture data. Spectrograms were generated using the ggplot2 package in RStudio (v4.3.1, R Project). The reflectance module was built under dark conditions, with two 90 w halogen lamps serving as the primary light source. A white polytetrafluoroethylene panel was used to calibrate the equipment. Results were analyzed by building spectrograms illustrating the percentage of reflectance of healthy and diseased plants. Reflectance data were collected in summer 2021, winter 2022, and autumn 2022.

4.10. Quantification of Benzoic Acid, Trans-Cinnamic Acid, and Salicylic Acid in Leaves Using GC-MS

To evaluate the hormonal response of healthy and HLB-diseased plants, the accumulation of salicylic acid (SA) and two precursors, benzoic acid (BA) and *trans*-cinnamic acid (*t*-CA), in leaves was quantified. Gas chromatography/mass spectrometry (GC-MS) was used at the end of the experiment (autumn 2022), following the methodology reported by Hijaz et al. [87] and modified by Nehela et al. [57]. Derivatization of the standards was performed using the same method. Calibration curves were generated by injecting 1 μL of pure standards at concentrations of 100, 50, 20, 10, 5, 2, and 1 ng μL^{-1} for BA, and 10, 5, 2, 1, 0.5, and 0.2 ng μL^{-1} for both SA and *t*-CA. The GC-MS was operated in SIM and Scan modes. Quantitative and qualitative ions were monitored for each standard. Results were compared with the NIST (National Institute of Standards and Technology) library using the MassHunter® program. Validation was conducted by assessing specificity, linearity, precision, accuracy, detection, and quantification limits, following Rawlinson et al.'s [88] criteria (Table S2).

4.11. Statistical Analysis

The analysis was conducted using RStudio version 4.3.1 [89]. The Shapiro–Wilk test assessed the normality of the study variables. The Box–Cox transformation was applied to variables that did not satisfy the normality assumption. Additionally, variables with repeated measurements over time were evaluated by adjusting the data to correlation structures based on the AIC (Akaike Information Criterion) [90]. Additionally, the residuals, kurtosis, and homogeneity of variance were visually inspected after the transformations to verify the proper fulfillment of statistical assumptions [91]. A mixed linear model was used to perform the analysis of variance (ANOVA). Also, Tukey's multiple comparison tests ($p \leq 0.05$) helped to determine differences between treatments. Data associated with levels of macro- and micronutrients were adjusted using a second-degree polynomial regression model, obtaining the 95% confidence intervals, R^2, R^{2Adj}, and the p-value from the F test

($p \leq 0.05$). Finally, principal component analysis (PCA) facilitated the exploration of the linear relationship among the study variables in both groups of plants.

5. Conclusions

CLas spreads more slowly in *C. latifolia* compared to susceptible citrus varieties, leading to a more gradual expression of symptoms. The disease affected plant development and physiological parameters related to photosynthesis, altering vascular structures and increasing starch accumulation in the leaf's midribs and mesophyll. Additionally, an increase in the endogenous content of salicylic acid and its biosynthesis precursors is confirmed in diseased plants compared to healthy ones. However, it was observed that the levels of benzoic acid in the leaves were higher, while the levels of salicylic acid were lower than in findings reported from susceptible citrus varieties. This finding may suggest a reduced conversion of benzoic acid to salicylic acid. Regarding the impact of combinatorial treatments using macro- and micronutrients, it was concluded that soil application of low doses of macronutrients in combination with micronutrients (Ma-1 + Mi-2) resulted in less damage to Persian lime plants affected by HLB. This treatment promoted more significant plant development, improved photosynthetic performance, and reduced metabolic alterations, such as starch accumulation in the leaves of diseased plants. Applying the fertilizer consistently and at lower doses may improve its utilization. In contrast, increasing the doses of macronutrients and simultaneously applying micronutrients to the soil and through foliar sprays does not seem advantageous, even for healthy Persian lime plants, under the conditions evaluated. Thus, careful fertilization management may optimize its use and help reduce the negative effects of HLB on diseased Persian lime plants. It is essential to reassess the timing and dosage of fertilization practices in regions where HLB has become prevalent, such as in the citrus-growing states of Mexico. This reevaluation could lead to potential benefits in environmental sustainability and production costs.

Supplementary Materials: The following supporting information can be downloaded here: https://www.mdpi.com/article/10.3390/plants14071086/s1. Figure S1: Leaf reflectance percentage at (A) 550 nm and (B) 790 nm for macro- and micronutrient levels. According to Tukey ($p < 0.05$), different letters represent significant differences. Values represent four replicates, mean $\pm$ standard deviation. Ma = macronutrients, Mi = micronutrients, ns= not significant. Figure S2: Effect of CLas infection and fertilization treatments on the content of (A) total chlorophyll, (B) sucrose, and (C) starch in leaves. The main effects, according to the general linear model (ANOVA), are indicated in graphs. According to Tukey ($p < 0.05$), different letters represent significant differences. The values represent four replicates, mean $\pm$ standard deviation. Ma = macronutrients, Mi = micronutrients. Significant p-values are shown in bold. Figure S3: Histological sections of leaf midribs show accumulation of starch grains in collenchyma, parenchyma, and pith cells in HLB-diseased plants vs. healthy plants. Diseased plants show anatomical changes in the vascular system, such as hyperplasia and increased phloem area. coll = collenchyma, par = parenchyma, Pf = phloem fibers, P = phloem, X = xylem, pi = pith, H = hyperplasia. Black arrows indicate the presence of starch. Figure S4: Effect of CLas infection and fertilization treatments on cumulative growth in (A) height, (B) canopy volume, and (C) trunk diameter. According to the general linear model (ANOVA), graphs indicate the main effects. According to Tukey ($p < 0.05$), different letters represent significant differences. Values represent four replicates, mean $\pm$ standard deviation. Ma = macronutrients, Mi = micronutrients. Significant p-values are shown in bold. Figure S5: Multivariate principal component analysis (biplot-PCA), including the study variables and their association with the groups of healthy and HLB-diseased Persian lime plants. Ht = height, CV = canopy volume, DT = trunk diameter, Chl = chlorophyll, Suc = sucrose, Glc = glucose, St = starch, tCA = *trans*-cinnamic acid, BA = benzoic acid, SA = salicylic acid, Refl = reflectance, Fv/Fm = photosystem II efficiency. Table S1: Fertilization treatments evaluated in healthy and HLB-diseased plants, consisting of three levels of macro- and micronutrients. Table S2: Validation of a chromatographic method for the quantification of BA, *t*-CA, and SA.

Author Contributions: Conceptualization, methodology, investigation, validation, and resources, L.A.P.-Z., J.A.V.-J., A.M.-H., F.O.-A., E.G.-P. and F.M-T.; formal analysis, L.A.P.-Z. and J.A.V.-J.; data curation, L.A.P.-Z.; writing—original draft preparation, L.A.P.-Z.; writing—review and editing, L.A.P.-Z., J.A.V.-J., A.M.-H., F.O.-A., E.G.-P. and F.M.-T.; visualization, L.A.P.-Z. and J.A.V.-J.; supervision, J.A.V.-J. and A.M.-H.; project administration, L.A.P.-Z. and J.A.V.-J.; funding acquisition, L.A.P.-Z. and J.A.V.-J. All authors have read and agreed to the published version of the manuscript.

Funding: Research funds were provided by Consejo Veracruzano de Investigación Científica y Desarrollo Tecnológico (COVEICYDET) (Project No. 15 1136/2021). L.A.P.Z. scholarship was provided by CONAHCyT.

Data Availability Statement: Data is contained within the article or Supplementary Material.

Conflicts of Interest: The authors declare no conflicts of interest.

References

1. da Graça, J.V.; Douhan, G.W.; Halbert, S.E.; Keremane, M.L.; Lee, R.F.; Vidalakis, G.; Zhao, H. Huanglongbing: An overview of a complex pathosystem ravaging the world's citrus: Citrus Huanglongbing. *J. Integr. Plant Biol.* **2016**, *58*, 373–387. [CrossRef] [PubMed]
2. Killiny, N. Made for each other: Vector–pathogen interfaces in the Huanglongbing pathosystem. *Phytopathology* **2022**, *112*, 26–43. [CrossRef]
3. Mora-Aguilera, G.; Robles-García, P.; López-Arroyo, J.I.; Flores-Sánchez, J.; Acevedo-Sánchez, G.; Domínguez-Monge, S.; Gutierrez-Espinosa, A.; Loeza-Kuk, E.; González-Gómez, R. Situación Actual y perspectivas del manejo del HLB de los cítricos. *Rev. Mex. Fitopatol.* **2014**, *32*, 108–119.
4. Farnsworth, D.; Grogan, K.A.; van Bruggen, A.H.C.; Moss, C. The potential economic cost and response to greening in Florida citrus. *Choices Mag. Food Farm Resour. Issues* **2014**, *29*, 188996.
5. Flores-Sánchez, J.L.; Mora-Aguilera, G.; Loeza-Kuk, E.; López-Arroyo, J.I.; Domínguez-Monge, S.; Acevedo-Sánchez, G.; Robles-García, P.; Flores-Sánchez, J.L.; Mora-Aguilera, G.; Loeza-Kuk, E.; et al. Pérdidas en producción inducidas por *Candidatus* Liberibacter asiaticus en limón Persa, en Yucatán México. *Rev. Mex. Fitopatol.* **2015**, *33*, 195–210.
6. Villar-Luna, H.; Santos-Cervantes, M.E.; Rodríguez-Negrete, E.A.; Méndez-Lozano, J.; Leyva-López, N.E. Economic and social impact of huanglongbing on the mexico citrus industry: A Review and future perspectives. *Horticulturae* **2024**, *10*, 481. [CrossRef]
7. Nehela, Y.; Killiny, N. Revisiting the complex pathosystem of huanglongbing: Deciphering the role of citrus metabolites in symptom development. *Metabolites* **2020**, *10*, 409. [CrossRef]
8. Ribeiro, C.; Stitt, M.; Hotta, C.T. How Stress affects your budget—Stress impacts on starch metabolism. *Front. Plant Sci.* **2022**, *13*, 774060. [CrossRef]
9. Kumar, N.; Kiran, F.; Etxeberria, E. Huanglongbing-Induced anatomical changes in citrus fibrous root orders. *HortScience* **2018**, *53*, 829–837. [CrossRef]
10. Deng, H.; Achor, D.; Exteberria, E.; Yu, Q.; Du, D.; Stanton, D.; Liang, G.; Gmitter, F.G., Jr. Phloem Regeneration is a mechanism for huanglongbing-tolerance of "Bearss" Lemon and "LB8-9" Sugar Belle® mandarin. *Front. Plant Sci.* **2019**, *10*, 277. [CrossRef]
11. Suh, J.H.; Tang, X.; Zhang, Y.; Gmitter, F.G.; Wang, Y. Metabolomic analysis provides new insight into tolerance of Huanglongbing in citrus. *Front. Plant Sci.* **2021**, *12*, 710598. [CrossRef] [PubMed]
12. Ma, W.; Pang, Z.; Huang, X.; Xu, J.; Pandey, S.S.; Li, J.; Achor, D.S.; Vasconcelos, F.N.C.; Hendrich, C.; Huang, Y.; et al. Citrus Huanglongbing is a pathogen-triggered immune disease that can be mitigated with antioxidants and gibberellin. *Nat. Commun.* **2022**, *13*, 529. [CrossRef]
13. Folimonova, S.Y.; Robertson, C.J.; Garnsey, S.M.; Gowda, S.; Dawson, W.O. Examination of the responses of different genotypes of citrus to Huanglongbing (citrus greening) under Different conditions. *Phytopathology* **2009**, *99*, 1346–1354. [CrossRef] [PubMed]
14. Sivager, G.; Calvez, L.; Bruyere, S.; Boisne-Noc, R.; Brat, P.; Gros, O.; Ollitrault, P.; Morillon, R. Specific physiological and anatomical traits associated with polyploidy and better detoxification processes contribute to improved Huanglongbing tolerance of the Persian lime compared with the Mexican lime. *Front. Plant Sci.* **2021**, *12*, 685679. [CrossRef]
15. Ortiz-Saavedra, S.; Domínguez-Monge, S.; Allende-Molar, R.; Mendoza-García, J.D.; Pérez-hernández, O.; Rodríguez-Quibrera, C.G.; Curti-Díaz, S.A.; Flores-Sánchez, J.L.; Sarmiento-Tejeda, R. Influencia de Huanglongbing en pérdidas de producción y calidad de frutos de limón Persa en Veracruz. *Supl. Rev. Mex. Fitopatol.* **2022**, *40*, S40.

16. Alquézar, B.; Carmona, L.; Bennici, S.; Miranda, M.P.; Bassanezi, R.B.; Peña, L. Cultural management of Huanglongbing: Current status and ongoing research. *Phytopathology®* **2022**, *112*, 11–25. [CrossRef] [PubMed]
17. Tripathi, R.; Tewari, R.; Singh, K.P.; Keswani, C.; Minkina, T.; Srivastava, A.K.; De Corato, U.; Sansinenea, E. Plant Mineral nutrition and disease resistance: A significant linkage for sustainable crop protection. *Front. Plant Sci.* **2022**, *13*, 883970. [CrossRef]
18. Ahmad, K.; Ismail, S.I. Enhanced nutritional programme: An innovative approach to controlling plant diseases in the tropics. In *Plant, Soil and Microbes: Volume 1: Implications in Crop Science*; Hakeem, K.R., Akhtar, M.S., Abdullah, S.N.A., Eds.; Springer International Publishing: Cham, Switzerland, 2016; pp. 225–251, ISBN 978-3-319-27455-3.
19. Ramirez, D.; Chaparro, J.; Wright, A.; Boman, B.; Gruber, B.; Ritenour, M.; Marino, S. Mitigation of Huanglongbing Effects on grapefruit trees using enhanced nutritional programs. *Proc. Fla. State Hortic. Soc.* **2016**, *129*, 51–55.
20. Nwugo, C.C.; Duan, Y.; Lin, H. Study on citrus response to Huanglongbing highlights a down-regulation of defense-related proteins in lemon plants upon 'Ca. Liberibacter Asiaticus' infection. *PLoS ONE* **2013**, *8*, e67442. [CrossRef]
21. Dong, Z.; Srivastava, A.K.; Liu, X.; Riaz, M.; Gao, Y.; Liang, X.; Tan, Q.; Sun, X.; Wu, S.; Hu, C. Interactions between nutrient and Huanglongbing pathogen in citrus: An overview and implications. *Sci. Hortic.* **2021**, *290*, 110511. [CrossRef]
22. Uthman, Q.O.; Kadyampakeni, D.M.; Nkedi-Kizza, P. Manganese Adsorption, availability, and uptake in citrus under microsprinkler irrigation. *Agrosystems Geosci. Environ.* **2020**, *3*, e20061. [CrossRef]
23. Atta, A.A.; Morgan, K.T.; Kadyampakeni, D.M.; Mahmoud, K.A. The effect of foliar and ground-applied essential nutrients on Huanglongbing-affected mature citrus trees. *Plants* **2021**, *10*, 925. [CrossRef] [PubMed]
24. da Silva, J.R.; de Alvarenga, F.V.; Boaretto, R.M.; Lopes, J.R.S.; Quaggio, J.A.; Coletta Filho, H.D.; Mattos, D. Following the Effects of micronutrient supply in HLB-infected trees: Plant responses and '*Candidatus* Liberibacter Asiaticus' acquisition by the Asian Citrus Psyllid. *Trop. Plant Pathol.* **2020**, *45*, 597–610. [CrossRef]
25. Atta, A.A.; Morgan, K.T.; Hamido, S.A.; Kadyampakeni, D.M.; Mahmoud, K.A. Water and soil nutrient dynamics of Huanglongbing-affected citrus trees as impacted by ground-applied nutrients. *Agronomy* **2020**, *10*, 1485. [CrossRef]
26. Zambon, F.T.; Kadyampakeni, D.M.; Grosser, J.W. Ground Application of overdoses of manganese have a therapeutic effect on sweet orange trees infected with *Candidatus* Liberibacter asiaticus. *HortScience* **2019**, *54*, 1077–1086. [CrossRef]
27. Morgan, K.T.; Rouse, R.E.; Ebel, R.C. Foliar applications of essential nutrients on growth and yield of 'Valencia' sweet orange infected with Huanglongbing. *HortScience* **2016**, *51*, 1482–1493. [CrossRef]
28. Vashisth, T.; Grosser, J. Comparison of controlled release fertilizer (CRF) for newly planted sweet orange trees under Huanglongbing prevalent conditions. *J. Hortic.* **2018**, *5*, 244. [CrossRef]
29. Ghimire, L.; Grosser, J.; Vashisth, T. Differences in nutrient uptake can influence the performance of citrus rootstocks under Huanglongbing conditions. *HortScience* **2023**, *58*, 40–46. [CrossRef]
30. Hallman, L.M.; Kadyampakeni, D.M.; Ferrarezi, R.S.; Wright, A.L.; Ritenour, M.A.; Rossi, L. Uptake of micronutrients in severely HLB-affected grapefruit trees grown on Florida Indian River flatwood soils. *J. Plant Nutr.* **2023**, *46*, 4110–4124. [CrossRef]
31. Zekri, M.; Obreza, T.A.; Koo, R.; Alferez, F.M. *Irrigation, Nutrition, and Citrus Fruit Quality*; University of Florida: Gainesville, FL, USA, 2018.
32. Kadyampakeni, D.M.; Chinyukwi, T. Are macronutrients and micronutrients therapeutic for restoring performance of trees affected by citrus greening? A discussion of current practices and future research opportunities. *J. Plant Nutr.* **2021**, *44*, 2949–2969. [CrossRef]
33. Atta, A.A.; Morgan, K.T.; Kadyampakeni, D.M.; Kamal, M.A. Effect of soil and/or foliar applied nutrients on leaf nutrient accumulation and water uptake on Huanglongbing affected 'Valencia' citrus trees. *Proc. Fla. State Hortic. Soc.* **2018**, *131*, 58–64.
34. Morgan, E.K.T.; Kadyampakeni, D.M. *Nutrition of Florida Citrus Trees*, 3rd ed.; University of Florida Institute of Food and Agricultural Sciences: Gainesville, FL, USA, 2020; p. 113.
35. Kwakye, S.; Kadyampakeni, D.M.; Morgan, K.; Vashisth, T.; Wright, A. Effects of iron rates on growth and development of young Huanglongbing-affected citrus trees in Florida. *HortScience* **2022**, *57*, 1092–1098. [CrossRef]
36. Kadyampakeni, D.M. Interaction of soil boron application with leaf B concentration, root length density, and canopy size of citrus affected by Huanglongbing. *J. Plant Nutr.* **2020**, *43*, 186–193. [CrossRef]
37. Kwakye, S.; Kadyampakeni, D.M.; Morgan, K.; Wright, A. Foliar Micronutrient applications enhance growth and yield of Huanglongbing (HLB)-affected sweet orange. *Soil Sci. Soc. Am. J.* **2022**, *87*, 365–377. [CrossRef]
38. Albrecht, U.; Bowman, K.D. Transcriptional response of susceptible and tolerant citrus to infection with *Candidatus* Liberibacter Asiaticus. *Plant Sci.* **2012**, *185–186*, 118–130. [CrossRef]
39. Sivager, G.; Calvez, L.; Bruyere, S.; Boisne-Noc, R.; Hufnagel, B.; Cebrian-Torrejon, G.; Doménech-Carbó, A.; Gros, O.; Ollitrault, P.; Morillon, R. Better tolerance to Huanglongbing Is conferred by tetraploid swingle citrumelo rootstock and is influenced by the ploidy of the scion. *Front. Plant Sci.* **2022**, *13*, 1030862. [CrossRef]

40. Flores-Sánchez, J.L.; Mora-Aguilera, G.; Loeza-Kuk, E.; López-Arroyo, J.I.; Gutiérrez-Espinosa, M.A.; Velázquez-Monreal, J.J.; Domínguez-Monge, S.; Bassanezi, R.B.; Acevedo-Sánchez, G.; Robles-García, P. Diffusion model for describing the regional spread of Huanglongbing from first-reported outbreaks and basing an area wide disease management strategy. *Plant Dis.* **2017**, *101*, 1119–1127. [CrossRef] [PubMed]
41. Cen, H.; Weng, H.; Yao, J.; He, M.; Lv, J.; Hua, S.; Li, H.; He, Y. Chlorophyll fluorescence imaging uncovers photosynthetic fingerprint of citrus huanglongbing. *Front. Plant Sci.* **2017**, *8*, 1509. [CrossRef]
42. Chen, Y.; Liu, L.; Guo, Q.; Zhu, Z.; Zhang, L. Effects of different water management options and fertilizer supply on photosynthesis, fluorescence parameters and water use efficiency of *Prunella vulgaris* seedlings. *Biol. Res.* **2016**, *49*, 12. [CrossRef]
43. Shahzad, F.; Chun, C.; Schumann, A.; Vashisth, T. Nutrient uptake in Huanglongbing-affected sweet orange: Transcriptomic and physiological analysis. *J. Am. Soc. Hortic. Sci.* **2020**, *145*, 349–362. [CrossRef]
44. Flores-de la Rosa, F.R.; Santillán-Mendoza, R.; Rodríguez-Quibrera, C.G.; Martínez-Ruiz, A.; Adame-García, J.; Luna-Rodriguez, M. Antioxidant gene expression, chlorophyll, and starch content in Persian lime (*Citrus latifolia* Tanaka Ex Q. Jiménez) trees with HLB by application of elicitors of plant resistance. *Mex. J. Biotechnol.* **2021**, *6*, 86–102. [CrossRef]
45. Smeekens, S. Sugar-induced signal transduction in plants. *Annu. Rev. Plant Physiol. Plant Mol. Biol.* **2000**, *51*, 49–81. [CrossRef]
46. Curtolo, M.; de Souza Pacheco, I.; Boava, L.P.; Takita, M.A.; Granato, L.M.; Galdeano, D.M.; de Souza, A.A.; Cristofani-Yaly, M.; Machado, M.A. Wide-Ranging transcriptomic analysis of *Poncirus trifoliata, Citrus sunki, Citrus sinensis* and Contrasting hybrids reveals HLB tolerance mechanisms. *Sci. Rep.* **2020**, *10*, 20865. [CrossRef] [PubMed]
47. Rosales, R.; Burns, J.K. Phytohormone changes and carbohydrate status in sweet orange fruit from Huanglongbing-Infected trees. *J. Plant Growth Regul.* **2011**, *30*, 312–321. [CrossRef]
48. Weng, H.; Lv, J.; Cen, H.; He, M.; Zeng, Y.; Hua, S.; Li, H.; Meng, Y.; Fang, H.; He, Y. Hyperspectral reflectance imaging combined with carbohydrate metabolism analysis for diagnosis of citrus Huanglongbing in different seasons and cultivars. *Sens. Actuators B Chem.* **2018**, *275*, 50–60. [CrossRef]
49. Peñuelas, J.; Filella, I. Visible and Near-infrared reflectance techniques for diagnosing plant physiological status. *Trends Plant Sci.* **1998**, *3*, 151–156. [CrossRef]
50. Martínez-Martínez, V.; Gomez-Gil, J.; Machado, M.L.; Pinto, F.A.C. Leaf and canopy reflectance spectrometry applied to the estimation of angular leaf spot disease severity of common bean crops. *PLoS ONE* **2018**, *13*, e0196072. [CrossRef]
51. Deng, X.; Huang, Z.; Zheng, Z.; Lan, Y.; Dai, F. Field detection and classification of citrus Huanglongbing based on hyperspectral reflectance. *Comput. Electron. Agric.* **2019**, *167*, 105006. [CrossRef]
52. He, C.; Li, X.; Liu, Y.; Yang, B.; Wu, Z.; Tan, S.; Ye, D.; Weng, H. Combining multicolor fluorescence imaging with multispectral reflectance imaging for rapid citrus Huanglongbing detection based on lightweight convolutional neural network using a handheld device. *Comput. Electron. Agric.* **2022**, *194*, 106808. [CrossRef]
53. Miles, G.P.; Stover, E.; Ramadugu, C.; Keremane, M.L.; Lee, R.F. Apparent tolerance to huanglongbing in citrus and citrus-related germplasm. *HortScience* **2017**, *52*, 31–39. [CrossRef]
54. Koh, E.-J.; Zhou, L.; Williams, D.S.; Park, J.; Ding, N.; Duan, Y.-P.; Kang, B.-H. Callose deposition in the phloem plasmodesmata and inhibition of phloem transport in citrus leaves infected with "*Candidatus* Liberibacter Asiaticus". *Protoplasma* **2012**, *249*, 687–697. [CrossRef]
55. Esquivel-Chávez, F.; Valdovinos-Ponce, G.; Mora-Aguilera, G.; Gómez-Jaimes, R.; Velázquez-Monreal, J.J.; Manzanilla-Ramírez, M.Á.; Flores-Sánchez, J.L.; López-Arroyo, J.I. Análisis histológico foliar de cítricos agrios y naranja dulce con síntomas ocasionados por *Candidatus* Liberibacter Asiaticus. *Agrociencia* **2012**, *46*, 769–782.
56. Gilroy, E.; Breen, S. Interplay between phytohormone signalling pathways in plant defence—Other than salicylic acid and jasmonic acid. *Essays Biochem.* **2022**, *66*, 657–671. [CrossRef]
57. Nehela, Y.; Hijaz, F.; Elzaawely, A.A.; El-Zahaby, H.M.; Killiny, N. Phytohormone profiling of the sweet orange (*Citrus Sinensis* (L.) Osbeck) leaves and roots using GC-MS-based method. *J. Plant Physiol.* **2016**, *199*, 12–17. [CrossRef]
58. Nehela, Y.; Hijaz, F.; Elzaawely, A.A.; El-Zahaby, H.M.; Killiny, N. Citrus Phytohormonal response to *Candidatus* Liberibacter asiaticus and its vector *Diaphorina citri. Physiol. Mol. Plant Pathol.* **2018**, *102*, 24–35. [CrossRef]
59. Zou, X.; Bai, X.; Wen, Q.; Xie, Z.; Wu, L.; Peng, A.; He, Y.; Xu, L.; Chen, S. Comparative analysis of tolerant and susceptible citrus reveals the role of methyl salicylate signaling in the response to Huanglongbing. *J. Plant Growth Regul.* **2019**, *38*, 1516–1528. [CrossRef]
60. Xu, M.; Li, Y.; Zheng, Z.; Dai, Z.; Tao, Y.; Deng, X. Transcriptional analyses of mandarins seriously infected by '*Candidatus* Liberibacter asiaticus'. *PLoS ONE* **2015**, *10*, e0133652. [CrossRef]
61. Nehela, Y.; Taha, N.A.; Elzaawely, A.A.; Xuan, T.D.; Amin, M.A.; Ahmed, M.E.; El-Nagar, A. Benzoic acid and its hydroxylated derivatives suppress early blight of tomato (*Alternaria solani*) via the induction of salicylic acid biosynthesis and enzymatic and nonenzymatic antioxidant defense machinery. *J. Fungi* **2021**, *7*, 663. [CrossRef]
62. Coquoz, J.-L.; Buchala, A.; Métraux, J.-P. The Biosynthesis of salicylic acid in potato plants. *Plant Physiol.* **1998**, *117*, 1095–1101. [CrossRef]

63. Nehela, Y.; Killiny, N. Gamma-aminobutyric acid supplementation boosts the phytohormonal profile in '*Candidatus* Liberibacter asiaticus'-infected citrus. *Plants* **2023**, *12*, 3647. [CrossRef]
64. Torres, L.M.F.; Olivas, A.F.; Fuentes, Y.M.O.; Arroyo, J.I.L.; Portugal, V.O.; Mendoza, A.B.; Morales, S.G.; Villa, V.M.Z. Comparison of Enzymes and phenolic compounds in three citrus species infected with *Candidatus* Liberibacter asiaticus. *Mex. J. Phytopathol.* **2017**, *35*, 314–325. [CrossRef]
65. Esteves, E.; Maltais-Landry, G.; Zambon, F.; Ferrarezi, R.S.; Kadyampakeni, D.M. Nitrogen, Calcium, and magnesium inconsistently affect tree growth, fruit yield, and juice quality of Huanglongbing-affected orange trees. *HortScience* **2021**, *56*, 1269–1277. [CrossRef]
66. Zhao, H.; Sun, R.; Albrecht, U.; Padmanabhan, C.; Wang, A.; Coffey, M.D.; Girke, T.; Wang, Z.; Close, T.J.; Roose, M.; et al. Small RNA profiling reveals phosphorus deficiency as a contributing factor in symptom expression for citrus Huanglongbing disease. *Mol. Plant* **2013**, *6*, 301–310. [CrossRef]
67. Cao, J.; Cheng, C.; Yang, J.; Wang, Q. Pathogen Infection drives patterns of nutrient resorption in citrus plants. *Sci. Rep.* **2015**, *5*, 14675. [CrossRef]
68. Gottwald, T.R.; Graham, J.H.; Irey, M.S.; McCollum, T.G.; Wood, B.W. Inconsequential effect of nutritional treatments on Huanglongbing control, fruit quality, bacterial titer and disease progress. *Crop Prot.* **2012**, *36*, 73–82. [CrossRef]
69. Li, S.; Wu, F.; Duan, Y.; Singerman, A.; Guan, Z. Citrus greening: Management strategies and their economic impact. *HortScience* **2020**, *55*, 604–612. [CrossRef]
70. Ozores-Hampton, M.; Roka, F.; Rouse, R.; Roberts, P. Dual treatment tested for HLB trees. *Citrus Ind. News*, May 2017; 28–31. Available online: https://crec.ifas.ufl.edu/media/crecifasufledu/extension/extension-publications/2017/2017_May_dual.pdf (accessed on 16 December 2024).
71. Bassanezi, R.B.; Primiano, I.V.; Vescove, H.V. Effect of enhanced nutritional programs and exogenous auxin spraying on huanglongbing severity, fruit drop, yield and economic profitability of orange orchards. *Crop Prot.* **2021**, *145*, 105609. [CrossRef]
72. Uthman, Q.O.; Kadyampakeni, D.M.; Nkedi-Kizza, P.; Kwakye, S.; Barlas, N.T. Boron, Manganese, and zinc sorption and leaf uptake on citrus cultivated on a sandy soil. *Plants* **2022**, *11*, 638. [CrossRef]
73. Curti-Díaz, S.A.; Loredo-Salazar, R.X.; Díaz-Zorrilla, U.; Sandoval-Rincón, J.A.; Hernández, H.J. *Tecnología para Producir Limón Persa.; INIFAP-CIRGOC. Campo Experimental Ixtacuaco*; Libro Técnico Núm. 8.: Veracruz, Mexico, 2000; 145p, ISBN 978-968-800-473-9.
74. Atta, A.A.; Morgan, K.T.; Hamido, S.A.; Kadyampakeni, D.M. Effect of essential nutrients on roots growth and lifespan of Huanglongbing affected citrus trees. *Plants* **2020**, *9*, 483. [CrossRef]
75. Gottwald, T.R.; da Graça, J.V.; Bassanezi, R.B. Citrus Huanglongbing: The Pathogen and Its Impact. Online. Plant Health Progress. Available online: https://apsjournals.apsnet.org/doi/10.1094/PHP-2007-0906-01-RV (accessed on 11 August 2024).
76. Shokrollah, H.; Lee Abdullah, T.; Sijam, K.; Akmar Abdullah, S.N. Potential use of selected citrus rootstocks and interstocks against HLB disease in Malaysia. *Crop Prot.* **2011**, *30*, 521–525. [CrossRef]
77. Doyle, J.J.; Doyle, J.L. A rapid DNA isolation procedure for small quantities of fresh leaf tissue. *Phytochem. Bull.* **1987**, *19*, 11–15.
78. Jagoueix, S.; Bove, J.M.; Garnier, M. The phloem-limited bacterium of greening disease of citrus is a member of the alpha subdivision of the proteobacteria. *Int. J. Syst. Bacteriol.* **1994**, *44*, 379–386. [CrossRef]
79. Li, W.; Hartung, J.S.; Levy, L. Quantitative real-time PCR for detection and identification of *Candidatus* Liberibacter species associated with citrus Huanglongbing. *J. Microbiol. Methods* **2006**, *66*, 104–115. [CrossRef]
80. Bao, M.; Zheng, Z.; Sun, X.; Chen, J.; Deng, X. Enhancing PCR capacity to detect *Candidatus* Liberibacter asiaticus utilizing whole genome sequence information. *Plant Dis.* **2020**, *104*, 527–532. [CrossRef]
81. Livak, K.J.; Schmittgen, T.D. Analysis of relative gene expression data using real-time quantitative PCR and the $2^{-\Delta\Delta CT}$ Method. *Methods* **2001**, *25*, 402–408. [CrossRef]
82. Kadyampakeni, D.M.; Morgan, K.T.; Schumann, A.W.; Nkedi-Kizza, P. Effect of irrigation pattern and timing on root density of young citrus trees infected with Huanglongbing disease. *HortTechnology* **2014**, *24*, 209–221. [CrossRef]
83. Das, D.; Seal, P.; Biswas, A.K. Influence of selenium on growth, antioxidants production and physiological parameters of rice (*Oryza Sativa* L.) seedlings and its possible reversal by coapplication of sulphate. *Am. J. Plant Sci.* **2019**, *10*, 2236–2278. [CrossRef]
84. Arnon, D.I. Copper enzymes in isolated chloroplasts. *Polyphenoloxidase in Beta vulgaris. Plant Physiol.* **1949**, *24*, 1–15. [CrossRef]
85. Fan, J.; Chen, C.; Brlansky, R.H.; Gmitter, F.G., Jr.; Li, Z.-G. Changes in carbohydrate metabolism in *Citrus Sinensis* infected with *Candidatus* Liberibacter Asiaticus. *Plant Pathol.* **2010**, *59*, 1037–1043. [CrossRef]
86. Zheng, Y.; Kumar, N.; Gonzalez, P.; Etxeberria, E. Strigolactones restore vegetative and reproductive developments in Huanglongbing (HLB) affected, greenhouse-grown citrus trees by modulating carbohydrate distribution. *Sci. Hortic.* **2018**, *237*, 89–95. [CrossRef]
87. Hijaz, F.; Killiny, N. Collection and chemical composition of phloem sap from *Citrus Sinensis* L. Osbeck (Sweet Orange). *PLoS ONE* **2014**, *9*, e101830. [CrossRef] [PubMed]

88. Rawlinson, C.; Kamphuis, L.G.; Gummer, J.P.A.; Singh, K.B.; Trengove, R.D. A rapid method for profiling of volatile and semi-volatile phytohormones using methyl chloroformate derivatisation and GC–MS. *Metabolomics* **2015**, *11*, 1922–1933. [CrossRef] [PubMed]
89. RStudio Team. RStudio: Integrated Development Enviroment for R (Version 4.3.1) R. RStudio PBC 2023. Available online: https://posit.co/download/rstudio-desktop/ (accessed on 20 February 2024).
90. Brochu, A.-S.; Durrivage, J.; Torres, D.; Pérez-López, E. Diet and injection, important recommendations to characterize *Clavibacter Michiganensis* –tomato interactions. *Plant Health Prog.* **2023**, *24*, 475–481. [CrossRef]
91. Kozak, M.; Piepho, H.-P. What's normal anyway? residual plots are more telling than significance tests when checking ANOVA assumptions. *J. Agron. Crop Sci.* **2018**, *204*, 86–98. [CrossRef]

plants

MDPI

Article

Fertilization Improves the Yield of *Sapindus saponaria* by Affecting Leaf–Soil–Microbial C–N–P Content and Stoichiometry

Juntao Liu [1,2,†], Hongbing Yang [3,†], Ling Zhou [1], Shangpeng Zhang [4], Jie Chen [5], Xu Wang [5], Shixiong Wu [1], Yingyun Gong [1], Guoqing Zhang [1], Weihua Zhang [2,*] and Liming Jia [1,*]

1. Key Laboratory of Silviculture and Conservation of the Ministry of Education, College of Forestry, Beijing Forestry University, Beijing 100083, China; ljt1120@bjfu.edu.cn (J.L.); zhouling89757@bjfu.edu.cn (L.Z.); wushixiong_1992@163.com (S.W.); oxalis623@163.com (Y.G.); zgq19890615@163.com (G.Z.)
2. Guangdong Provincial Key Laboratory of Silviculture, Protection and Utilization/Guangdong Academy of Forestry, Guangzhou 510520, China
3. Key Laboratory of Forest Ecology and Environment of National Forestry and Grassland Administration, Ecology and Nature Conservation Institute, Chinese Academy of Forestry, Beijing 100091, China; philyhb@163.com
4. Beijing Normal University, Beijing 100091, China; zhangsp@mail.bnu.edu.cn
5. Research Institute of Tropical Forestry, Chinese Academy of Forestry, Guangzhou 510520, China; chenjiecaf@hotmail.com (J.C.); cafwangxu111@126.com (X.W.)

* Correspondence: zwh523@sinogaf.cn (W.Z.); jlm@bjfu.edu.cn (L.J.)

† These authors contributed equally to this work.

Abstract: The purpose of this study was to evaluate the effects of different nitrogen (N), phosphorus (P), and potassium (K) fertilization ratios on the carbon (C), N, and P contents and their ecological stoichiometric characteristics in the leaf–soil–microbial system of *Sapindus saponaria* and elucidate their relationship with yield. A "3414" experimental design was employed in a 6-year-old *Sapindus saponaria* woodland located in Fujian Province of China. Fourteen N–P–K fertilization treatments with three replicates were established. Leaf, soil, and microbial samples were collected and analyzed for C, N, and P contents. Redundancy Analysis (RDA), Partial Least Squares Path Modeling (PLS–PM), and the entropy-weighted technique of ranking preferences by similarity to optimal solutions (TOPSIS) were utilized to assess the relationships among variables and determine optimal fertilization strategies. It was found through research that different fertilization treatment methods have a significant impact on both the soil nutrient content and the C, N, and P contents of soil microorganisms. Compared with the control group, soil organic C, total N, and total P, and microbial C, N, and P contents increased by 14.25% to 52.61%, 3.90% to 39.84%, 9.52% to 150%, 6.65% to 47.45%, 11.84% to 46.50%, and 14.91% to 201.98%, respectively. Results from Redundancy Analysis (RDA) indicated that soil organic C, total N, and total P exerted a significant influence on the leaf nutrients. PLS-PM demonstrated that fertilization indirectly affected leaf nutrient accumulation and yield by altering soil properties, with soil total phosphorus and leaf phosphorus being key determinants of yield. Additionally, soil microbial entropy impacted yield by regulating microbial biomass stoichiometric ratios. The entropy-weighted TOPSIS model identified the $N_2P_2K_2$ treatment (600 kg/ha N, 500 kg/ha P, and 400 kg/ha K) as the most effective fertilization strategy. Optimizing N–P–K fertilization ratios significantly enhances leaf nutrient content and soil microbial biomass C, N, and P, thereby increasing *Sapindus saponaria* yield. This research clarifies the underlying mechanisms through which fertilization exerts an impact on the C–N–P stoichiometry within the leaf–soil–microbial system. Moreover, it furnishes a scientific foundation for the optimization of fertilization management strategies in *Sapindus saponaria* plantations.

https://doi.org/10.3390/plants14091360

Keywords: *Sapindus saponaria*; fertilization; microbial nutrient limitation; C–N–P stoichiometry; TOPSIS

1. Introduction

As global climate change intensifies and the issue of ecosystem degradation becomes increasingly severe, the efficient management of nutrient cycling within ecosystems has emerged as a crucial area of research. Specifically, the stoichiometric equilibrium of the three key elements—carbon (C), nitrogen (N), and phosphorus (P)—has become a central focus [1–3]. In forest ecosystems worldwide, the ecological stoichiometry associated with plants, soils, and microorganisms provides essential insights into ecosystem processes and functions [4,5]. According to the dynamic balance mechanism and growth rate theory, the stoichiometric characteristics of C, N, and P are extensively utilized in assessing plant nutrient balance, allocation of limiting elements, and decomposition processes of litter [6–8]. Grasping the correlations and equilibrium conditions of C, N, and P elements in soil and microorganisms is of vital importance. This understanding is key to clarifying the mechanisms of nutrient cycling, particularly in terms of how varying fertilization ratios of N, P, and potassium (K) influence the dynamics of soil nutrients [9,10]. Soil microbial biomass changes are sensitive indicators of soil nutrient alterations and serve as effective indicators for assessing the intensity of soil organic matter metabolism [11]. Previous studies have demonstrated that the utilization of N, P, and K fertilizers not only enhances leaf nutrient content and the survival environment of soil microorganisms but also significantly increases soil microbial biomass and overall soil nutrient levels, thereby affecting the composition structure and variety of microbial communities [12–14]. However, the specific roles of different fertilization practices in influencing the ecological stoichiometric properties of C, N, and P in leaf tissues and soil microbiomass remain underexplored.

It has been shown that the C, N, and P stoichiometric traits and plant yields in the plant–soil–microbe system are chiefly impacted by natural factors such as climatic factors, topographic features, altitude, soil properties, and plant traits, and human factors such as fertilizer application, mulching, pruning, and so on [15–18], with the application of fertilizer having a notably significant influence on the stoichiometric ratios of N, P, and K present in the leaves, soil, and microbial biomass of the plants [15]. Attention should be drawn to the fact that over fertilization reduces the efficiency of fertilizer utilization and affects plant leaves, soil, and microbial biomass C (MBC), N (MBN), P (MBP), and stoichiometric proportions, whereas lesser use of fertilizers or irrational ratios will result in an imbalance of nutrients in plant leaves, soil, and microbials, leading to the reduction in soil fertility, and this subsequently influences plant growth [15,19–22]. Scientific studies have confirmed that balanced fertilization helps to promote plant leaf, soil, and microbial nutrient balance and improve soil quality, which in turn increases crop yields and improves fruit quality [23]. While high P or K input significantly impacted the diversity in combination with the composition of soil bacterial communities while decreasing differences in soil C, N, and P and their stoichiometric ratios, high N input significantly increased MBC and MBC:MBN when compared to low N input over time under various fertilization conditions [24]. A number of studies have discovered that the addition of P significantly decreases the N:P and C:N ratios of plant, soil, and microbial biomass, with little effect on the C:N ratio [7]. In addition, N nitrogen addition significantly reduced plant and soil C:N but increased plant N:P ratios and soil microbial carbon and phosphorus limiting [3]. In addition, soil nutrient status has important direct effects on plant leaf C, N, and P content and stoichiometry [25]. Meanwhile, plant leaf C and N content and C:N were significantly correlated with soil C and N content [26]. Reasonable fertilization management can maximize crop yield [23]. In actual production, due to the lack of scientific rationing, ratios of N, P, and K are often unbalanced, which affects plant nutrient uptake and utilization, reduces yield and quality, and increases the risk of nutrient loss and environmental pollution. Therefore, reasonable N, P, and K ratios can significantly promote plant fruit yield, and there is an urgent need to explore scientific fertilization rationing strategies, which are essential for enhancing sustainable plant development.

In recent years, the methods currently commonly used for fertilizer evaluation include the affiliation function value method and the principal component analysis method [27]. The TOPSIS method is a multi-attribute decision-making method of ideal objective similarity and approximation of ideal solution ranking method, while the entropy-weighted TOPSIS model is an improved model combining the entropy-weighted method and the TOPSIS model, which is the most recent research result in the comprehensive evaluation method [27–29]. It has been applied to the fertilization evaluation of different plants such as *Panax notoginseng* [28] and *Mangifera indica* [30], but we have not yet seen any report on the evaluation of the C, N, and P content of leaf–soil–microbial C, N, and P content and stoichiometric ratio of *Sapindus saprophyllus*, as well as the yield, by using the entropy-weighted TOPSIS method. Therefore, an entropy-weighted TOPSIS comprehensive evaluation system was constructed to systematically quantify the C, N, and P coupling characteristics and yield response mechanism in the leaf–soil–microbe system of *Sapindus mukorossi* and to break through the limitations of the traditional single-factor evaluation through the optimization of multi-dimensional index weights and relative proximity ranking to provide scientific bases for *Sapindus mukorossi*'s ecocultivation and nutrient precision management. It provides a scientific basis for the ecological cultivation and precise management of nutrients in *Sapindus indica*.

Sapindus saponaria is an energy plant extensively spread across the tropical and subtropical areas of China, with multiple functions such as biomass energy production, ecological restoration, and landscaping [27,31–33]. In Fujian Province, one of its primary cultivation areas, the planting area has reached 20,000 hectares. Traditional fertilization practices, characterized by irrational fertilizer structures and unscientific fertilizer mixing, often lead to severe environmental problems such as soil nutrient imbalance, acidification, and groundwater pollution, which seriously affect *Sapindus saponaria* yield and sustainable development [27]. Although previous studies have shown the effects of N, P, and K fertilization on *Sapindus saponaria* growth, photosynthetic physiology, and soil nutrient balance, systematic research is lacking about the combined effects of different fertilization treatments on yield concerning leaf, soil, and microbial mass C, N, and P, and stoichiometric traits. In this context, the present study employs a "3414" experimental design to explore the effects of varying N–P–K fertilization ratios on the C, N, and P contents and their stoichiometric characteristics in the leaf–soil–microbial system of a 6-year-old *Sapindus saponaria* plantation. Utilizing Redundancy Analysis (RDA), Partial Least Squares Path Modeling (PLS–PM), and entropy-weighted TOPSIS, this study aims to (1) quantify the effects of various N–P–K ratios on the C, N, and P contents of *Sapindus saponaria* leaves, soil, and microorganisms; (2) reveal the relationship regarding the relationship among soil and microbial biomass N, P, and K content and stoichiometric characteristics with leaf nutrient profiles; (3) verify whether fertilization indirectly affects microbial biomass by altering soil nutrients, thereby improving *Sapindus saponaria* yield; and (4) develop a scientific nutrient management plan for *Sapindus saponaria* plantations to promote sustainable development and productivity. These objectives aim to provide comprehensive insights into optimizing fertilization strategies for *Sapindus saponaria*, enhancing yield, and maintaining ecosystem health.

2. Results

2.1. Effects of Different Fertilization Treatments on Carbon, Nitrogen, Phosphorus, and Stoichiometry of *Sapindus saponaria* Leaves

Compared with the CK treatment group, different fertilization treatments significantly increased leaf C, N, and P contents (Figure 1), which took values ranging from 2.74 to 7.21%, −14.44 to 13.16%, and −5 to 41.67%, respectively, and leaf C and P contents under each fertilization treatment showed a tendency of increasing and then decreasing with the additions of applied N, P, and K at the levels of P_2K_2, N_2K_2, or N_2P_2; all of them in the leaf C and P content reached the highest in the $N_2P_2K_2$ treatment. Compared with $N_0P_2K_2$, $N_2P_0K_2$, and $N_2P_2K_0$ treatments, the leaf organic C and total P contents of treatment $N_2P_2K_2$ were increased by 3.19% and 44.92%, 3.79% and 19.58%, and 2.87% and 11.76%, respectively, and there was a significant difference between treatments $N_2P_2K_2$ and $N_0P_2K_2$ and $N_2P_0K_2$ and $N_2P_2K_0$. Moreover, one-way ANOVA showed that different fertilization treatments had a significant effect on leaf C:P and N:P stoichiometry, while there was no significant effect on its leaf C:N.

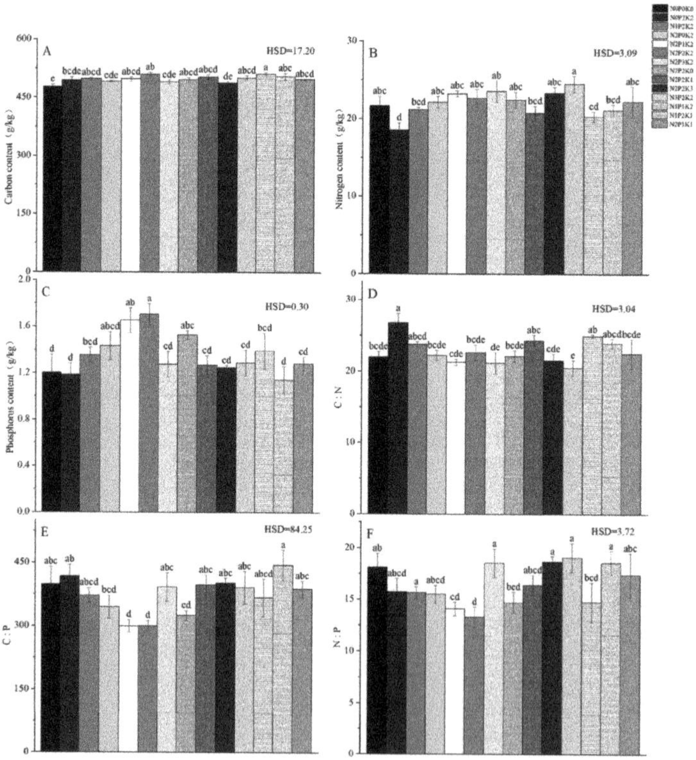

Figure 1. Characteristics of organic C, total N, and total P contents and their stoichiometric ratios of *Sapindus saponaria* soils under different N–P–K fertilization treatments: (**A**) leaf C content, (**B**) leaf N content, (**C**) leaf P content, (**D**) LC:LN, (**E**) LC:LP, and (**F**) LN:LP. Different lowercase letters indicate significant differences between fertilization treatments at the 0.05 statistical significance level.

2.2. Soil Carbon, Nitrogen, and Phosphorus Contents, and Stoichiometric Ratios Under Different Fertilization Treatments

Compared to CK, soil organic carbon (SOC), total nitrogen (TN), and total phosphorus (TP) contents increased by 14.25% to 52.61%, 3.90% to 39.84%, and 9.52% to 150% given different fertilization strategies (Figure 2). In 0–20 cm and 20–40 cm soil layers, SOC and

TN contents presented a trend of growth followed by decline with the increase in N, P, and K application. The increases in SOC, TN, and TP were more obvious in the 0–20 cm layer than in the 20–40 cm layer, with P content showing the greatest increase across soil layers.

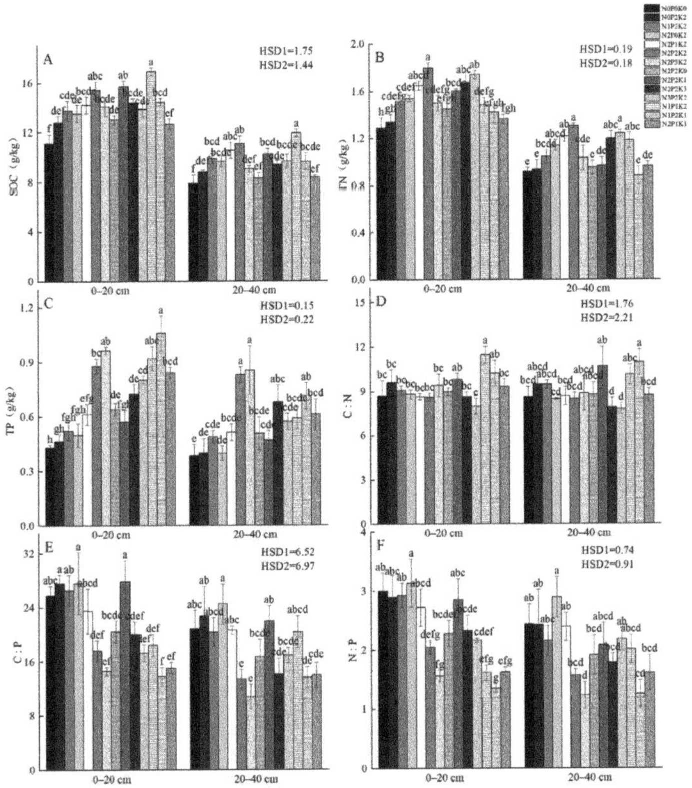

Figure 2. Characteristics of SOC, TN, and TP contents and their stoichiometric ratios of *Sapindus saponaria* soils under different N–P–K fertilization treatments. (**A**) SOC, organic carbon; (**B**) TN, total nitrogen; (**C**) TP, total phosphorus; (**D**) C:N, carbon/nitrogen ratio; (**E**) C:P, carbon/phosphorus ratio; (**F**) N:P, nitrogen/phosphorus ratio. HSD1: 0–20 cm. HSD2: 20–40 cm. A significant difference at the 0.05 level exists between fertilization treatments denoted by different lowercase letters.

Soil C:N, C:P, and N:P ratios under different fertilization treatments ranged from 7.97 to 11.44, 13.76 to 27.81, and 1.35 to 3.13, respectively (Figure 2). Compared to CK, the soil C:N ratio tended to increase in all treatments except $N_2P_1K_2$, $N_2P_2K_2$, $N_2P_2K_3$, and $N_3P_2K_2$. Soil N:P ratios showed significant differences among treatments, while N:P ratios showed no significant differences.

2.3. Microbial Carbon, Nitrogen, and Phosphorus Contents, and Stoichiometric Ratios Under Different Fertilization Treatments

Compared to the control treatment (CK), soil MBC, MBN, and MBP contents increased by 6.65–47.45%, 11.84–46.50%, and 14.91–201.98%, respectively, under fertilization treatments. At P_2K_2 and N_2P_2 levels, soil MBN and MBP contents initially increased and then decreased with increasing N and K application, while MBP content steadily increased with increasing P application in the N_2K_2 treatment. Soil microbial biomass ratios (MBC:MBN, MBC:MBP, and MBN:MBP) ranged from 9.33 to 14.89, 12.19 to 27.81, and 0.82 to 2.44, respectively (Figure 3). One-way ANOVA indicated significant differences in soil MBC,

MBN, and MBP contents among treatments ($p < 0.05$), but no significant differences in their stoichiometric ratios.

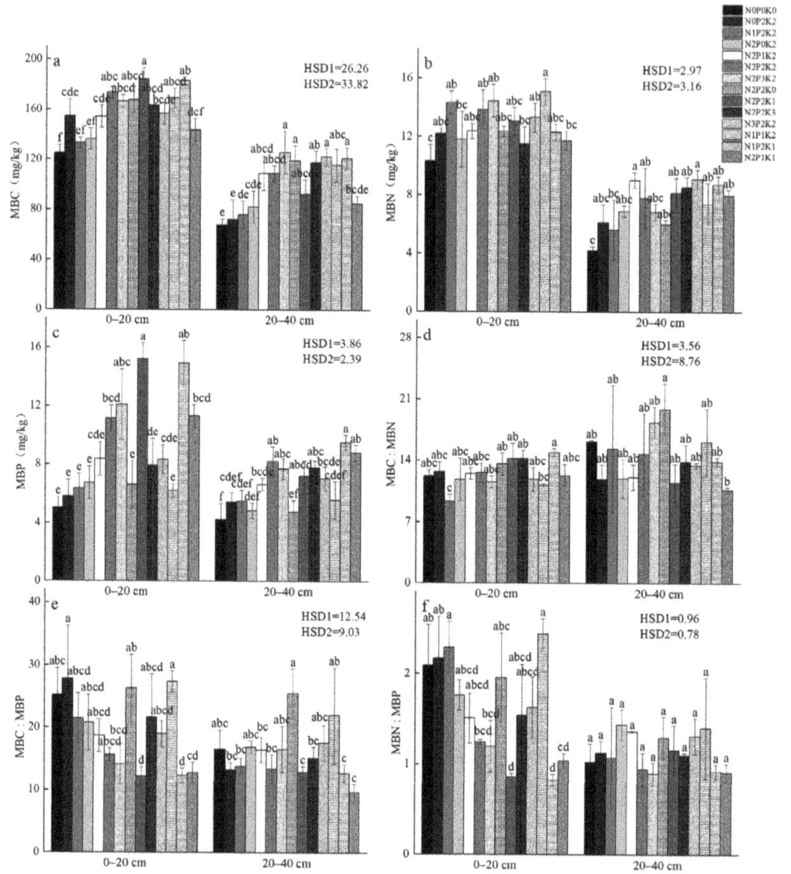

Figure 3. Characteristics of MBC, MBN, and MBP and stoichiometric ratio of *Sapindus saponaria* soil under different N–P–K fertilization treatments. (**a**) MBC: microbial carbon content; (**b**) MBN: microbial nitrogen content; (**c**) MBP: microbial phosphorus content; (**d**) MBC:MBN: microbial carbon to nitrogen ratio; (**e**) MBC:MBP, microbial carbon to phosphorus ratio; (**f**) MBN:MBP, microbial nitrogen to phosphorus ratio. At the 0.05 significance level, distinct lowercase letters denote substantial differences among fertilization treatments.

2.4. Leaf–Soil–Microbial Biomass C, N, P Contents, and Stoichiometric Characteristics

In the 0–20 cm soil layer, the first axis and second RDA axis explained 22.67% and 20.92% of the variation in leaf nutrient characteristics, respectively, collectively accounting for 43.59% of the variation (Figure 4a). Leaf N content showed a positive relationship to leaf P content and leaf N:P ratio and soil N and P contents and had a negative correlation with soil N:P ratio and leaf C:N ratio. Leaf C:P ratio was positively correlated with MBP content and MBC:MBN, whereas leaf N:P ratio was negatively associated with MBC:MBP.

Within the soil layer from 20 to 40 cm, the two RDA axes explained 38.32% of the variation in leaf nutrient characteristics (Figure 4b). Soil TP, TN, and SOC contents significantly affected leaf nutrient characteristics ($p < 0.05$), with SOC having the greatest impact (Table S1).

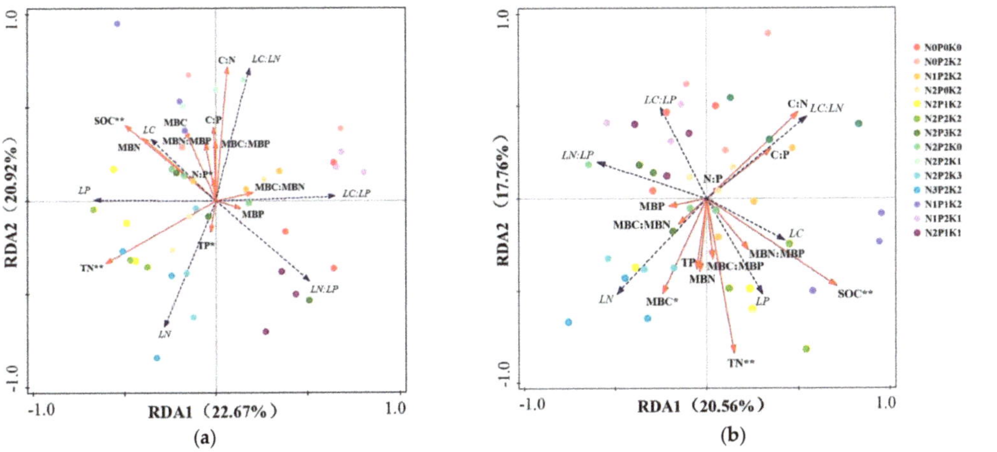

Figure 4. RDA of soil and microbial biomass C, N, P, and stoichiometric ratios of *Sapindus saponaria* under different fertilization treatments: (**a**) 0 to 20 cm soil layer and (**b**) 20 to 40 cm soil layer.

As can be seen from Figure 5, in 0–20 cm and 20–40 cm soil layers, leaf C content showed highly significant positive correlations with SOC, TP, and soil MBC, MBN, MBP, and soil C:N, and significant negative correlations with soil N:P and leaf LN:LP. The leaf LC:LN exhibited a significant positive correlation with the soil C:N. Meanwhile, it showed a significant negative correlation with the leaf N content. Soil N:P was significantly positively correlated with MBN:MBP and soil C:P and significantly negatively correlated with soil TP, MBC content, and soil C:N. Highly significant positive correlations were detected among the soil MBC, MBN, and MBP contents.

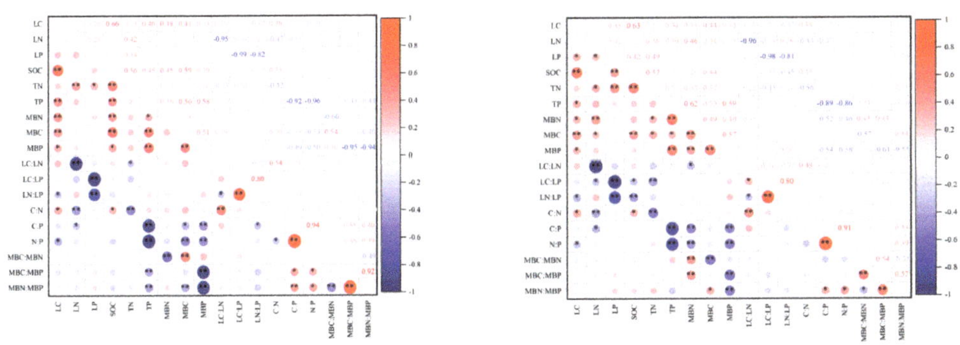

(**a**) 0–20 cm soil layer (**b**) 20–40 cm soil layer

Figure 5. Correlations between carbon, nitrogen, and phosphorus contents of leaves, soil, and microbiomass of *Sapindus saponaria* in 0–20 cm (**a**) and 20–40 cm (**b**) soil layers. LC: leaf blade carbon content; LN: leaf blade nitrogen content; LP: leaf blade phosphorus content; SOC: organic carbon; TN: total nitrogen; TP: total phosphorus; C:N: SOC to TN ratio; C:P: SOC to TP ratio; N:P: TN to TP ratio; LC:LN: leaf blade carbon to nitrogen ratio; LC:LP: leaf blade carbon to phosphorus ratio; LN:LP: leaf blade nitrogen to phosphorus ratio. MBC: microbial biomass carbon content; MBN: microbial biomass nitrogen content; MBP: microbial biomass phosphorus content; MBC:MBN: microbial biomass carbon to nitrogen ratio; MBC:MBP: microbial biomass carbon to phosphorus ratio; MBN:MBP: microbial biomass nitrogen to phosphorus ratio.

2.5. Multi-Objective Decision-Making and Evaluation Processes Leveraging the Entropy-Weighted TOPSIS Approach

The entropy-weighted TOPSIS model was used to comprehensively evaluate each fertilization treatment based on multiple indicators (Table 1). In the 0–20 cm soil layer, the order of advantages and disadvantages of the top 5 treatments of *Sapindus saponaria* obtained by TOPSIS was $N_2P_2K_2 > N_1P_2K_1 > N_2P_3K_2 > N_2P_2K_0 > N_2P_1K_2$, of which the best comprehensive evaluation of the soil layer of 20–40 cm and 0–20 cm was the $N_2P_2K_2$ treatment. Therefore, the N, P, and K fertilization of 600 kg/ha, 500 kg/ha, and 400 kg/ha in the field can effectively improve the comprehensive indexes of *Sapindus saponaria*, respectively.

Table 1. Comprehensive evaluation results of *Sapindus saponaria* under different fertilization treatments.

Treatment No.	0–20 cm Euclidean Space Distance D^+	Euclidean Space Distance D^-	C_i	Ranking	Treatment No.	20–40 cm Euclidean Space Distance D^+	Euclidean Space Distance D^-	C_i	Ranking
$N_0P_0K_0$	0.32	0.03	0.09	14	$N_0P_0K_0$	0.32	0.04	0.10	14
$N_0P_2K_2$	0.29	0.07	0.19	13	$N_0P_2K_2$	0.29	0.05	0.16	13
$N_1P_2K_2$	0.25	0.12	0.34	11	$N_1P_2K_2$	0.23	0.11	0.34	12
$N_2P_0K_2$	0.24	0.11	0.33	12	$N_2P_0K_2$	0.24	0.13	0.36	11
$N_2P_1K_2$	0.17	0.20	0.53	5	$N_2P_1K_2$	0.14	0.22	0.61	2
$N_2P_2K_2$	0.09	0.27	0.75	1	$N_2P_2K_2$	0.05	0.30	0.84	1
$N_2P_3K_2$	0.15	0.21	0.58	3	$N_2P_3K_2$	0.17	0.23	0.57	3
$N_2P_2K_1$	0.21	0.15	0.42	10	$N_2P_2K_1$	0.23	0.15	0.40	9
$N_2P_2K_0$	0.18	0.23	0.56	4	$N_2P_2K_0$	0.22	0.14	0.38	10
$N_2P_2K_3$	0.21	0.16	0.42	9	$N_2P_2K_3$	0.16	0.20	0.56	6
$N_3P_2K_2$	0.18	0.19	0.50	7	$N_3P_2K_2$	0.17	0.21	0.56	5
$N_1P_1K_2$	0.19	0.20	0.51	6	$N_1P_1K_2$	0.16	0.21	0.57	4
$N_1P_2K_1$	0.18	0.25	0.58	2	$N_1P_2K_1$	0.21	0.22	0.50	7
$N_2P_1K_1$	0.19	0.16	0.45	8	$N_2P_1K_1$	0.22	0.15	0.42	8

Note: D^+ and D^- represent the distance between the evaluation object and the positive and negative ideal solution, respectively; Ci is the comprehensive evaluation value.

2.6. Analysis Using Partial Least Squares Path Model (PLS–PM)

From the PLS–PM about the yield of *Sapindus saponaria*, it was observed that the model fitness of soil layers at depths of 0–20 cm (topsoil) and 20–40 cm (subsoil) met the acceptable requirements and explained 58.7% and 60.5% of the variation in the yield of *Sapindus saponaria*, respectively (Figure 6), with fertilizer, LP, LC:P, and MBX:MBP directly and significantly affecting its yield. It was easy to find that leaf C:P affected the highest total effect on *Sapindus saponaria* yield and leaf P content had the greatest direct or indirect effect on *Sapindus saponaria* yield. In the topsoil case, there were two distinct pathways affecting the yield of its saprophytes (Figure 6a; Table S2) as follows: ① fertilizer application directly positively affects the yield, and ② fertilizer application ultimately negatively affects the yield by directly positively affecting the TP content, which in turn positively affects the MBP content, which in turn negatively affects the MBC:MBP and MBN:MBP. In addition to this, fertilization also affects the leaf P content and its C:P by influencing the TP content, which in turn positively affects the yield. In the subsoil case (Figure 6b; Table S2), there were four distinct pathways affecting the yield of its saprophytes as follows: (1) fertilizer application positively affects its yield by positively affecting TP, which in turn positively affects LP; (2) fertilizer application negatively affects its yield by positively affecting TP, which in turn negatively affects C:P and N:P; (3) fertilizer application negatively affected QN by affecting, in turn, MBN content, which in turn positively affected yield; and (4) fertilizer application affected MBN content by positively affecting TP content, which in turn positively affected yield.

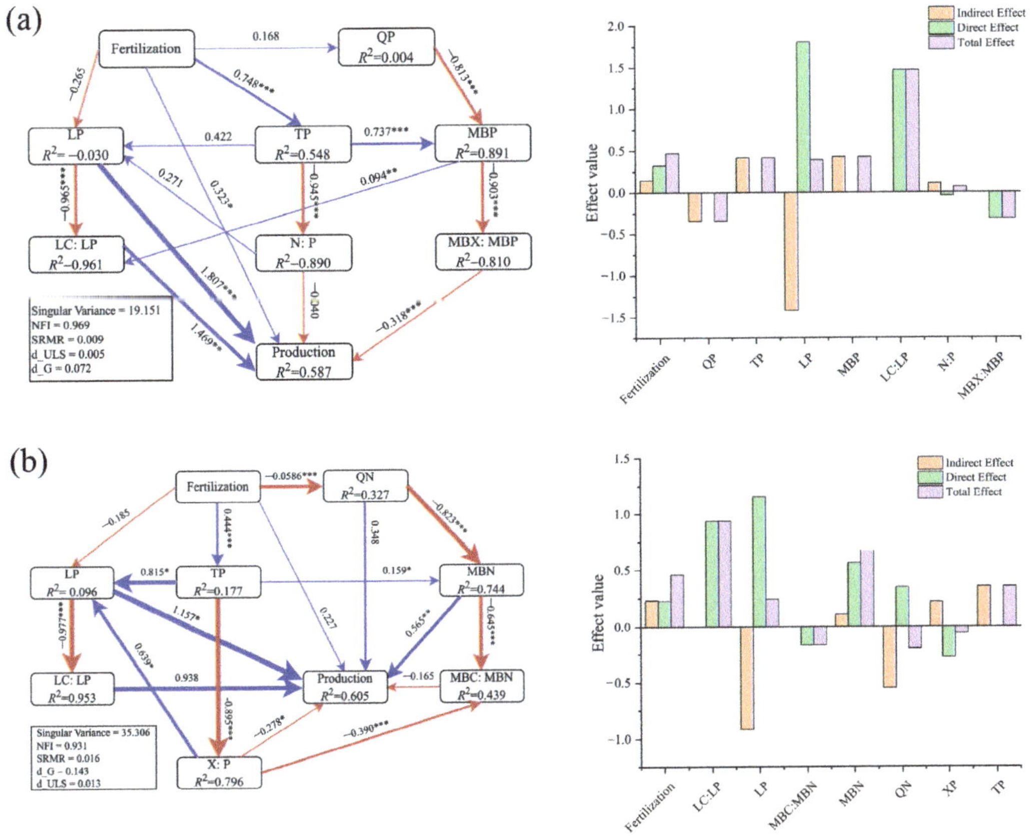

Figure 6. PLS–PM analysis for 0 to 20 cm and 20 to 40 cm soil layers: (**a**) 0–20 cm layer soil and (**b**) 20–40 cm layer soil. Fertilization denotes the amount of fertilizer applied. QN and QP represent microbial entropy nitrogen and phosphorus; MBX:MBP, MBC:MBP, and MBN:MBP ratios; and X:P: C:P, and N:P ratios. Production denotes the yield of the fruit. Blue lines indicate positive impacts, and red lines indicate negative impacts. Straight line thickness indicates relative magnitude of impact. The table of numbers on the lines were path coefficients. *, $p < 0.05$; **, $p < 0.01$; ***, $p < 0.001$. R^2 denotes adjusted fit.

In addition, the yield of *Sapindus saponaria* is regulated by several factors, and long-term input of inorganic fertilizer directly reduces its yield by decreasing leaf P content on the one hand, and suitable fertilization facilitates the increase in soil TP content in order to increase leaf P content, which significantly improves the yield of *Sapindus saponaria.* On the other hand, by affecting microbial entropy to influence the microbial biomass content and its stoichiometric ratio, which directly or indirectly acted on the yield.

3. Discussion

3.1. Characterization of C, N, and P Contents and Their Stoichiometric Ratios in Leaves

Numerous studies indicate that N and P stand out as the two primary limiting factors constraining plant growth. Concurrently, the leaf N:P ratio is extensively adopted as a crucial indicator for gauging plant nutrient limitations. The magnitude of plant leaves' N:P ratio reflects the relative limitation of N or P at the community level. Changes in this ratio have the potential to initiate modifications in plant functional traits, vegetation composition, and plant diversity [34–36]. There is a general consensus that plant leaves with N:P ratios lower than 14 indicate that the plant may be N-limited; N:P ratios higher than 16 indicate P-limitation; and if the N:P ratios of plant leaves are between 14 and 16, the plant may be limited by both N and P elements [37]. Also, high C:N ratios reflect that N may be a limiting factor, while low C:N ratios indicate relative N abundance, which may affect plant C balance [38,39]. In the context of this research, the mean value of the N:P ratio of *Sapindus saponaria* leaves was significantly higher than the mean value of 16.30 for Chinese plants and the global mean range of 12.7–13.8 [40,41], and the overall P-limiting characteristics were demonstrated, which indicated that the P-limiting intensity of the *Sapindus saponaria* leaves was greater in this forest stand.

Even though utilizing P fertilizer can boost plant leaf P uptake and enhance the resistance to P stress, it cannot offset the negative effects of inefficient P fertilizer utilization, as well as the limited ability to effectively mitigate the intensity of P limitation, which is closely bound up with the specific types of fertilizer application, plant characteristics, regional differences, etc. [42–44]. Specifically, leaf N and P contents in the study area were 18.52–23.55 and 1.14–1.71, respectively (Figure 1). These findings are consistent with the analysis of 753 species by Han et al. [37] and further confirm the higher degree of P-limitation exhibited by plant leaves in the subtropics. This effect of high leaf N:P ratio can be attributed to the low natural P content and strong N deposition in the soils of subtropical regions of China [37]. At the ecosystem scale, there is a coupling relationship between plant leaf P content and soil P content [45,46], which agrees with the outcomes presented herein. In response to leaf P limitation, soil P content can be effectively increased by fertilization, which in turn increases plant leaf P content, thereby reducing the leaf N:P ratios and mitigating the intensity of P limitation [47]. In conclusion, moderate fertilization can significantly increase leaf C, N, and P contents to alleviate plant leaf P-limitation and better maintain plant growth function.

3.2. Characterization of Soil–Microbial C, N, and P Contents and Stoichiometric Ratios

The findings of this research demonstrated that fertilization significantly influenced the nutrient levels and fertility augmentation of *Sapindus saponaria* growing soils. Specifically, fertilization significantly increased soil C, N, and P contents and effectively improved soil fertility. The result here is in accordance with those of other tree species [23,48]. The reasons for the efficiency of fertilizer application (increased yield, improved fertility, etc.) may lie in the fact that fertilizer application directly increases soil C, N, and P contents, and applying fertilizer stimulates plant growth and boosts the amount of plant residues incorporated into the soil, thus indirectly increasing the accumulation of soil nutrients [49]. Compared with the no fertilization treatment, TP content displayed an upward trend with the increase in P fertilizer at different levels of P fertilizer addition. This is due to the fact that P easily combines with Al^{3+} and Fe elements triggered by soil acidification and then transforms into low-soluble compounds, which can effectively increase soil effective P content [50]. Furthermore, findings indicated that SOC content increased at first and later decreased when N, P, and K were applied to the soil. According to the "stoichiometric decomposition theory" formulated by Hessen et al. [51], the imbalance of stoichiometric ratios in the soil was alleviated by N inputs, and microbial activity could be increased, thus accelerating SOC decomposition. Decomposition rates and microbial activity were highest when C and N inputs corresponded to microbial stoichiometric C:N ratios. When fertilizer was applied in excess, it is likely that it exerted an influence on the richness and activity of the soil microbial community. This, in turn, had an impact on the biodegradation process of SOC sources, ultimately leading to a decreased SOC content [52]. This indicates the importance of moderate fertilization in maintaining the balance of SOC content.

C and nutrient stoichiometric ratios (C:N, C:P, and N:P) in soil act as vital indices of the efficiency of soil nutrients [53,54]. The results of this study yielded that the C:N, C:P, and N:P of saprophytic soils under different fertilization treatments ranged from 7.97 to 11.44, 13.76 to 27.81, and 1.35 to 3.13, respectively, and the average C:N ratio of the soils in the study area, which was 9.21, was notably lower than the corresponding ratio of soils across China, which stood at 11.9. This reflects the higher rate of decomposition of soil organic matter as well as the higher potential for N accumulation in the investigated region. The average C:P value of the soil was 21.13, which was considerably below the average level (136) across China [55]. This suggests that microbial mineralization of soil organic matter has a high potential for releasing P, which is conducive to promoting P uptake by plants [56]. In addition, the average soil N:P ratio was measured at 2.31, which was substantially lower than both the global average ratio of 5.9 and the average ratio of 5.2 observed in China [55], which also suggests that soils in this region are more significantly P-limited.

In addition, fertilization significantly affected soil microbial biomass and its stoichiometric ratio characteristics, and soil TP and SOC contents were the primary determinants contributing to the variations in soil microbial biomass. The mean ratios of MBC:MBN, MBC:MBP, and MBN:MBP in *Sapindus saponaria* soils under various fertilization treatments were 12.49, 19.63, and 1.61, respectively. Specifically, the MBC:MBN ratio significantly exceeded the global average (typically, 8.3–10) [57], while the MBN:MBP ratio fell below the global average level (typically, 3.5–8.0) [5]. These imbalances suggest a fungal-dominated microbial community structure in the study area, as fungi generally exhibit higher C:N and lower N:P ratios compared to bacteria [5,57]. Since a higher MBC:MBN ratio usually reflects a fungal-dominated community structure, fungi tend to grow and metabolize in high C:N ratio environments more than bacteria, which also suggests that microorganisms are somewhat P-limited in terms of nutrient limitation [5,57,58]. Overall, under various fertilization treatments, the P content of saprophytic soils experienced a more significant increase compared to other indicators, thus indicating that the nutrient limiting factor for soil microorganisms is P content [5,58]. Given that the analysis within this study did not include the soil enzyme activities and the traits of microbial community structure, the intrinsic regulatory mechanisms between the soil–microbial biomass and its stoichiometric ratio characteristics and its other microbial properties still need to be explored in depth in the follow-up.

3.3. *Influence of Fertilization Regimes on Fruit Yield of Sapindus saponaria*

The effectiveness of N, P, and K fertilizers in increasing crop yields is in general agreement with the conclusions reached in most studies [23,59]. This is due to the fact that balanced fertilization enhances soil nutrient availability and improves plant growth conditions. The findings of this study indicated that the C, N, and P contents present in both the plants and soil microorganisms were better in the $N_2P_2K_2$ treatment using the entropy-weighted TOPSIS model (Table 1). It was also found that $N_2P_2K_2$ treatment was the optimal fertilization rate for fruit yield enhancement in *Sapindus saponaria* (Figure S1), but excessive fertilization led to yield decline instead, probably because excess nutrients may lead to soil compaction and nutrient imbalance when fertilization exceeds the threshold, thus inhibiting the plant root growth and nutrient uptake, which ultimately led to yield decline, and this phenomenon is applicable to other crops as well [60]. Therefore, when fertilizer is applied within reasonable limits, the supply of N, P, and K in the soil increases, which in turn promotes higher fruit yields. Additionally, moderate fertilization helps to enhance plant resistance, reduce pests and diseases, and further increase yield.

Long-term balanced fertilization (such as $N_2P_2K_2$) significantly improves the structure of the microbial community by increasing the contents of SOC, N, and P. It also enhances MBC and metabolic activity while reducing metabolic entropy (CO_2 emission per unit of biomass). This optimization can promote soil nutrient mineralization and the efficiency of plant nutrient uptake, thus supporting high fruit yields [61]. Some studies have emphasized that in calcareous alluvial soils deficient in P, rational application of P is a prerequisite for improving the utilization efficiency of N fertilizers [62]. This is consistent with the contribution of P to yield in the $N_2P_2K_2$ treatment of this study, and it explains that excessive application of N may lead to a relative deficiency of P, inhibiting the symbiosis of arbuscular mycorrhizal fungi (AMF) and further limiting the nutrient uptake by the roots. In addition, the suitable soil pH value for plants of the *Sapindus* is 5–6. If excessive fertilization leads to a decrease in the soil pH value in the experimental area, thus resulting in severe soil acidification, it may directly inhibit root development and microbial activities, resulting in a decline in yield [63]. The results of multiple studies have shown that the combined application of organic fertilizers can enhance the stability of soil aggregates and the complexity of the microbial network, which may further improve the sustainability of yield [61,62]. Based on the limitations of current research, on the basis of optimizing existing N–P–K inorganic fertilizer ratios, research will be conducted on the coupled application of organic and biological fertilizers. The focus is on analyzing soil–plant system nutrient cycling characteristics, microbial community structure responses, and synergistic yield-quality improvement mechanisms under fertilizer synergism. By constructing a three-in-one "inorganic–organic–biological" fertilization management model, this study aims to provide multi-dimensional solutions for the sustainable development of *Sapindus*.

3.4. Plant–Soil–Microbe Coupling

From the results of RDA and PLS-PM, it was concluded that the TP and SOC contents of the soil substantially influenced the yield of *Sapindus saponaria*, MBN or MBP content, and microbial activity. One possible explanation for this occurrence is the rapid escalation of the soil's TP levels and SOC content after fertilization, which can provide raw materials for soil microbial activities and accelerate the absorption and conversion of soil nutrients. The correlation analysis presented evidence that the MBC, MBN, and MBP of *Sapindus saponaria* soil exhibited a highly significant positive correlation with SOC, TN, and TP contents ($p < 0.01$). This result corresponds with the conclusions drawn from most investigations into woody plants, where fertilization promoted soil nutrient accumulation in the 0–20 cm and 20–40 cm soil layers, affecting *Sapindus saponaria* yield mainly through changes in soil P content and its N:P, leaf P content and its C:P ratio, and microbial entropy N or P (Figure 6. When compared to the 0–20 cm soil layer, the fluctuations in soil nutrient content and its stoichiometric ratios within the 20–40 cm soil layer, along with the variations in leaf P content and the C:N ratio of leaves corresponding to the 20–40 cm soil layer, provided a more substantial explanation for the changes in the yield of *Sapindus saponaria*, but the difference in the degree of explanation between the two layers was not significant. This may be due to the fact that *Sapindus saponaria* has a deeper root system, so that the main nutrients absorbed by it mainly come from the deeper soil; soil nutrient status of the 20–40 cm soil layer may better reflect the long-term nutrient supply status of the plant, thus exerting a more pronounced influence on the leaf nutrient status and yield; 20–40 cm soil layer is more capable of regulating the rate of nutrient release due to the deep regulation of soil stoichiometric equilibrium; soil enzyme activities show differences in different soil layers [64,65], but the activities of P-related metabolizing enzymes in their two soil horizons may not be so different as to affect soil P levels to a similar degree and have a similar effect on yields.

In conclusion, P content in soil and leaves serves as a restricting factor in determining crop yield [66–68], and the key reason for the changes in crop yields lies in the soil nutrient status, especially the P content within the soil, while soil microbial biomass affects yield indirectly by influencing soil enzyme activity and nutrient levels and indirectly affects yield [69]. However, exorbitant fertilization practices have the potential to trigger severe soil degradation, thereby throwing the soil's nutrient structure out of balance and subsequently impeding the growth and development of the plant root system, reducing nutrient and water uptake capacity, and indirectly inhibiting yield. Soil MBN and MBP content are key factors in regulating the main pathway of *Sapindus saponaria* yield, and TP content had a significant direct effect on soil microbiomass (Figure 6), suggesting that soil environmental factors jointly driving the shift of MBC, MBN, and MBP were generated under the distinct fertilization regimens. This aligns generally with the results of investigations conducted by other scholars [70–72].

4. Materials and Methods

4.1. Overview of the Experimental Site

Our experimental site was in Jianning County, Fujian Province (116°47′20″ E, 26°40′3″ N), which belongs to the subtropical monsoon climate zone, with a cold climate in winter and a large temperature difference between day and night in summer. The soil type of the experimental site was sandy clay loam. The values of soil physicochemical properties in the background are presented in an article published by our team [27]. The average plant height and ground diameter within the stand of the asexual line 'Yuanhua' were 2.38 m and 6.97 cm. The location of the experimental site and experimental design of different N–P–K treatments are presented in the Supplementary Materials (Figure S2).

4.2. Experimental Design

The experimental materials were planted in 6 years old Sapindus "Yuanhua" asexual forest stands. A "3414" randomized block design was used, including 3 factors (N, P, and K), 4 levels for each factor, a total of 14 combinations of treatments, 3 replications, and a total of 42 plots. Here, 0 level was for no fertilizer application and served as a control. Fertilizer application rates are listed in Table 2. The treatment plots were separated by isolation rows. Fertilizer was applied three times a year, on April 10 (during flowering, 30% of the total fertilizer application), July 20 (during fruiting, 30% of the total fertilizer application), and November 1 (after harvest, 40% of the total fertilizer application).

Table 2. Combination of factors and levels in each treatment.

Processing Number	Fertilization Levels ($kg{\cdot}ha^{-1}$)		
	N	P	K
$N_0P_0K_0$	0	0	0
$N_0P_2K_2$	0	500	400
$N_1P_2K_2$	300	500	400
$N_2P_0K_2$	600	0	400
$N_2P_1K_2$	600	250	400
$N_2P_2K_2$	600	500	400
$N_2P_3K_2$	600	750	400
$N_2P_2K_0$	600	500	0
$N_2P_2K_1$	600	500	200
$N_2P_2K_3$	600	500	600
$N_3P_2K_2$	900	500	400
$N_1P_1K_2$	300	250	400
$N_1P_2K_1$	300	500	200
$N_2P_1K_1$	600	250	200

Fertilizer was applied by furrow application, mixing the fertilizer well into the soil and covering it immediately. Maintenance and management were the same for all plots except for the fertilizer treatment. The test fertilizers were urea (46.0% N), calcium superphosphate (12.0% P_2O_5), and potassium sulfate (60.0% K_2O).

4.3. Sample Collection and Measurement

Soil sampling and determination: Three sampling points were randomly selected in each plot during the fruit expansion period of *Sapindus saponaria*. Soil samples were collected from 0 to 20 cm and 20 to 40 cm soil layers at each sampling site. After removing impurities and sapodilla root residues from the collected soil samples, they were divided into two parts. One part was put into a self-sealing bag and brought back to the laboratory to dry naturally and pass through a 2 mm sieve for the determination of soil chemical properties; the other part was put into a 50 mL centrifuge tube, and the soil samples were preserved in an icebox and brought back to the laboratory at 4 °C for low-temperature storage for the determination of soil microbial biomass. SOC was determined by the potassium dichromate oxidation–external heating method [73]; soil nitrogen was determined by the Kjeldahl method; and soil total phosphorus was determined by the molybdenum-antimony colorimetric method. MBC and MBN were determined by chloroform fumigation–K_2SO_4 extraction, and soil MBPs was determined by chloroform fumigation–$NaHCO_3$ leaching [49,74]. Yield: after harvesting at maturity, the plants were weighed on an electronic scale, and the average yield per plant was calculated.

4.4. Calculation of Leaf–Soil–Microbe-Related Indicators

The C:N ratio is the ratio of SOC content to TN content, the soil C:P ratio is the ratio of SOC content to TP content, and the soil N:P ratio is the ratio of TN content to TP content. Leaf carbon to nitrogen ratio (LC:LN) is the ratio of organic C content to total N content, carbon to phosphorus ratio (LC:LP) is the ratio of organic C content to total P content, and nitrogen to phosphorus ratio is the ratio of leaf total N content to total P content. Soil microbial entropy carbon (QC) is calculated as SOC content over MBC content, microbial entropy nitrogen (QN) is calculated as TN content over MBN content, and microbial entropy phosphorus (QP) is calculated as TP content over MBP content [75].

4.5. Multi-Objective Decision Making and Evaluation Based on Entropy-Weighted TOPSIS Approach

The entropy-weighted TOPSIS method involves both the entropy-weighted method and the TOPSIS method; the entropy-weighted method first standardizes each index and calculates the weight value of each evaluation index, and then further calculates the weight value of entropy weight and uses the weight value of entropy weight to multiply by the standardized original data to obtain the new data value. Using the new data values for the TOPSIS method for calculation, the Ci value of each ideal solution of the proximity procedure of each evaluation object is finally obtained so as to judge and measure the superiority and inferiority ranking of the evaluation object. The specific calculation process can be referred to in the literature [29,30]. The analysis process is as follows:

(1) An evaluation index matrix of leaf and soil C, N, and P contents, as well as yield attributes of *Sapindus* indica under different fertilization treatments, was established as follows.

$$X = \begin{bmatrix} X_{11} & \cdots & X_{1j} & \cdots & X_{1m} \\ \cdots & & \cdots & & \cdots \\ X_{i1} & \cdots & X_{ij} & \cdots & X_{im} \\ \cdots & & \cdots & & \cdots \\ X_{n1} & \cdots & X_{nj} & \cdots & X_{nm} \end{bmatrix} \quad (1)$$

where X_{ij} is the j indicator value of the i treatment of the original data (i = 1, . . ., n; j = 1, . . ., m), n is the number of treatments (n = 14), and m is the number of evaluation indicators (m = 10).

(2) Standardization of evaluation indicators to harmonize the types and dimensions of each indicator, with the following formula.

For positive indicators, use the following formula:

$$X'_{ij} = \frac{X_{ij} - \max(X_{1j}, X_{2j}, \ldots, X_{ij})}{\max(X_{1j}, X_{2j}, \ldots, X_{ij}) - \min(X_{1j}, X_{2j}, \ldots, X_{ij})} + 1 \quad (2)$$

For negative indicators, use the following formula:

$$X'_{ij} = \frac{\max(X_{ij}, X_{2j}, \ldots, X_{ij}) - X_{ij}}{\max(X_{ij}, X_{ij}, \ldots, X_{ij}) - \min(X_{1j}, X_{2j}, \ldots, X_{ij})} + 1 \quad (3)$$

where X_{ij} is the j indicator value for the i treatment (i = 1, . . ., n; j = 1, . . ., m) and X'_{ij} is the normalized X_{ij}.

(3) The proportion of the j indicator represented by the i treatment (P_{ij}) is calculated as follows:

$$P_{ij} = \frac{X'_{ij}}{\sum_{i=1}^{n} X_{ij}} \quad (4)$$

(4) The entropy value e_j of the j indicator is calculated as follows:

$$e_j = -\frac{\sum\limits_{i-1}^{n} P_{ij} \ln(P_{ij})}{\ln n} \quad (5)$$

(5) The coefficient of variation g_j for the j indicator is calculated as follows:

$$g_i = 1 - e_j \quad (6)$$

(6) The weight W_j of the j indicator is calculated as follows:

$$W_j = \frac{g_i}{\sum_{i=1}^{m} g_i} \quad (7)$$

(7) A weighted normalized decision matrix (R) is formed from the normalized decision matrix $X = (X'_{ij})14 \times 10$ and the weight vector W = (w1, w2, w3, . . ., w14).

$$R = (R_{ij})_{m \times n} = (W_j X_{ij})_{m \times n} \quad (8)$$

(8) Determine the optimal solution Z_{ij}^+ and the worst solution Z_{ij}^- to form the optimal vector Z^+ and the worst vector Z^-, respectively.

$$Z_{ij}^+ = (\max R_{i1}^+, \max R_{i2}^+, \max R_{i3}^+, \ldots, \max R_{i10}^+) \quad (9)$$

$$Z_{ij}^- = (\max R_{i1}^-, \max R_{i2}^-, \max R_{i3}^-, \ldots, \max R_{i10}^-) \quad (10)$$

(9) Determination of Euclidean spatial distances D^+ and D^- between 14 fertilization treatments and the worst solution.

$$D_j^+ = \sqrt{\sum_{j=1}^{m}\left[w_j \times (r_{ij} - Z_{ij}^+)\right]^2} \tag{11}$$

$$D_j^- = \sqrt{\sum_{j=1}^{m}\left[w_j \times (r_{ij} - Z_{ij}^-)\right]^2} \tag{12}$$

(10) Calculate the comprehensive evaluation value C_i of each treatment, i.e., calculate the closeness of the evaluation object to the optimal program as follows:

$$C_i = \frac{D_i^-}{D_i^+ + D_i^-} \tag{13}$$

4.6. Statistics and Analysis of Data

Data were organized using Microsoft Office 2010 and analyzed using IBM SPSS Statistics 21.0. One-way Analysis of Variance (ANOVA) and Least Significant Difference (LSD) tests or Tukey's Honestly Significant Difference (HSD) tests were performed to compare indicators between different treatment groups, with comprehensive analysis and visualization implemented in Python 3. Pearson correlation analysis was used to assess the associations among leaf, soil, and microbial C, N, and P contents and their stoichiometric ratios. Redundancy analysis (RDA, the gradient length < 3) was performed using Canoco5 to screen key soil factors that affect leaf nutrient factors. The PLS–PM method was used to evaluate factors influencing the yield of *Sapindus saponaria* by the plsdepot (version 0.2.0) and seminr (version 2.3.4) packages in R (version 4.3). The model's goodness of fit was assessed using the following established thresholds: normed fit index (NFI) > 0.90 and standardized root mean square residual (SRMR) < 0.08, in accordance with Hu and Bentler's [76] dual-criteria approach. Additionally, the stability of path coefficients was evaluated through a bootstrap analysis with 1000 resamples, employing 95% bias-corrected confidence intervals to test statistical significance.

5. Conclusions

This research delved into how balanced nitrogen–phosphorus–potassium (N–P–K) fertilization impacted the nutrient changes in the leaves, the soil, and the soil microbial biomass, along with the yield of the 'Yuanhua' variety of *Sapindus saponaria*. The results demonstrated that different fertilization treatments significantly enhanced leaf and soil nutrient contents and soil microbial biomass C, N, and P levels. Utilizing the entropy-weighted TOPSIS model, the $N_2P_2K_2$ treatment (600 kg·ha^{-1} N, 500 kg·ha^{-1} P, and 400 kg·ha^{-1} K) was identified as the optimal fertilization strategy for maximizing both nutrient content and yield. The best fertilization strategy for this area is to apply N, P, and K to the soil. Applying fertilizers with appropriate N–P–K ratios can remarkably elevate C, N, and P contents of leaf-soil microbes in *Sapindus saponaria* stands and boost the yields of *Sapindus saponaria*. Soil TP, TN, and SOC contents significantly affected leaf nutrient characteristics. Fertilizer application indirectly affected leaf nutrient content by changing soil properties, which in turn had a significant effect on *Sapindus saponaria* yield, especially total soil and leaf P content, which were the main contributors to the yield of *Sapindus saponaria*, and the microbial entropy of soil microorganisms regulated microbial biomass stoichiometric ratios, which further affected *Sapindus saponaria* yield either directly or indirectly. We propose that in the future, investigators should carry out a further examination of the variations in the functional aspects of the microbial community that occur due to the various fertilization treatments and their potential mechanisms of influence on *Sapindus saponaria* yield. Considering that fertilizer application affects soil properties to significantly influence microbial characteristics, which in turn exacerbates or alleviates P limitation and affects plant growth and development, it is reasonable to scientifically formulate fertilizer application strategies to improve soil quality and reduce environmental pollution.

Supplementary Materials: The following supporting information can be downloaded at: https://www.mdpi.com/article/10.3390/plants14091360/s1, Figure S1: Effect of different fertilization treatments on fruit yield of *Sapindus indicus*; Figure S2: Location of the experimental site and experimental design of different NPK treatments; Table S1: RDA of leaf nutrient content influenced by soil properties in different soil layers under different fertilization treatments; Table S2: The contents of soil QC, QN, and QP under different fertilization treatments and soil layers.

Author Contributions: J.L. and L.J. originated the research concept and crafted the experimental framework. J.L., L.J. and H.Y. wrote and completed the manuscript. J.L., Y.G., W.Z. and L.Z. organized and analyzed the data. S.W., S.Z. and X.W. performed and conducted the experiments. J.L., L.J., G.Z. and J.C. modified the language, polished the grammar, and revised the manuscript. All those listed as authors participated in the development of the article and consented to the published form of the final manuscript. All authors have read and agreed to the published version of the manuscript.

Funding: This study was supported by the National Natural Science Foundation of China (No. 32071793) and the Special Foundation for National Science and Technology Basic Research Program of China (No. 2019FY100803).

Data Availability Statement: The original contributions presented in this study are included in the article/Supplementary Materials. Further inquiries can be directed to the corresponding author(s).

Acknowledgments: We thank the original contributions put forward in this study, which have been incorporated into the article's materials; further inquiries can be directed to the corresponding author/s.

Conflicts of Interest: The research was not conducted with any commercial or financial relationships that could be construed as a potential conflict of interest. We express our heartfelt gratitude to Shuijing Luo for his tending of the plants.

References

1. Carnicer, J.; Sardans, J.; Stefanescu, C.; Ubach, A.; Bartrons, M.; Asensio, D.; Peñuelas, J. Global Biodiversity, Stoichiometry and Ecosystem Function Responses to Human-Induced C–N–P Imbalances. *J. Plant Physiol.* **2015**, *172*, 82–91. [CrossRef] [PubMed]
2. Qi, D.; Feng, F.; Lu, C.; Fu, Y. C:N:P Stoichiometry of Different Soil Components after the Transition of Temperate Primary Coniferous and Broad-Leaved Mixed Forests to Secondary Forests. *Soil Tillage Res.* **2022**, *216*, 105260. [CrossRef]
3. Xu, H.; Qu, Q.; Li, G.; Liu, G.; Geissen, V.; Ritsema, C.J.; Xue, S. Impact of Nitrogen Addition on Plant-Soil-Enzyme C–N–P Stoichiometry and Microbial Nutrient Limitation. *Soil Biol. Biochem.* **2022**, *170*, 108714. [CrossRef]
4. Ding, D.; Arif, M.; Liu, M.; Li, J.; Hu, X.; Geng, Q.; Yin, F.; Li, C. Plant-Soil Interactions and C:N:P Stoichiometric Homeostasis of Plant Organs in Riparian Plantation. *Front. Plant Sci.* **2022**, *13*, 979023. [CrossRef]
5. Gao, D.; Bai, E.; Wang, S.; Zong, S.; Liu, Z.; Fan, X.; Zhao, C.; Hagedorn, F. Three-dimensional Mapping of Carbon, Nitrogen, and Phosphorus in Soil Microbial Biomass and Their Stoichiometry at the Global Scale. *Glob. Change Biol.* **2022**, *28*, 6728–6740. [CrossRef]
6. Chen, X.; Chen, H.Y.H. Plant Mixture Balances Terrestrial Ecosystem C:N:P Stoichiometry. *Nat. Commun.* **2021**, *12*, 4562. [CrossRef]
7. Sun, Y.; Wang, C.; Chen, X.; Liu, S.; Lu, X.; Chen, H.Y.H.; Ruan, H. Phosphorus Additions Imbalance Terrestrial Ecosystem C:N:P Stoichiometry. *Glob. Change Biol.* **2022**, *28*, 7353–7365. [CrossRef]
8. Zhang, P.; Lü, X.-T.; Li, M.-H.; Wu, T.; Jin, G. N Limitation Increases along a Temperate Forest Succession: Evidences from Leaf Stoichiometry and Nutrient Resorption. *J. Plant Ecol.* **2022**, *15*, 1021–1035. [CrossRef]
9. Hu, B.; Xie, M.; Li, H.; Zhao, W.; Hu, J.; Jiang, Y.; Ji, W.; Li, S.; Hong, Y.; Yang, M.; et al. Stoichiometry of Soil Carbon, Nitrogen, and Phosphorus in Farmland Soils in Southern China: Spatial Pattern and Related Dominates. *Catena* **2022**, *217*, 106468. [CrossRef]
10. Liu, Y.; Li, C.; Cai, G.; Sauheitl, L.; Xiao, M.; Shibistova, O.; Ge, T.; Guggenberger, G. Meta-Analysis on the Effects of Types and Levels of N, P, and K Fertilization on Organic Carbon in Cropland Soils. *Geoderma* **2023**, *437*, 116580. [CrossRef]

11. Han, B.; Yao, Y.; Wang, Y.; Su, X.; Ma, L.; Chen, X.; Li, Z. Microbial Traits Dictate Soil Necromass Accumulation Coefficient: A Global Synthesis. *Glob. Ecol. Biogeogr.* **2023**, *33*, 151–161. [CrossRef]
12. Barłóg, P.; Grzebisz, W.; Łukowiak, R. Fertilizers and Fertilization Strategies Mitigating Soil Factors Constraining Efficiency of Nitrogen in Plant Production. *Plants* **2022**, *11*, 1855. [CrossRef] [PubMed]
13. Dincă, L.C.; Grenni, P.; Onet, C.; Onet, A. Fertilization and Soil Microbial Community: A Review. *Appl. Sci.* **2022**, *12*, 1198. [CrossRef]
14. Rutkowski, K.; Łysiak, G.P. Effect of Nitrogen Fertilization on Tree Growth and Nutrient Content in Soil and Cherry Leaves (*Prunus cerasus* L.). *Agriculture* **2023**, *13*, 578. [CrossRef]
15. Fang, Z.; Yu, H.; Li, C.; Wang, B.; Huang, J. Soil Microbial Biomass C:N:P Stoichiometry Is Driven More by Climate, Soil Properties and Plant Traits than by N Enrichment in a Desert Steppe. *Catena* **2022**, *216*, 106402. [CrossRef]
16. Jia, Y.; Cheng, Z.; Peng, Y.; Yang, G. Microplastics Alter the Equilibrium of Plant-Soil-Microbial System: A Meta-Analysis. *Ecotoxicol. Environ. Saf.* **2024**, *272*, 116082. [CrossRef]
17. Song, S.; Xiong, K.; Chi, Y. Ecological Stoichiometric Characteristics of Plant–Soil–Microorganism of Grassland Ecosystems under Different Restoration Modes in the Karst Desertification Area. *Agronomy* **2023**, *13*, 2016. [CrossRef]
18. Song, Z.; Zuo, X.; Zhao, X.; Qiao, J.; Ya, H.; Li, X.; Yue, P.; Chen, M.; Wang, S.; Medina-Roldán, E. Plant Functional Traits Mediate the Response Magnitude of Plant-Litter-Soil Microbial C:N:P Stoichiometry to Nitrogen Addition in a Desert Steppe. *Sci. Total Environ.* **2024**, *915*, 169915. [CrossRef]
19. Li, C.; Zhao, C.; Zhao, X.; Wang, Y.; Lv, X.; Zhu, X.; Song, X. Beneficial Effects of Biochar Application with Nitrogen Fertilizer on Soil Nitrogen Retention, Absorption and Utilization in Maize Production. *Agronomy* **2023**, *13*, 113. [CrossRef]
20. Liu, Q.; Xu, H.; Yi, H. Impact of Fertilizer on Crop Yield and C:N:P Stoichiometry in Arid and Semi-Arid Soil. *Int. J. Environ. Res. Public Health* **2021**, *18*, 4341. [CrossRef]
21. Sardans, J.; Rivas-Ubach, A.; Peñuelas, J. The C:N:P Stoichiometry of Organisms and Ecosystems in a Changing World: A Review and Perspectives. *Perspect. Plant Ecol. Evol. Syst.* **2012**, *14*, 33–47. [CrossRef]
22. Zhang, Y.; Ye, C.; Su, Y.; Peng, W.; Lu, R.; Liu, Y.; Huang, H.; He, X.; Yang, M.; Zhu, S. Soil Acidification Caused by Excessive Application of Nitrogen Fertilizer Aggravates Soil-Borne Diseases: Evidence from Literature Review and Field Trials. *Agric. Ecosyst. Environ.* **2022**, *340*, 108176. [CrossRef]
23. Li, Z.; Zhang, R.; Xia, S.; Wang, L.; Liu, C.; Zhang, R.; Fan, Z.; Chen, F.; Liu, Y. Interactions between N, P and K Fertilizers Affect the Environment and the Yield and Quality of Satsumas. *Glob. Ecol. Conserv.* **2019**, *19*, e00663. [CrossRef]
24. Zhou, J.; Jiang, X.; Wei, D.; Zhao, B.; Ma, M.; Chen, S.; Cao, F.; Shen, D.; Guan, D.; Li, J. Consistent Effects of Nitrogen Fertilization on Soil Bacterial Communities in Black Soils for Two Crop Seasons in China. *Sci. Rep.* **2017**, *7*, 3267. [CrossRef]
25. Lin, Y.; Lai, Y.; Tang, S.; Qin, Z.; Liu, J.; Kang, F.; Kuang, Y. Climatic and Edaphic Variables Determine Leaf C, N, P Stoichiometry of Deciduous Quercus Species. *Plant Soil* **2022**, *474*, 383–394. [CrossRef]
26. Fan, H.; Wu, J.; Liu, W.; Yuan, Y.; Hu, L.; Cai, Q. Linkages of Plant and Soil C:N:P Stoichiometry and Their Relationships to Forest Growth in Subtropical Plantations. *Plant Soil* **2015**, *392*, 127–138. [CrossRef]
27. Liu, J.; Wang, D.; Yan, X.; Jia, L.; Chen, N.; Liu, J.; Zhao, P.; Zhou, L.; Cao, Q. Effect of Nitrogen, Phosphorus and Potassium Fertilization Management on Soil Properties and Leaf Traits and Yield of *Sapindus mukorossi*. *Front. Plant Sci.* **2024**, *15*, 1300683. [CrossRef]
28. Tuo, Y.; Wang, Z.; Zheng, Y.; Shi, X.; Liu, X.; Ding, M.; Yang, Q. Effect of Water and Fertilizer Regulation on the Soil Microbial Biomass Carbon and Nitrogen, Enzyme Activity, and Saponin Content of Panax Notoginseng. *Agric. Water Manag.* **2023**, *278*, 108145. [CrossRef]
29. Yan, F.; Liu, X.; Bai, W.; Fan, J.; Zhang, F.; Xiang, Y.; Hou, X.; Pei, S.; Dai, Y.; Zeng, H.; et al. Multi-Objective Optimization of Water and Nitrogen Regimes for Drip-Fertigated Sugar Beet in a Desert Climate. *Field Crops Res.* **2022**, *288*, 108703. [CrossRef]
30. Liu, X.; Peng, Y.; Yang, Q.; Wang, X.; Cui, N. Determining Optimal Deficit Irrigation and Fertilization to Increase Mango Yield, Quality, and WUE in a Dry Hot Environment Based on TOPSIS. *Agric. Water Manag.* **2021**, *245*, 106650. [CrossRef]
31. Han, P.; Zhang, W.; Wang, G.; Sun, W.; Huang, Y. Changes in Soil Organic Carbon in Croplands Subjected to Fertilizer Management: A Global Meta-Analysis. *Sci. Rep.* **2016**, *6*, 27199. [CrossRef]
32. Liu, J.; Zhou, L.; Wang, D.; Gong, Y.; Yan, X.; Cao, Q.; Wu, S.; Weng, J.; Zhang, G.; Jia, L. The Effect of Balanced N, P and K Fertilization on Fine Root Traits and Soil Properties in *Sapindus mukorossi*. *Forests* **2024**, *15*, 94. [CrossRef]
33. Zhao, G.; Gao, Y.; Gao, S.; Xu, Y.; Liu, J.; Sun, C.; Gao, Y.; Liu, S.; Chen, Z.; Jia, L. The Phenological Growth Stages of *Sapindus mukorossi* According to BBCH Scale. *Forests* **2019**, *10*, 462. [CrossRef]
34. Ågren, G.I.; Wetterstedt, J.Å.M.; Billberger, M.F.K. Nutrient Limitation on Terrestrial Plant Growth—Modeling the Interaction between Nitrogen and Phosphorus. *New Phytol.* **2012**, *194*, 953–960. [CrossRef]
35. Li, Y.; Niu, S.; Yu, G. Aggravated Phosphorus Limitation on Biomass Production under Increasing Nitrogen Loading: A Meta-analysis. *Glob. Change Biol.* **2016**, *22*, 934–943. [CrossRef]

36. Tessier, J.T.; Raynal, D.J. Use of Nitrogen to Phosphorus Ratios in Plant Tissue as an Indicator of Nutrient Limitation and Nitrogen Saturation. *J. Appl. Ecol.* **2003**, *40*, 523–534. [CrossRef]
37. Han, W.; Fang, J.; Guo, D.; Zhang, Y. Leaf Nitrogen and Phosphorus Stoichiometry across 753 Terrestrial Plant Species in China. *New Phytol.* **2005**, *168*, 377–385. [CrossRef]
38. Raven, J.A.; Handley, L.L.; Andrews, M. Global Aspects of C/N Interactions Determining Plant-Environment Interactions. *J. Exp. Bot.* **2003**, *55*, 11–25. [CrossRef]
39. Thuille, A.; Laufer, J.; Höhl, C.; Gleixner, G. Carbon Quality Affects the Nitrogen Partitioning between Plants and Soil Microorganisms. *Soil Biol. Biochem.* **2015**, *81*, 266–274. [CrossRef]
40. Elser, J.J.; Fagan, W.F.; Denno, R.F.; Dobberfuhl, D.R.; Folarin, A.; Huberty, A.; Interlandi, S.; Kilham, S.S.; McCauley, E.; Schulz, K.L.; et al. Nutritional Constraints in Terrestrial and Freshwater Food Webs. *Nature* **2000**, *408*, 578–580. [CrossRef]
41. Reich, P.B.; Oleksyn, J. Global Patterns of Plant Leaf N and P in Relation to Temperature and Latitude. *Proc. Natl. Acad. Sci. USA* **2004**, *101*, 11001–11006. [CrossRef] [PubMed]
42. Bindraban, P.S.; Dimkpa, C.O.; Pandey, R. Exploring Phosphorus Fertilizers and Fertilization Strategies for Improved Human and Environmental Health. *Biol. Fertil. Soils* **2020**, *56*, 299–317. [CrossRef]
43. Feng, J.; Chen, L.; Xia, T.; Ruan, Y.; Sun, X.; Wu, T.; Zhong, Y.; Shao, X.; Tang, Z. Microbial Fertilizer Regulates C:N:P Stoichiometry and Alleviates Phosphorus Limitation in Flue-Cured Tobacco Planting Soil. *Sci. Rep.* **2023**, *13*, 10276. [CrossRef] [PubMed]
44. Francis, B.; Aravindakumar, C.T.; Brewer, P.B.; Simon, S. Plant Nutrient Stress Adaptation: A Prospect for Fertilizer Limited Agriculture. *Environ. Exp. Bot.* **2023**, *213*, 105431. [CrossRef]
45. He, M.; Dijkstra, F.A.; Zhang, K.; Li, X.; Tan, H.; Gao, Y.; Li, G. Leaf Nitrogen and Phosphorus of Temperate Desert Plants in Response to Climate and Soil Nutrient Availability. *Sci. Rep.* **2014**, *4*, 6932. [CrossRef]
46. Tang, Z.; Xu, W.; Zhou, G.; Bai, Y.; Li, J.; Tang, X.; Chen, D.; Liu, Q.; Ma, W.; Xiong, G.; et al. Patterns of Plant Carbon, Nitrogen, and Phosphorus Concentration in Relation to Productivity in China's Terrestrial Ecosystems. *Proc. Natl. Acad. Sci. USA* **2018**, *115*, 4033–4038. [CrossRef]
47. Maity, A.; Marathe, R.A.; Sarkar, A.; Basak, B.B. Phosphorus and Potassium Supplementing Bio-Mineral Fertilizer Augments Soil Fertility and Improves Fruit Yield and Quality of Pomegranate. *Sci. Hortic.* **2022**, *303*, 111234. [CrossRef]
48. Zhang, H.; Gong, L.; Ding, Z.; Wu, X. Fine-Root Soil Stoichiometry of *Picea* Schrenkiana Fisch. et Mey. and Its Correlation with Soil Environmental Factors under Different Nitrogen Input Levels in the Tianshan Mountains, Xinjiang. *Forests* **2023**, *14*, 426. [CrossRef]
49. Dong, R.; Yang, S.; Wang, X.; Xie, L.; Ma, Y.; Wang, Y.; Zhang, L.; Zhang, M.; Qin, J. C:N:P Stoichiometry in Plant, Soil and Microbe in Sophora Moorcroftiana Shrubs across Three Sandy Dune Types in the Middle Reaches of the Yarlung Zangbo River. *Front. Plant Sci.* **2023**, *13*, 1060686. [CrossRef]
50. Xu, M.; Zhu, Y.; Zhang, S.; Feng, Y.; Zhang, W.; Han, X. Global Scaling the Leaf Nitrogen and Phosphorus Resorption of Woody Species: Revisiting Some Commonly Held Views. *Sci. Total Environ.* **2021**, *788*, 147807. [CrossRef]
51. Hessen, D.O.; Ågren, G.I.; Anderson, T.R.; Elser, J.J.; De Ruiter, P.C. Carbon sequestration in ecosystems: The role of stoichiometry. *Ecology* **2004**, *85*, 1179–1192. [CrossRef]
52. Chen, C.; Hall, S.J.; Coward, E.; Thompson, A. Iron-Mediated Organic Matter Decomposition in Humid Soils Can Counteract Protection. *Nat. Commun.* **2020**, *11*, 2255. [CrossRef] [PubMed]
53. Kirkby, C.A.; Richardson, A.E.; Wade, L.J.; Batten, G.D.; Blanchard, C.; Kirkegaard, J.A. Carbon-Nutrient Stoichiometry to Increase Soil Carbon Sequestration. *Soil Biol.* **2013**, *60*, 77–86. [CrossRef]
54. Tipping, E.; Somerville, C.J.; Luster, J. The C:N:P:S Stoichiometry of Soil Organic Matter. *Biogeochemistry* **2016**, *130*, 117–131. [CrossRef]
55. Tian, H.; Chen, G.; Zhang, C.; Melillo, J.M.; Hall, C.A.S. Pattern and Variation of C:N:P Ratios in China's Soils: A Synthesis of Observational Data. *Biogeochemistry* **2010**, *98*, 139–151. [CrossRef]
56. Li, Y.; Wu, J.; Liu, S.; Shen, J.; Huang, D.; Su, Y.; Wei, W.; Syers, J.K. Is the C:N:P Stoichiometry in Soil and Soil Microbial Biomass Related to the Landscape and Land Use in Southern Subtropical China? *Glob. Biogeochem. Cycles* **2012**, *26*, 4399. [CrossRef]
57. Cleveland, C.C.; Liptzin, D. C:N:P Stoichiometry in Soil: Is There a "Redfield Ratio" for the Microbial Biomass? *Biogeochemistry* **2007**, *85*, 235–252. [CrossRef]
58. Xu, X.; Thornton, P.E.; Post, W.M. A Global Analysis of Soil Microbial Biomass Carbon, Nitrogen and Phosphorus in Terrestrial Ecosystems. *Glob. Ecol. Biogeogr.* **2013**, *22*, 737–749. [CrossRef]
59. Zhang, X.; Li, S.; An, X.; Song, Z.; Zhu, Y.; Tan, Y.; Guo, X.; Wang, D. Effects of Nitrogen, Phosphorus and Potassium Formula Fertilization on the Yield and Berry Quality of Blueberry. *PLoS ONE* **2023**, *18*, e0283137. [CrossRef]
60. Chen, K.; Ma, L.; Chen, C.; Liu, N.; Wang, B.; Bao, Y.; Liu, Z.; Zhou, G. Long-Term Impact of N, P, K Fertilizers in Different Rates on Yield and Quality of *Anisodus tanguticus* (Maxinowicz) Pascher. *Plants* **2023**, *12*, 2102. [CrossRef]
61. Zhu, L.; Luan, L.; Chen, Y.; Wang, X.; Zhou, S.; Zou, W.; Han, X.; Duan, Y.; Zhu, B.; Li, Y.; et al. Community Assembly of Organisms Regulates Soil Microbial Functional Potential through Dual Mechanisms. *Glob. Change Biol.* **2024**, *30*, e17160. [CrossRef] [PubMed]

62. Lin, X.G.; Feng, Y.Z. Research Progresses of Farmland Microorganisms in Fluvo-Aquic Soil of China. *Chin. J. Eco-Agric.* **2016**, *24*, 416–434. [CrossRef]
63. Liu, J.; Wang, L.; Sun, C.; Xi, B.; Li, D.; Chen, Z.; He, Q.; Weng, X.; Jia, L. Global Distribution of Soapberries (*Sapindus* L.) Habitats under Current and Future Climate Scenarios. *Sci. Rep.* **2021**, *11*, 19740. [CrossRef] [PubMed]
64. Piotrowska-Długosz, A.; Długosz, J.; Frąc, M.; Gryta, A.; Breza-Boruta, B. Enzymatic Activity and Functional Diversity of Soil Microorganisms along the Soil Profile—A Matter of Soil Depth and Soil-Forming Processes. *Geoderma* **2022**, *416*, 115779. [CrossRef]
65. Zhao, Y.; Han, Z.; Zhang, G.; Chen, D.; Zang, L.; Liu, Q.; Guo, Y.; Xie, P.; Chen, H.; He, Y. Variability of Soil Enzyme Activities and Nutrients with Forest Gap Renewal Interacting with Soil Depths in Degraded Karst Forests. *Ecol. Indic.* **2024**, *166*, 112332. [CrossRef]
66. Harish, M.N.; Choudhary, A.K.; Bhupenchandra, I.; Dass, A.; Rajanna, G.A.; Singh, V.K.; Bana, R.S.; Varatharajan, T.; Verma, P.; George, S.; et al. Double Zero-Tillage and Foliar-P Nutrition Coupled with Bio-Inoculants Enhance Physiological Photosynthetic Characteristics and Resilience to Nutritional and Environmental Stresses in Maize–Wheat Rotation. *Front. Plant Sci.* **2022**, *13*, 959541. [CrossRef]
67. Khan, F.; Siddique, A.B.; Shabala, S.; Zhou, M.; Zhao, C. Phosphorus Plays Key Roles in Regulating Plants' Physiological Responses to Abiotic Stresses. *Plants* **2023**, *12*, 2861. [CrossRef]
68. Van Der Velde, M.; Folberth, C.; Balkovič, J.; Ciais, P.; Fritz, S.; Janssens, I.A.; Obersteiner, M.; See, L.; Skalský, R.; Xiong, W.; et al. African Crop Yield Reductions Due to Increasingly Unbalanced Nitrogen and Phosphorus Consumption. *Glob. Change Biol.* **2014**, *20*, 1278–1288. [CrossRef]
69. Liu, J.; Shu, A.; Song, W.; Shi, W.; Li, M.; Zhang, W.; Li, Z.; Liu, G.; Yuan, F.; Zhang, S.; et al. Long-Term Organic Fertilizer Substitution Increases Rice Yield by Improving Soil Properties and Regulating Soil Bacteria. *Geoderma* **2021**, *404*, 115287. [CrossRef]
70. Li, P.; Yang, Y.; Han, W.; Fang, J. Global Patterns of Soil Microbial Nitrogen and Phosphorus Stoichiometry in Forest Ecosystems. *Glob. Ecol. Biogeogr.* **2014**, *23*, 979–987. [CrossRef]
71. Luo, G.; Xue, C.; Jiang, Q.; Xiao, Y.; Zhang, F.; Guo, S.; Shen, Q.; Ling, N. Soil Carbon, Nitrogen, and Phosphorus Cycling Microbial Populations and Their Resistance to Global Change Depend on Soil C:N:P Stoichiometry. *mSystems* **2020**, *5*, e00162-20. [CrossRef] [PubMed]
72. Tang, S.; Ma, Q.; Marsden, K.A.; Chadwick, D.R.; Luo, Y.; Kuzyakov, Y.; Wu, L.; Jones, D.L. Microbial Community Succession in Soil Is Mainly Driven by Carbon and Nitrogen Contents Rather than Phosphorus and Sulphur Contents. *Soil Biol. Biochem.* **2023**, *180*, 11. [CrossRef]
73. Liu, K.-L.; Chen, B.-Y.; Zhang, B.; Wang, R.-H.; Wang, C.-S. Understory Vegetation Diversity, Soil Properties and Microbial Community Response to Different Thinning Intensities in *Cryptomeria japonica* var. *sinensis* plantations. *Front. Microbiol.* **2023**, *11*, 1117384. [CrossRef]
74. Wang, G.; Xiao, H.; Xin, Z.; Luo, F.; Jin, Y.; Liu, M.; Li, J. Changes in Plant-Soil-Microbe C-N-P Contents and Stoichiometry during Poplar Shelterbelt Degradation. *Catena* **2024**, *243*, 108227. [CrossRef]
75. Chi, Y.; Song, S.; Xiong, K.; Albasher, G.; Fang, J. Responses of Soil Microbial Biomass, Microbial Entropy and Soil-Microorganism Stoichiometry Imbalance to Different Utilization Patterns in the Artificial Grassland of Karst Desertification Area. *Front. Microbiol.* **2023**, *14*, 129353. [CrossRef]
76. Hu, L.; Bentler, P.M. Cutoff Criteria for Fit Indexes in Covariance Structure Analysis: Conventional Criteria versus New Alternatives. *Struct. Equ. Model. A Multidiscip. J.* **1999**, *6*, 1–55. [CrossRef]

plants

MDPI

Article

Optimization of Hydroponic Wheat Sprouts as an Alternative Livestock Feed: Yield and Biochemical Composition Under Different Fertilization Regimes

Andrius Grigas [1,*], Dainius Steponavičius [1], Indrė Bručienė [1], Ričardas Krikštolaitis [2], Tomas Krilavičius [2], Aušra Steponavičienė [3] and Dainius Savickas [1]

[1] Faculty of Engineering, Vytautas Magnus University Agriculture Academy, Studentų St. 15A, Akademija, LT-53362 Kaunas, Lithuania; dainius.steponavicius@vdu.lt (D.S.); indre.bruciene@vdu.lt (I.B.); dainius.savickas@vdu.lt (D.S.)

[2] Faculty of Informatics, Vytautas Magnus University, LT-44248 Kaunas, Lithuania; ricardas.krikstolaitis@vdu.lt (R.K.); tomas.krilavicius@vdu.lt (T.K.)

[3] Catering Department, Kaunas University of Applied Sciences, Pramonės Ave. 22, LT-50387 Kaunas, Lithuania; steponaviciene.ausra@gmail.com

* Correspondence: andrius.grigas@vdu.lt

Abstract: This study investigated the effects of macronutrient type and concentration on the biomass yield and biochemical composition of hydroponically grown wheat sprouts (HWS), with the aim of identifying fertilization strategies that optimize both productivity and feed quality. HWS were cultivated using a nutrient film technique over a 7-day period under controlled environmental conditions, with treatments including calcium nitrate (CN1–CN3), potassium phosphate (CP1–CP3), potassium sulfate (CK1–CK2), and a balanced NPK 20–20–20 fertilizer (NPK1–NPK3), each applied at three increasing concentrations. The quantitative parameters assessed included biomass yield per unit of dry seed (DP, kg kg^{-1}) and dry matter content (DM, %), while qualitative traits included crude protein (CP), ether extract (EE), crude fiber (CF), and ash content. Results indicated that balanced NPK fertilization significantly enhanced performance, with NPK3 achieving the highest biomass yield (6.39 kg kg^{-1}), CP (24.26%), CF (5.63%), and ash (16.0%) content. In contrast, CN3 treatments reduced yield (4.84 kg kg^{-1}) despite increasing CP (19.65%), indicating trade-offs between nitrogen enrichment and vegetative expansion. Phosphorus-based treatments (CP2–CP3) improved nutrient density without suppressing yield. Regression analyses revealed strong correlations between DM and both CF (R^2 = 0.81) and ash (R^2 = 0.71), supporting their utility as indirect indicators of feed quality. EE content remained stable (2.07–2.67%) across all treatments, suggesting its limited responsiveness to macronutrient manipulation. These findings highlight the importance of nutrient synergy in hydroponic systems and provide a practical framework for tailoring fertilization regimes to meet specific agronomic and nutritional objectives in precision livestock feeding and provide practical guidance for optimizing hydroponic livestock feed production.

Keywords: hydroponic forage production; precision livestock feeding; nutrient optimization; feed quality; alternative feed systems

1. Introduction

Feeding the world's growing population is becoming an increasingly complex challenge. As demand for meat, dairy, and other animal products rises, pressure increases on agricultural systems to deliver not just more feed, but feed that is sustainable, nutritious, and produced with fewer natural resources. Traditional forage crops, such as alfalfa and maize silage, continue to play a major role in livestock diets, but their reliance on fertile land, intensive water use, and seasonal conditions makes them vulnerable to supply fluctuations and environmental concerns [1,2]. At the same time, the carbon footprint associated with large-scale forage production remains a significant issue [3], and as climate change accelerates, agriculture must adapt [4].

One promising approach that has gained attention in recent years is hydroponic cultivation [5]. Unlike conventional farming, hydroponics allows plants to grow without soil, offering a controlled environment where water and nutrient use can be carefully managed. Studies show hydroponic systems can decrease water usage by up to 90% while drastically reducing land requirements [6,7]. This technology enables efficient fodder production in environments where conventional agriculture is constrained—such as arid regions with limited water availability or space-restricted settings like peri-urban or controlled environment systems [8,9].

Among the crops suited for hydroponic production, wheat sprouts (HWS) stand out. They grow quickly, reaching harvest in just 7 to 10 days, and offer a dense package of crude protein, minerals, antioxidants, and bioactive compounds [10,11]. Their nutritional profile, combined with high digestibility, has made HWS a strong candidate as an alternative forage, particularly for ruminants. Research also points to HWS improving rumen efficiency and reducing methane emissions, adding environmental benefits to their list of advantages [12,13].

However, in hydroponic systems, plants are entirely dependent on the nutrient solution supplied for their growth and development. This makes the composition of that solution critical—not only for maximizing biomass but also for tailoring the nutritional qualities of the final product. Key macronutrients like nitrogen (N), phosphorus (P), and potassium (K) are fundamental to plant function. N is vital for protein and amino acid synthesis; P is involved in energy metabolism and structural integrity; and K supports enzyme activity, osmotic regulation, and stress tolerance [14,15].

Despite these well-known roles, relatively little research has been performed to explore how varying the balance and concentration of N, P, and K specifically affects both the yield and the feed quality of HWS. Most previous studies have looked either at biomass alone or at the effect of individual nutrients, rather than considering their interactions [16,17]. In practical animal nutrition, the value of a feed source is determined not solely by its biomass yield, but also by its internal composition, specifically the concentrations of crude protein, fiber, minerals, and its overall digestibility. Also, future studies will include detailed mineral and ionic composition analysis to better assess feed quality.

An additional complexity arises from the fact that the augmentation of a single nutrient can result in trade-offs, potentially enhancing certain plant traits while adversely affecting others. For instance, adding more nitrogen might boost protein content, but could also suppress lipid accumulation. Similarly, an excess of potassium can raise fiber content, which might hurt digestibility if it goes too far [18,19]. These interactions mean that simply maximizing one nutrient will not necessarily produce the best overall feed.

Considering this, the current study was undertaken to evaluate how different fertilization regimes, applying nitrogen, phosphorus, and potassium alone and in combination, influence the biomass production and biochemical composition of HWS. It was hypothesized that both the type and concentration of macronutrient fertilizers would have a significant impact on the quantitative yield and composite quality of HWS. Furthermore, the study investigated whether the relationships between nutrient concentrations and plant responses exhibited linear or non-linear trends, with the objective of identifying threshold levels beyond which additional fertilizer inputs fail to yield significant benefits and may even negatively impact feed quality.

The main hypothesis of this study was that both the type and concentration of macronutrient fertilizers would have a significant impact on the quantitative yield and composite quality of HWS. It was also aimed to identify and demonstrate that targeted nutrient modulation can be used to direct plant development towards desirable traits, such as higher protein, fiber content, or increased yield. These traits are particularly important in precision agriculture and livestock systems, which require both efficiency and nutritional adequacy. By addressing these questions, this research seeks to support the development of nutritionally optimized hydroponic feed systems that align with both economic and environmental goals in modern livestock production.

2. Materials and Methods

The studies were conducted in 2023–2025 at the Agricultural Machines Technological Process Research Laboratory of Vytautas Magnus University Agriculture Academy, Lithuania.

2.1. Hydroponic System Setup

Scientific research was carried out using specialized equipment adapted to hydroponic cultivation and scientific research—HWS cultivation stand HAS1 (Figure 1).

(a) (b)

Figure 1. (**a**) HWS cultivation stand HAS1 and (**b**) HWS during cultivation.

Its structure (2150 × 1400 × 1400 mm) was constructed from extruded aluminum profiles. The interior of the bench houses a unit for HWS cultivation and measurement, featuring four growth trays (Figure 2), each sized 1000 × 225 × 70 mm. Each tray offers a growing space of 0.9 m^2. The nutrient film technique (NFT) was chosen as the nutrient solution supply system. The core mechanism of the NFT involves technologically inclining the growth trays at a predetermined angle, denoted as α, to facilitate the delivery of the nutrient solution to the plants. This configuration leverages the disparity in potential energy to induce a gravitational flow of the nutrient solution across the plant roots arrayed along the inclined plane. This ensures an efficient distribution of humidity, oxygen, and nutrients essential for plant growth. Subsequently, the solution flows to the lowest part of the shelf, where it is either directed towards waste management systems or subjected to purification and recycling processes for further use. In this empirical study, the nutrient solution that passed through the roots was discharged into the sewer and not reused to reduce as much as possible the growth of various pathogens and the imbalance of nutrients. This hydroponic cultivation methodology is extensively utilized for a wide spectrum of agricultural crops, demonstrating pronounced effectiveness for species necessitating shorter growth periods, such as lettuce and various leafy vegetables [20]. The NFT system is flexible and beneficial for the environment, making it great for innovative farming methods. It helps save water and uses nutrients wisely, promoting water conservation and nutrient efficiency.

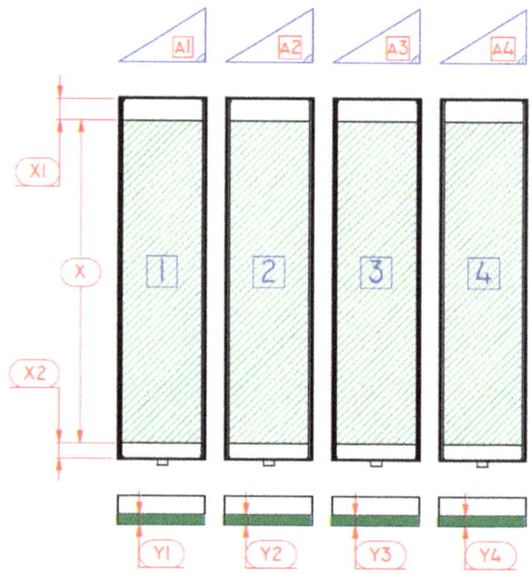

Figure 2. Growth trays and seed spreading parameters in the HAS1 stand. A1–A4—slope angle α (%), Y1–Y4—thickness of spread seeds (mm), X—length of spread seeds (mm), X1 and X2—distance from the edges of the growth trays (mm).

The slope angle α of the HAS1 stand growth trays was set at 6.5% [21]. Other parameters indicated in Figure 2 were chosen as follows: the thickness of the spread seeds Y1–Y4—30 mm, the length X—825 mm, and the distances from the edges of the growth trays X1 and X2—85 mm each (Figure 2).

2.2. *Hydroponic Wheat Sprouts Growth Parameters*

To accurately assess the effects of different fertilization strategies on HWS, it is essential to establish a controlled and standardized growth environment. The following section outlines the key cultivation parameters maintained during the experiment, ensuring reproducibility and minimizing external variability that could influence biomass yield or biochemical composition.

2.2.1. Microclimate

The temperature throughout the growth cycle was dependent on the temperature of the laboratory room containing the HAS1 stand. The relative humidity of the environment in the growth chamber was controlled (<60%) with an installed air extraction and intake system and monitored. The concentration of CO_2 in the HAS1 stand was continuously monitored throughout the growth cycle to observe plant respiratory activity. No external CO_2 supplementation was applied, and ambient levels were maintained without artificial control. This allowed the study to capture natural variations in CO_2 concentration resulting from plant metabolism. All microclimatic parameters are presented in Table 1 that presents the average environmental parameters recorded during the growth cycle for each treatment, including temperature, relative humidity, CO_2 concentration, and vapor pressure deficit (VPD), all of which remained within controlled ranges, with slight variations due to treatment-specific irrigation activity and plant transpiration rates.

Table 1. Dynamic parameters of empirical research.

	Temperature, °C	Humidity, %	CO_2, ppm	VPD *
Control	21.87 ± 0.05	37.88 ± 0.26	591.58 ± 2.26	1.616 ± 0.008
CN1	21.69 ± 0.05	47.56 ± 0.24	605.96 ± 2.54	1.349 ± 0.008
CN2	21.26 ± 0.05	44.2 ± 0.21	601.91 ± 2.68	1.396 ± 0.007
CN3	20.95 ± 0.05	34.84 ± 0.38	616.35 ± 3.02	1.602 ± 0.011
CP1	20.49 ± 0.05	45.73 ± 0.59	609.26 ± 2.78	1.31 ± 0.005
CP2	21.18 ± 0.05	48.78 ± 0.52	642.97 ± 2.77	1.29 ± 0.008
CP3	20.67 ± 0.05	49.78 ± 0.26	636.68 ± 3.78	1.23 ± 0.006
CK1	21.13 ± 0.05	41.82 ± 0.35	589.69 ± 3.01	1.46 ± 0.006
CK2	21.14 ± 0.05	47.62 ± 0.62	630.42 ± 2.12	1.31 ± 0.007
CK3	21.47 ± 0.05	42.62 ± 0.57	633.76 ± 3.59	1.47 ± 0.008
NPK1	20.57 ± 0.05	56.96 ± 0.28	624.91 ± 3.16	1.04 ± 0.008
NPK2	20.46 ± 0.05	40.76 ± 0.71	648.26 ± 3.43	1.43 ± 0.008
NPK3	21.18 ± 0.05	43.86 ± 0.42	591.43 ± 2.23	1.41 ± 0.005

* Vapor pressure deficit.

2.2.2. Lighting, Photoperiod, and Irrigation Strategy

The HWS growing stand HAS1 is equipped with a smart light emitting diode (LED) plant lighting system, which can create up to 1500 (μmol m^{-2}) s^{-1} of photosynthetically active radiation flux density (PAR), which can be adjusted, as well as changing the light spectrum and duration (photoperiod). The intensity and color spectrum of the light could be modified to accommodate the specific needs of a plant variety, its stage of development, and desired growth outcomes. LEDs offer notable benefits over conventional plant growth lighting options, such as fluorescent lights, including higher efficiency in turning electricity into usable light and emitting less heat [22]. Since this study did not analyze various recipes of the light spectrum and, in general, the influence of light on the cultivation of HWS, a full spectrum (380–780 nm) light was chosen. Regarding the PAR and lighting duration, a light flux of 250 (μmol m^{-2}) s^{-1} and a 12 h photoperiod were selected (from 8:00 a.m. till 8:00 p.m.).

Regarding the irrigation of HWS, a flow rate of 2 L min^{-1} of the nutrient solution was chosen so that the HWS and seeds receive a sufficient amount of water, nutrients, and oxygen, but at the same time, due to the high flow rate, especially in the first days of growth, the seeds would not be washed to the lowest part of the growth tray. Since the supply of nutrient solution was in all cases 8 times day^{-1} and irrigated for 60 s, the total amount of water per growth tray day^{-1} was 16 L.

2.2.3. Preparation of Seeds and Cultivation of Hydroponic Wheat Sprouts

Spring wheat (*Triticum aestivum* L.) seeds of the Scirocco variety were used for an empirical study. Prior to testing, seed germination was determined using the moistened tissue method [23], and it was 98.1 ± 2.2%. Before starting each growth period, it is important to clean the seeds and the HAS1 stand to reduce the spread of diseases as much as possible. Sodium hypochlorite (NaOCl) solution was used to disinfect both the seeds and the surfaces of the HAS1 system. For seed disinfection, a 2% NaOCl solution was used, with seeds soaked for 10 min. For surface disinfection of the HAS1 system, a stronger 20%

NaOCl solution was applied with a shortened exposure time of 5 min. After disinfection, the seeds and equipment surfaces were thoroughly rinsed with clean water to remove all the disinfectants. Disinfection of the seeds was followed by their soaking phase. During the seed soaking process, the seeds were immersed in tap water for a period of 24 h, allowing them to absorb moisture and begin the germination process. This procedure activates enzymes in the seeds that break down complex nutrients into simpler forms that are more easily absorbed by plants. In addition, soaking helps to soften the seed coat, promoting faster and more uniform germination. It also helps identify which seeds are likely to germinate because healthy seeds sink in water, while bad ones tend to float. Soaking ensures all seeds begin sprouting at the same time, leading to even growth.

After disinfection and soaking, the wheat seeds were sown manually onto each growth tray. A total of 1.55 kg of soaked seeds, equivalent to 1.00 kg of dry seeds, was used per tray. This corresponds to a seeding density of approximately 1.72 kg/m^2 (dry seed basis), given each tray's growing area of 0.9 m^2. The soaked seed mass was determined based on pre-trial water absorption tests and standardized to ensure consistent germination and sprouting. The seed layer thickness was approximately 30 mm, as defined by the cultivation protocol. This seeding rate allowed uniform distribution and optimal conditions for hydroponic sprout development.

2.2.4. Nutrient Solution

To evaluate the influence of different fertilizer types and concentrations on the quantitative and qualitative characteristics of HWS, a series of nutrient solutions were prepared using four common fertilizers: calcium nitrate ($Ca(NO_3)_2$), potassium phosphate (KH_2PO_4), potassium sulfate (K_2SO_4), and a balanced NPK 20-20-20 compound fertilizer. These compounds were selected based on their widespread use in hydroponic nutrient formulations, including Hoagland-type solutions, to isolate and compare the effects of nitrogen, phosphorus, potassium, calcium, and sulfur on plant development. Each fertilizer was tested at three concentration levels, coded as CN1–CN3, CP1–CP3, CK1–CK3, and NPK1–NPK3, respectively. These concentrations were selected based on published ranges used in hydroponic fodder cultivation and our preliminary optimization trials.

The nutrient solutions were prepared using 800 L of tap water per growth cycle. Prior to fertilizer addition, the physicochemical properties of the water were assessed to ensure compatibility with hydroponic systems. Tap water had a pH range of 7.3–7.7 and an electrical conductivity of 0.643 ± 0.015 μS/cm (Table 2). Key ions included chloride (16–18 mg/L), sulfate (24.9 mg/L), and sodium (11.0 mg/L), while nitrate and ammonium levels were below quantification limits. Although deionized water is recommended for controlled nutrition studies, untreated tap water was used to reflect practical, real-world hydroponic conditions. The low baseline ion concentrations were accounted for in the treatment design. These values complied with EU drinking water standards and did not necessitate pre-treatment. However, to optimize nutrient availability in the hydroponic environment, the pH of the water was adjusted from its natural range to 5.8–6.0 using nitric acid before the preparation of nutrient solutions. Fresh nutrient solutions were prepared prior to each growth cycle to prevent nutrient degradation or cross-contamination between treatments.

The calcium nitrate treatments (CN1–CN3) supplied nitrogen in the form of nitrate (NO_3^-) alongside calcium (Ca^{2+}), with concentrations ranging from 100 to 300 mg L^{-1} of nitrogen and 143 to 429 mg L^{-1} of calcium. The potassium phosphate treatments (CP1–CP3) delivered phosphorus ($H_2PO_4^-$) and potassium (K^+) in concentrations ranging from 100 to 300 mg L^{-1} of phosphorus, corresponding to 126 to 378 mg L^{-1} of potassium. Potassium sulfate treatments (CK1–CK3) provided potassium and sulfur (SO_4^{2-}), with potassium ranging from 100 to 300 mg L^{-1} and sulfur from 18 to 54 mg L^{-1}. Lastly, the NPK compound fertilizer treatments (NPK1–NPK3) included a mixture of nitrogen forms (NO_3^-, NH_4^+, and NH_2), phosphorus, and potassium, at respective concentrations of 100 to 300 mg L^{-1} nitrogen, 43.6 to 130.8 mg L^{-1} phosphorus, and 83 to 249 mg L^{-1} potassium. The pH of the nutrient solution was adjusted and maintained between 5.8 and 6.2 throughout the growing cycle to ensure optimal nutrient availability.

Table 2. Nutrient concentrations in solution and physicochemical properties of fertilizer treatments.

		Nutrients Concentration in Solution, mg L^{-1}			EC	pH
	Control *	–	–		0.643 ± 0.015	5.8–6.0
$Ca(NO_3)^2$		N (NO_{3^-})	Ca^{2+}			
	CN1	100	143		1.316 ± 0.061	5.8–6.0
	CN2	200	286		2.070 ± 0.220	5.8–6.0
	CN3	300	429		2.928 ± 0.086	5.8–6.0
KH_2PO_4		P ($H_2PO_{4^-}$)	K^+			
	CP1	100	126		0.782 ± 0.029	5.8–6.0
	CP2	200	252		0.913 ± 0.059	5.8–6.0
	CP3	300	378		1.045 ± 0.088	5.8–6.0
K_2SO_4		K^+	S ($SO_{4^{2-}}$)			
	CK1	100	18		0.940 ± 0.060	5.8–6.0
	CK2	200	36		1.228 ± 0.120	5.8–6.0
	CK3	300	54		1.518 ± 0.175	5.8–6.0
NPK 20-20-20		N (NO_{3^-}, NH_{4^+}, NH_2)	P ($H_2PO_{4^-}$)	K^+		
	NPK1	100	43.6	83	1.600 ± 0.124	5.8–6.0
	NPK2	200	87.2	166	2.551 ± 0.248	5.8–6.0
	NPK3	300	130.8	249	3.489 ± 0.371	5.8–6.0

* Tap water.

2.2.5. Investigated Quantitative and Qualitative Parameters

Quantitative and qualitative parameters were observed and recorded in this study. As for the quantitative parameters, the main ones were the yield potential of germinated seeds DP (kg kg^{-1}) (i.e., the yield from one kilogram of dry seeds) and the proportion of dry matter (LST ISO 712:2010 [24]) in the feed (%) after a 7-day growth cycle. The DP was recorded using a Flintec PB7.5 (C3 accuracy (0.023%) class) load cells (Flintec Inc., Hudson, MA, USA). The sensor's signals were amplified using Laumas TLB485 digital-analog weight transmitter (24-bit analog to digital converter, 4.8 kHz) (Laumas, Parma, Italy).

The following criteria were used to evaluate the qualitative parameters of HWS: crude protein (%)—ISO 20483:2006 [25], crude fiber (%)—ISO 6865:2000 [26], ash (%)—ISO 5984:2022(E) [27], ether extract content (%)—AOAC 920.39-1920/LST ISO 1443 [28].

2.3. Statistical Analysis

Each treatment was conducted with four replicates (n = 4) in a completely randomized design (CRD), with tray positions randomly assigned within the growth chamber to minimize environmental variability, and the duration of each growth cycle was 7 days. Samples were collected at the end of each cycle between 8:00 and 9:00 a.m. to minimize diurnal variation. Data was analyzed using analysis of variance (ANOVA) with Statistica 10.0 statistical software. The post hoc test for the significant difference ($R_{0.05}$) was used to compare the arithmetic means of the data. $R_{0.05}$ was calculated with a 95% level of confidence [29].

3. Results and Discussion

The results revealed that nutrient composition, not just concentration, was the main determinant of productivity and compositional outcomes. High levels of nitrogen or calcium did not necessarily improve biomass yield, and in some cases led to physiological stress or suboptimal resource allocation. These findings highlight the importance of balanced fertilization strategies and highlight the risks of overly simplistic nutrient supplementation, especially in closed-loop hydroponic systems where the system's ability to resist changes in nutrient concentration, pH, or ionic balance when external inputs are added is limited.

Although HWS cultivation has been widely explored in recent years, few studies have simultaneously evaluated such a broad spectrum of physiological and biochemical parameters under controlled conditions.

3.1. Biomass Yield and Dry Matter Content Response to Fertilization

After extensive experimental studies, it became evident that the nutrient composition played a pivotal role in shaping both biomass yield and dry matter content. The control treatment (tap water), which received no supplemental macronutrients, produced 5.18 ± 0.22 kg kg^{-1} DP, with a DM content of 13.30%. The relatively high biomass yield in the control group can be explained using endogenous seed reserves, which support initial sprout development even in the absence of external macronutrient supplementation. This baseline allowed for comparative evaluation of the various fertilization regimes.

Among the tested treatments, calcium nitrate application revealed a distinct non-linear effect on growth. While the lowest concentration (CN1) slightly enhanced biomass production, higher concentrations (CN2 and CN3) led to significant reductions, accompanied by elevated DM content (Figure 3). This pattern suggests that moderate nitrate levels support vegetative expansion [30], but excessive concentrations may impose osmotic or oxidative stress, potentially due to nitrate accumulation and the ATP-intensive demands of nitrate reduction [31]. Importantly, this growth reduction is more plausibly attributed to nitrate-specific stress—including osmotic load and energy demand—rather than to a general nutrient deficiency. In this short-cycle hydroponic system, the seedlings rely primarily on internal seed reserves, and the solution was purposefully unbalanced to isolate the effects of individual macronutrients. Thus, the observed effects reflect excess nitrate, not the absence of other nutrients, and demonstrate the limits of applying Liebig's Law in such early growth contexts. In addition, increased calcium availability may inhibit cell elongation by reinforcing pectin cross-linking within the cell wall matrix [32], thereby limiting volumetric growth despite higher tissue density.

In contrast, treatments with potassium phosphate (CP1–CP3) consistently promoted biomass accumulation across all tested concentrations, while maintaining relatively stable DM percentages (Figure 3). This uniform increase implies that biomass gains stemmed from effective tissue accumulation rather than water-driven expansion [33]. The synergistic effects of phosphorus, essential for ATP synthesis and nucleic acid formation, osmotic regulation, and enzymatic activation, led to sustainable and efficient biomass growth [34].

Potassium sulfate (CK1–CK3) treatments presented a more moderate response. Biomass increased by up to 8.3% under CK2, with minimal change in DM%, indicating a subtler but positive metabolic enhancement. Sulfur, incorporated into essential amino acids such as methionine and cysteine, contributes to protein synthesis and cellular function, while potassium continues to support energy-efficient carbon assimilation [35]. Although less dramatic than NPK or phosphate treatments, these findings highlight how potassium and sulfur can fine-tune metabolic efficiency, especially under specific environmental or developmental conditions.

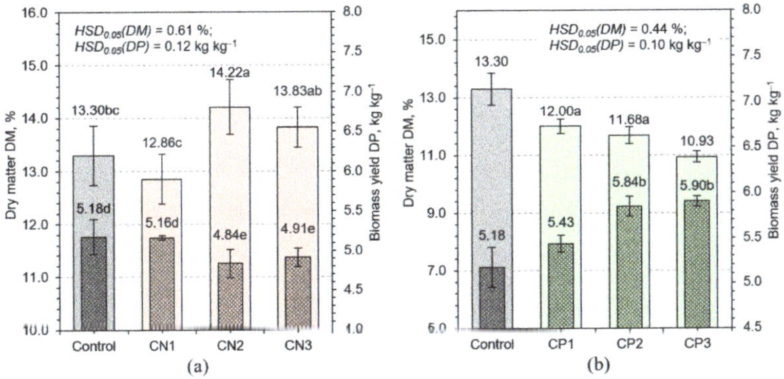

Figure 3. Effect of different calcium nitrate (**a**) and potassium phosphate (**b**) concentrations on biomass yield DP and dry matter content DM of HWS. Any two samples with a common letter are not significantly different ($p < 0.05$), as assessed using the least significant difference.

Notably, the most pronounced increases in DP were recorded under NPK fertilization. NPK2 and NPK3 treatments enhanced biomass by 16.0% and 19.7%, respectively, with only modest increases in DM% (Figure 4). These results underscore the critical role of nutrient synergy, where simultaneous provision of nitrogen, phosphorus, and potassium supports both expansive growth and compositional integrity. The observed results likely reflect coordinated physiological processes such as increased nutrient uptake, hormonal balance, and synchronized cell division and elongation, culminating in high productivity without the typical trade-offs associated with dilution effects.

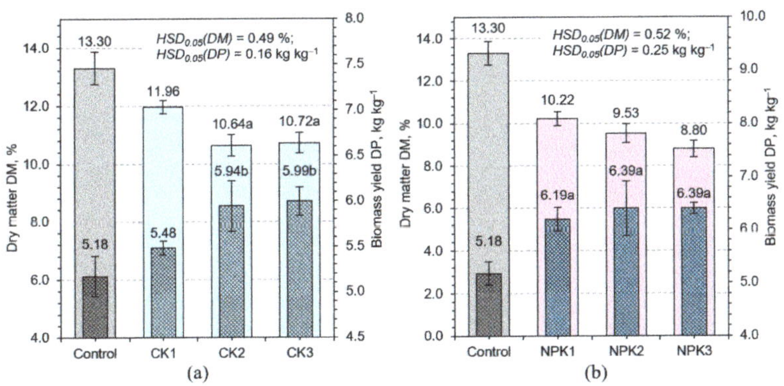

Figure 4. Effect of different potassium sulfate (**a**) and NPK 20–20–20 (**b**) concentrations on biomass yield, DP, and dry matter content, DM, of HWS. Any two samples with a common letter are not significantly different ($p < 0.05$), as assessed using the least significant difference.

Regression analysis further highlighted these dynamics. A strong inverse correlation between biomass yield and DM% (R^2 = 0.86) was observed (Figure 5), aligning with the well-established dilution effect in rapidly expanding tissues. NPK treatments deviated from this trend, suggesting that balanced nutrition can decouple this relationship to some extent, enabling robust growth without compromising tissue density. Treatments that induced physiological stress, such as CN3, exhibited reduced biomass yet elevated DM%, likely reflecting increased osmolyte or structural carbohydrate accumulation (Figure 3).

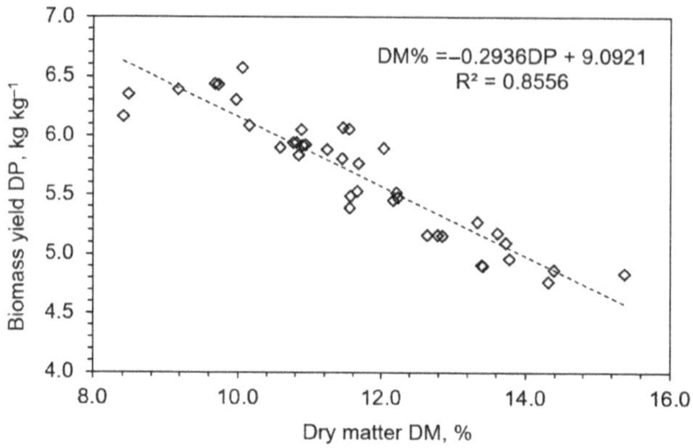

Figure 5. Linear regression between dry matter content (DM, %) and biomass yield (DP, kg kg^{-1}) across all fertilizer treatments.

These physiological insights hold practical significance for feed production systems. High-yield scenarios appear best supported by balanced fertilization strategies, particularly those incorporating complete NPK blends (e.g., NPK3). However, treatments such as CP2 and CK2 may offer strategic advantages when the production goal shifts toward improving nutrient density or digestibility per unit mass. Thus, the choice of fertilization regime can be tailored to align with distinct agronomic or nutritional objectives. Importantly, the observed trade-offs reinforce the value of adaptive nutrient management practices that integrate ongoing compositional monitoring to optimize both productivity and feed quality.

3.2. *Effect of Fertilization on Feed Quality Parameters*

Beyond DP (kg kg^{-1}) and DM (%), the nutritional value of HWS is defined by its compositional profile. This empirical study focused on four primary feed quality metrics: crude protein (CP), ether extract (EE), crude fiber (CF), and ash. Each of these parameters responds differently to nutrient availability, reflecting distinct metabolic and developmental pathways. Understanding their response to fertilization is crucial for designing nutrient strategies that optimize both growth and feed quality.

3.2.1. Crude Protein and Ether Extract

Crude protein is a key metric for evaluating the nutritional quality of HWS, as it reflects the plant's capacity to assimilate nitrogen and synthesize amino acids and structural proteins [36]. In the control group, CP content was 16.33%, which corresponds to baseline levels observed in unfertilized HWS (Figure 6). The application of calcium nitrate significantly increased CP, peaking at 19.65% in CN3. However, this rise was accompanied by a reduction in biomass, indicating a negative trade-off between nitrogen loading and overall growth. This phenomenon is likely a consequence of metabolic imbalance: excessive nitrate can induce energy-demanding reduction processes, resulting in an accumulation of unused amino acids, increased oxidative stress, and reduced allocation of resources to structural expansion [37].

Regression analysis revealed a curvilinear relationship between CP and biomass yield (DP), with CP initially increasing alongside biomass, followed by a plateau and eventual decline at the highest yield levels (Figure 7). This pattern confirms the classical 'dilution effect', where the protein concentration decreases due to disproportionate accumulation of water and carbohydrates in rapidly expanding tissues [38]. The mechanism underlying this effect involves carbon–nitrogen metabolic uncoupling under accelerated growth, where structural and storage proteins cannot be synthesized at a rate sufficient to maintain proportional increases in CP.

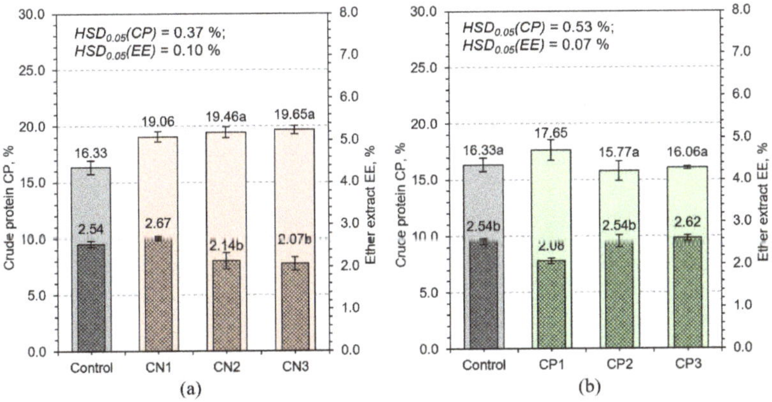

Figure 6. Effects of different calcium nitrate (**a**) and potassium phosphate (**b**) concentrations on crude protein CP and ether extract EE content of HWS. Any two samples with a common letter are not significantly different ($p < 0.05$), as assessed using the least significant difference.

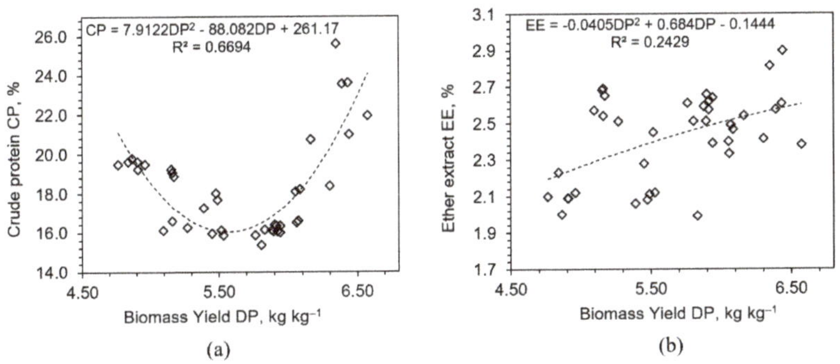

Figure 7. Relationships between biomass yield (DP, kg kg^{-1}) and (**a**) crude protein (CP, %) and (**b**) ether extract (EE, %) contents across all fertilizer treatments.

NPK3 exhibited the highest CP content (24.26%) while also supporting maximal biomass production, illustrating that nutrient synergy is essential for the simultaneous optimization of yield and compositional quality (Figure 8). Phosphorus enhances energy metabolism via ATP synthesis, enabling efficient protein assembly and nucleic acid replication [39]. Potassium plays a pivotal role in maintaining turgor pressure, enzyme activation, and regulation of pH and ion gradients, all of which support sustained metabolic throughput and nitrogen assimilation [40].

The intermediate variants CP2 and NPK2 showed high CP levels with moderate biomass, resulting in dense, protein-rich tissues (Figures 6 and 8). These treatments present a valuable compromise for systems where feed quality is prioritized over sheer yield. From a nutritional standpoint, such profiles may improve feed conversion efficiency in livestock by providing concentrated protein per unit of dry matter. These results suggest that CP should be interpreted in context, considering mass and density, rather than as a stand-alone metric. Further physiological insights can be drawn by considering root architecture and nutrient uptake dynamics. It is likely that balanced fertilization not only improved internal biochemical processes but also supported root surface area development and ion transporter activity. Such mechanisms would further enhance nitrogen uptake and assimilation efficiency, explaining the superior CP outcomes in NPK treatments (Figure 8).

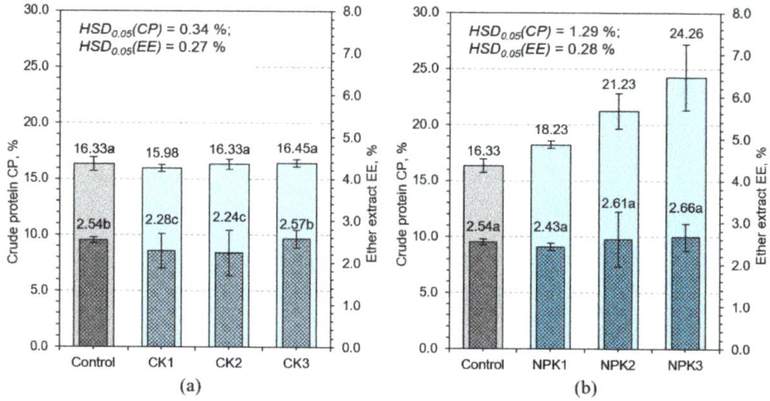

Figure 8. Effect of different potassium sulfate (**a**) and NPK 20–20–20 (**b**) concentrations on crude protein CP and ether extract EE content of HWS. Any two samples with a common letter are not significantly different ($p < 0.05$), as assessed using the least significant difference.

Ether extract, representing lipid content, remained relatively stable across treatments (2.07–2.67%), and this lack of variability suggests that lipid metabolism in early vegetative stages of wheat is tightly regulated and not easily influenced by external nutrient availability. Cereals typically synthesize lipids during reproductive phases [41]; therefore, the early-stage hydroponic environment may not provide the developmental triggers necessary for lipid accumulation. This hypothesis is supported by weak correlations between EE and both biomass and dry matter content ($R^2 < 0.25$). Despite the low responsiveness, EE remains an important metric for energy value in feed. Its stability ensures a predictable contribution to feed energy balance, which is beneficial for ration formulation. However, any attempts to enhance lipid content through macronutrient fertilization are unlikely to succeed without modifying environmental factors, growth duration, or introducing genetic traits associated with higher baseline lipid production. Strategies such as mild salinity, altered photoperiods, or cultivar selection may offer alternative routes for manipulating EE content.

In conclusion, CP responded dynamically to nutrient composition and concentration, confirming its role as a sensitive and integrative indicator of physiological status under hydroponic conditions. On the other hand, EE remained almost unchanged despite the different fertilizers applied. This indicates that the amount of lipids is biologically inert; it remains stable, as it depends largely on the growth phase and genetics, and not on external fertilizers or environmental conditions. Together, these metrics highlight the importance of understanding both flexible and stable traits in feed optimization.

3.2.2. Crude Fiber and Crude Ash

Crude fiber content is indicative of plant tissue maturity and structural development. It encompasses cellulose, hemicellulose, and lignin—compounds that form the primary and secondary cell walls [42]. In the control group, CF was 1.91%, reflecting minimal cell wall thickening typical of early-stage wheat (Figure 9). Fertilization has induced significant increases in CF across all treatments, with marked differences depending on nutrient type. Calcium nitrate treatments resulted in modest CF increases (e.g., CN3–2.53%), reflecting nitrogen's role in supporting cell elongation but not secondary wall deposition. Nitrogen promotes protein synthesis and cellular expansion, but without adequate phosphorus and potassium, the synthesis of wall polymers remains constrained [43]. This limitation results in tissue that expands volumetrically but lacks commensurate structural reinforcement.

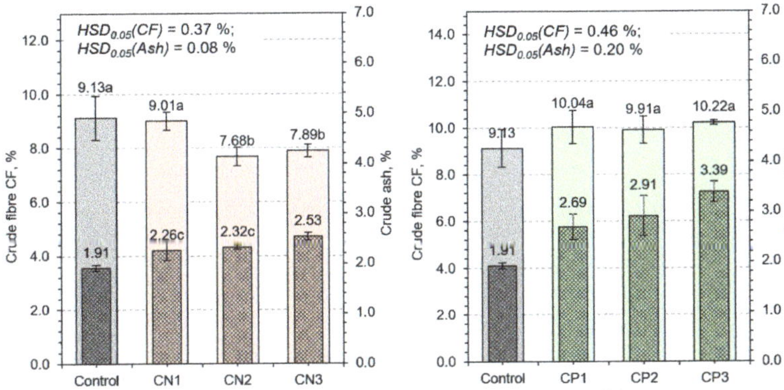

(a) (b)

Figure 9. Effect of different calcium nitrate (**a**) and potassium phosphate (**b**) concentrations on crude fiber CF and ash content of HWS. Any two samples with a common letter are not significantly different ($p < 0.05$), as assessed using the least significant difference.

Treatments with elevated phosphorus concentrations, particularly CP3 (3.39%), demonstrated enhanced CF accumulation (Figure 9). Phosphorus is crucial for ATP production, which drives biosynthesis of complex carbohydrates such as hemicellulose and pectin [44]. Additionally, phosphorus supports cell division in meristematic zones, thereby promoting tissue differentiation and wall maturation [45].

Potassium sulfate further increased the fiber content (CK3–4.05%), highlighting potassium's regulatory role in carbohydrate partitioning, enzymatic activity, and osmotic balance [46] (Figure 10). The contribution of sulfur may also be significant, as it is involved in the biosynthesis of structural amino acids like cysteine and methionine, which play roles in lignification and cross-linking of wall proteins.

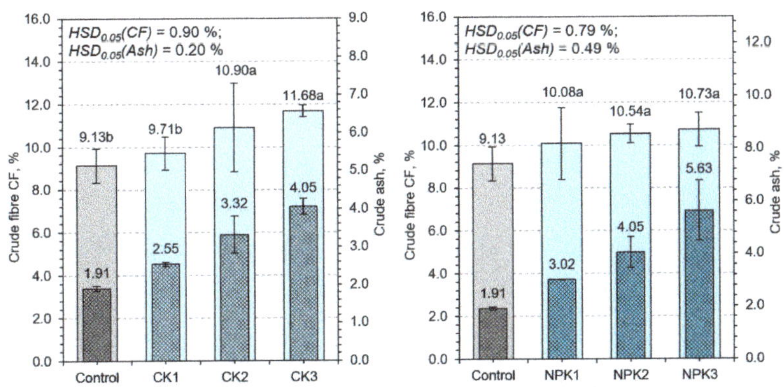

(a) (b)

Figure 10. Effect of different potassium sulfate (**a**) and NPK 20–20–20 (**b**) concentrations on crude fiber CF and ash content of HWS. Any two samples with a common letter are not significantly different ($p < 0.05$), as assessed using the least significant difference.

The highest CF value was observed in NPK3 (5.63%), clearly demonstrating the synergistic action of all three macronutrients. This result indicates that only under balanced nutrient supply can the plant simultaneously expand and fortify its tissues. Regression analysis confirmed a strong positive relationship between CF and DM% ($R^2 = 0.81$), suggesting that fiber content serves as a reliable proxy for physiological maturity in HWS.

Ash content, representing total inorganic residue post-combustion, is another critical indicator of nutritional value, reflecting mineral accumulation within plant tissues. In the control group, the ash content was 9.13%. CN treatments reduced ash levels (e.g., CN3 = 7.68%), potentially due to ionic imbalance or competitive inhibition of mineral uptake by excess nitrate. In contrast, phosphorus and potassium treatments elevated ash content to over 10%, with the highest values in NPK3 (16.0%). The increase in ash content under NPK and CP treatments likely reflects improved root function, transporter activity, and membrane permeability, all of which enhance mineral uptake. Potassium facilitates the uptake of calcium, magnesium, and micronutrients by stabilizing membrane potential and activating proton pumps. These physiological improvements result in not only greater mineral content but also improved distribution within tissues.

Regression between ash and DM% revealed a non-linear trend ($R^2 = 0.71$), with mineral accumulation plateauing at high DM values (Figure 11). This suggests a physiological ceiling beyond which further mineral uptake is constrained by internal regulatory mechanisms, possibly related to vacuolar storage capacity [47], phloem transport limits [48], or ion toxicity thresholds. Moreover, the concurrent increase in both CF and ash suggests coordinated metabolic prioritization during tissue maturation. As the plant shifts from rapid expansion to strengthening and nutrient loading, structural and mineral constituents are deposited simultaneously. This coordination enhances both mechanical stability and nutritional density of the feed, making such biomass more suitable for animals requiring mineral-rich diets.

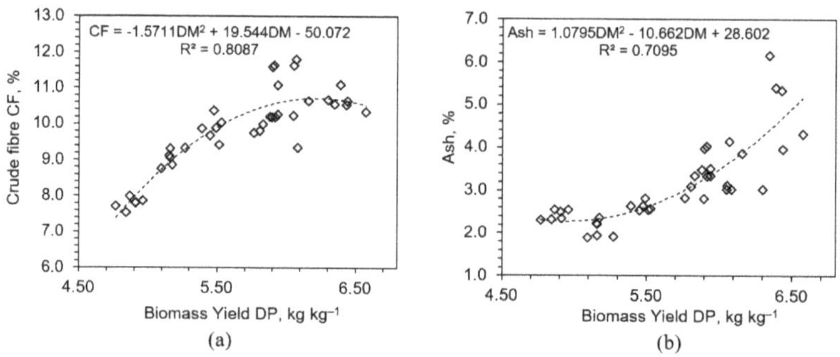

Figure 11. Relationships between biomass yield (DP, kg kg^{-1}) and (**a**) crude fiber (CF, %) and (**b**) crude ash (%) contents across all fertilizer treatments.

In summary, the CF and ash content are both highly responsive to fertilization, particularly under balanced nutrient regimes. They offer complementary insights into the structural and mineral maturation of HWS. Their correlations with DM% and distinct patterns across treatments underscore their value as diagnostic indicators for both physiological assessment and quality control in precision agriculture systems.

3.3. Practical Implications for Fertilization Strategy

The comprehensive analysis of biomass yield, dry matter accumulation, and feed quality parameters under varying fertilization regimes reveals insights for optimizing HWS production. The findings indicate that no single macronutrient guarantees both high biomass and nutritional quality. Instead, optimal outcomes are achieved through the combination and concentration of nitrogen, phosphorus, and potassium.

For producers aiming to maximize biomass yield, the use of balanced NPK fertilization, particularly at moderate to high concentrations (e.g., NPK2 and NPK3) is recommended. In contrast, if the production goal is to enhance nutrient density, especially crude protein or mineral content, phosphorus-dominant regimes (e.g., CP2 or CP3) offer a valuable alternative. These variants achieved significant improvements in CP, CF, and ash content while avoiding the dilution effect commonly seen in high-yield systems. Also, the low responsiveness of the ether extract to fertilization implies that strategies aimed at increasing lipid content should involve alternative approaches, such as cultivar selection, modification of growth duration, or controlled stress induction, rather than nutrient manipulation. Importantly, the strong regression patterns observed between dry matter and both CF and ash suggest that simple indirect measurements (e.g., DM%) can serve as reliable indicators of HWS quality traits. This opens the possibility for streamlined monitoring protocols in commercial systems.

It is worth noting that the biochemical composition of the hydroponic wheat sprouts obtained in this study suggests specific applications for different livestock categories. From a nutritional standpoint, the enhanced crude protein (up to 24.26%) and ash content (up to 16.0%) observed under NPK3 treatment suggest that hydroponic wheat sprouts can serve as a protein- and mineral-rich supplement in ruminant diets, potentially reducing the need for additional concentrates or mineral premixes. Moreover, the relatively low crude fiber content ensures good digestibility, making the sprouts suitable not only for cattle and sheep but also potentially for monogastric animals when included in balanced rations. Compared to conventional forages like alfalfa (~18–20% CP) or maize silage (~8–10% CP), the protein concentration in HWS under optimized fertilization is highly competitive. These compositional traits highlight the potential of HWS as a high-value, space- and resource-efficient feed component in precision livestock nutrition systems.

3.4. Limitations of the Study

While the study presents robust evidence on the influence of macronutrient composition on biomass and feed quality traits in HWS, certain limitations should be acknowledged. Firstly, the short cultivation period (7 days) may have constrained the full expression of slower-developing traits, particularly those related to lipid accumulation and secondary metabolite pathways. Longer growth durations might reveal additional nutrient effects or temporal shifts in compositional traits such as ether extract or complex structural carbohydrates.

Secondly, the ash content was assessed as a total inorganic residue, without elemental differentiation. As such, the specific contributions of individual macro- and micronutrients (e.g., calcium, magnesium, iron, and zinc) remain unresolved. Future studies using inductively coupled plasma optical emission spectroscopy (ICP-OES) or similar elemental analysis techniques could clarify the precise mineral dynamics and potential antagonisms or synergies among elements.

Thirdly, this study was conducted under highly controlled conditions in a growth chamber, which, while eliminating environmental noise, limits the extrapolation of findings to commercial-scale systems that involve variable humidity, temperature, and light quality. Additionally, potential cultivar-specific responses were not addressed, as only one wheat genotype was tested. Genetic variation may influence nutrient responsiveness, and incorporating multiple cultivars would enhance the generalizability of findings. Despite these limitations, the study provides a solid foundation for further research and optimization in controlled environment HWS systems.

4. Conclusions

This study demonstrated that both the type and concentration of macronutrient fertilizers significantly influence the biomass yield and nutritional composition of HWS. The highest biomass yield (6.39 kg kg^{-1} DP) was achieved under NPK3 treatment, which also resulted in the highest crude protein (24.26%) and crude fiber (5.63%) content, indicating that balanced fertilization with nitrogen, phosphorus, and potassium promotes both productivity and feed quality. Excessive calcium nitrate (CN3) led to a reduced yield (4.84 kg kg^{-1} DP) despite elevated protein levels (19.65%), highlighting a trade-off between nitrogen-induced protein synthesis and overall vegetative growth. Potassium phosphate treatments, particularly CP2 and CP3, improved both protein (up to 17.65%) and ash content (10.22%) without causing yield suppression, suggesting these are effective strategies for enhancing nutrient density. Regression analyses revealed strong correlations between biomass and dry matter ($R^2 = 0.86$), and between dry matter and both fiber ($R^2 = 0.81$) and ash content ($R^2 = 0.71$), indicating these parameters are physiologically linked and can be used for predictive quality assessments. The relatively unchanged ether extract content (2.07–2.67%) across all treatments underscores its developmental regulation and limited responsiveness to macronutrient manipulation. Among all treatments, NPK3 demonstrated the best overall performance, achieving the highest biomass yield and superior nutritional quality, making it the most suitable fertilization regime for hydroponic wheat sprout production.

This study contributes a comprehensive perspective by integrating statistical correlations with mechanistic physiological interpretations, providing new insights into nutrient use efficiency and compositional optimization. It also demonstrates how regression-based modeling can help predict critical quality traits, enabling data-driven decisions in modern HWS production systems. Collectively, the findings underscore the importance of nutrient synergy and targeted fertilization strategies to optimize both yield and compositional quality in controlled environment HWS systems, providing practical guidance for sustainable feed production in precision agriculture.

Author Contributions: Conceptualization, A.G.; methodology, A.G. and D.S. (Dainius Savickas); validation, D.S. (Dainius Steponavičius), R.K. and T.K.; investigation, A.G. and D.S. (Dainius Savickas); writing—original draft preparation, A.G., I.B. and D.S. (Dainius Savickas); writing—review and editing, A.S.; visualization, A.G. and D.S. (Dainius Savickas); supervision, D.S. (Dainius Steponavičius). All authors have read and agreed to the published version of the manuscript.

Funding: This project has received funding from the Research Council of Lithuania (LMTLT), agreement No. S-ITP-24-10.

Data Availability Statement: The original contributions presented in this study are included in the article. Further inquiries can be directed to the corresponding author.

Conflicts of Interest: The research was not conducted with any commercial or financial relationships that could be construed as a potential conflict of interest.

References

1. Neto, C.; Catarino, A.; Sobreiro, J.; das Dores, J.; Patanita, M.; Tomaz, A.; Palma, P. Effect of Different Irrigated Crop Successions on Soil Carbon and Nitrogen–Phosphorus–Potassium Budget under Mediterranean Conditions. *Agriculture* **2024**, *14*, 1908. [CrossRef]
2. Rencricca, G.; Froldi, F.; Moschini, M.; Trevisan, M.; Ghnimi, S.; Lamastra, L. The Environmental Impact of Permanent Meadows-Based Farms: A Comparison among Different Dairy Farm Management Systems of an Italian Cheese. *Sustain. Prod. Consum.* **2023**, *37*, 53–64. [CrossRef]
3. Yang, S.; Mahmood, M.; Baral, R.; Wu, H.; Almloff, M.; Stanton, L.E.; Min, D.; Smiley, B.K.; Iiams, J.C.; Yu, J. Forage Conservation Is a Neglected Nitrous Oxide Source. *PNAS Nexus* **2024**, *3*, 373. [CrossRef]
4. FAO. *The State of Food and Agriculture 2021: Making Agrifood Systems More Resilient to Shocks and Stresses*; FAO: Rome, Italy, 2021.

5. Grigas, A.; Steponavičius, D.; Kemzūraitė, A.; Tarasevičienė, Ž.; Domeika, R. Spatial Heterogeneity in the Properties of Hydroponic Wheat Fodder and Its Sustainability. *Sci. Rep.* **2024**, *14*, 19312. [CrossRef]
6. Bihari, C.; Ahamad, S.; Kumar, M.; Kumar, A.; Kamboj, A.D.; Singh, S.; Srivastava, V.; Gautam, P. Innovative Soilless Culture Techniques for Horticultural Crops: A Comprehensive Review. *Int. J. Environ. Clim. Change* **2023**, *13*, 4071–4084. [CrossRef]
7. Dutta, M.; Gupta, D.; Sahu, S.; Limkar, S.; Singh, P.; Mishra, A.; Kumar, M.; Mutlu, R. Evaluation of Growth Responses of Lettuce and Energy Efficiency of the Substrate and Smart Hydroponics Cropping System. *Sensors* **2023**, *23*, 1875. [CrossRef]
8. Ahamed, M.S.; Sultan, M.; Shamshiri, R.R.; Rahman, M.M.; Aleem, M.; Balasundram, S.K. Present Status and Challenges of Fodder Production in Controlled Environments: A Review. *Smart Agric. Technol.* **2023**, *3*, 100080. [CrossRef]
9. Gwynn-Jones, D.; Dunne, H.; Donnison, I.; Robson, P.; Sanfratello, G.M.; Schlarb-Ridley, B.; Hughes, K.; Convey, P. Can the Optimisation of Pop-up Agriculture in Remote Communities Help. Feed. the World? *Glob. Food Sec.* **2018**, *18*, 35–43. [CrossRef]
10. Czarnek, K.; Tatarczak-Michalewska, M.; Dreher, P.; Rajput, V.D.; Wójcik, G.; Gierut-Kot, A.; Szopa, A.; Blicharska, E. UV-C Seed Surface Sterilization and Fe, Zn, Mg, Cr Biofortification of Wheat Sprouts as an Effective Strategy of Bioelement Supplementation. *Int. J. Mol. Sci.* **2023**, *24*, 10367. [CrossRef]
11. Zada, T.; Khan, S.; Islam, Z.; Sultan, A.; Khan, K.; Albekairi, T.H.; Khan, A.A.; Aziz, T.; Khan, R.U. Growth Performance, Serum Biochemical Profile and Liver Histopathology of Broiler Chickens Supplemented with Graded Level of Dried Sprouted Wheat Grains. *Ital. J. Anim. Sci.* **2024**, *23*, 880–886. [CrossRef]
12. Mpanza, T.D.E.; Dhlamini, T.C.; Pierneef, R.E.; Mbatha, K.R. Enteric Methane Emission, Rumen Fermentation and Microbial Profiles of Meat-Master Lambs Supplemented with Barley Fodder Sprouts. *Fermentation* **2022**, *8*, 434. [CrossRef]
13. Zang, Y.; Richards, A.T.; Seneviratne, N.; Gutierrez-Oviedo, F.A.; Harding, R.; Ranathunga, S.; McFadden, J.W. Replacing Conventional Concentrates with Sprouted Barley or Wheat: Effects on Lactational Performance, Nutrient Digestibility, and Milk Fatty Acid Profile in Dairy Cows. *J. Dairy Sci.* **2024**, *107*, 5529–5541. [CrossRef]
14. Tripathi, R.; Tewari, R.; Singh, K.P.; Keswani, C.; Minkina, T.; Srivastava, A.K.; De Corato, U.; Sansinenea, E. Plant Mineral Nutrition and Disease Resistance: A Significant Linkage for Sustainable Crop Protection. *Front. Plant Sci.* **2022**, *13*, 883970. [CrossRef]
15. Tariq, A.; Zeng, F.; Graciano, C.; Ullah, A.; Sadia, S.; Ahmed, Z.; Murtaza, G.; Ismoilov, K.; Zhang, Z. Regulation of Metabolites by Nutrients in Plants. In *Plant Ionomics: Sensing, Signaling, and Regulation*; Wiley: Hoboken, NJ, USA, 2023; pp. 1–18.
16. Li, T.; Lalk, G.T.; Arthur, J.D.; Johnson, M.H.; Bi, G. Shoot Production and Mineral Nutrients of Five Microgreens as Affected by Hydroponic Substrate Type and Post-Emergent Fertilization. *Horticulturae* **2021**, *7*, 129. [CrossRef]
17. Pastorelli, G.; Serra, V.; Turin, L.; Attard, E. Hydroponic Fodders for Livestock Production—A Review. *Ann. Anim. Sci.* **2024**, *24*, 645–656. [CrossRef]
18. Sriagtula, R.; Martaguri, I.; Sowmen, S. Evaluation of Nutrient Solution Dose and Harvest Time on Forage Sorghum (*Sorghum bicolor*, L. Moench) in Hydroponic Fodder System. In *IOP Conference Series: Earth and Environmental Science*; IOP Publishing: Bandung, Indonesia, 2021; Volume 888, p. 012068.
19. Zörb, C.; Senbayram, M.; Peiter, E. Potassium in Agriculture–Status and Perspectives. *J. Plant Physiol.* **2014**, *171*, 656–669. [CrossRef]
20. Mohammed, S.B.; Sookoo, R. Nutrient Film Technique for Commercial Production. *Agric. Sci. Res. J.* **2016**, *6*, 269–274.
21. Grigas, A.; Kemzūraitė, A.; Steponavičius, D.; Steponavičienė, A.; Domeika, R. Impact of Slope of Growing Trays on Productivity of Wheat Green Fodder by a Nutrient Film Technique System. *Water* **2020**, *12*, 3009. [CrossRef]
22. Cocetta, G.; Casciani, D.; Bulgari, R.; Musante, F.; Kołton, A.; Rossi, M.; Ferrante, A. Light Use Efficiency for Vegetables Production in Protected and Indoor Environments. *Eur. Phys. J. Plus* **2017**, *132*, 43. [CrossRef]
23. Zeng, Y.J.; Qi, X.N.; Li, Y.Z.; Wang, Y.Q.; Wang, Y.R. Optimising the Preparation Method of Seeds of Agriophyllum Squarrosum Prior to Staining in Tetrazolium Solution. *Seed Sci. Technol.* **2014**, *42*, 300–306. [CrossRef]
24. *LST ISO 712:2010*; Grains and Their Products, Determination of Moisture Content, Reference Method. Lithuanian Standardization Department: Vilnius, Lithuania, 2010.
25. *ISO 20483: 2006*; Cereals and Pulses-Determination of the Nitrogen Content and Calculation of the Crude Protein Content-Kjeldahl Method. International Organization for Standardization: Geneva, Switzerland, 2006.
26. *ISO 6865: 2000*; Animal Feeding Stuffs—Determination of Crude Fiber Content—Method with Intermediate Filtration. ISO (International Organization for Standardization): Geneva, Switzerland, 2000.
27. *ISO 5984:2022(E)*; Animal Feeding Stuffs—Determination of Crude Ash. International Organization for Standardization: Geneva, Switzerland, 2022.
28. *LST ISO 1443:1973*; Meat and Meat Products—Determination of Total Fat Content. Lithuanian Standardization Department: Vilnius, Lithuania, 1973.
29. Olsson, U.; Engstrand, U.; Rupšys, P. *Statistical Methods SAS and MINITAB*; LŽŪU Press Center: Akademija, Lithuanian, 2000. (In Lithuanian)

30. Devienne, F.; Mary, B. Influence of Constant Nitrate Concentration on N Uptake and Growth of Wheat. In *Structure and Function of Roots, Proceedings of the Fourth International Symposium on Structure and Function of Roots, Stará Lesná, Slovakia, 20–26 June 1993*; Springer: Berlin/Heidelberg, Germany, 1995; pp. 141–147.
31. Zhang, X.; Franzisky, B.L.; Eigner, L.; Geilfus, C.; Zörb, C. Antagonism of Chloride and Nitrate Inhibits Nitrate Reductase Activity in Chloride-Stressed Maize. *Plant Growth Regul.* **2021**, *93*, 279–289. [CrossRef]
32. Chebli, Y.; Geitmann, A. Cellular Growth in Plants Requires Regulation of Cell Wall Biochemistry. *Curr. Opin. Cell Biol.* **2017**, *44*, 28–35. [CrossRef] [PubMed]
33. Hilty, J.; Muller, B.; Pantin, F.; Leuzinger, S. Plant Growth: The What, the How, and the Why. *New Phytol.* **2021**, *232*, 25–41. [CrossRef] [PubMed]
34. Khan, F.; Siddique, A.B.; Shabala, S.; Zhou, M.; Zhao, C. Phosphorus Plays Key Roles in Regulating Plants' Physiological Responses to Abiotic Stresses. *Plants* **2023**, *12*, 2861. [CrossRef]
35. Vuppaladadiyam, A.K.; Yao, J.G.; Florin, N.; George, A.; Wang, X.; Labeeuw, L.; Jiang, Y.; Davis, R.W.; Abbas, A.; Ralph, P.; et al. Impact of Flue Gas Compounds on Microalgae and Mechanisms for Carbon Assimilation and Utilization. *ChemSusChem* **2018**, *11*, 334–355. [CrossRef]
36. Souza, S.R.; Stark, E.M.L.M.; Fernandes, M.S.; Magalhaes, J.R. Effects of Supplemental Nitrogen on Nitrogen-assimilation Enzymes, Free Amino Nitrogen, Soluble Sugars, and Crude Protein of Rice. *Commun. Soil Sci. Plant Anal.* **1999**, *30*, 711–724. [CrossRef]
37. Tsouvaltzis, P.; Kasampalis, D.S.; Aktsoglou, D.-C.; Barbayiannis, N.; Siomos, A.S. Effect of Reduced Nitrogen and Supplemented Amino Acids Nutrient Solution on the Nutritional Quality of Baby Green and Red Lettuce Grown in a Floating System. *Agronomy* **2020**, *10*, 922. [CrossRef]
38. Smith, M.R.; Reis Hodecker, B.E.; Fuentes, D.; Merchant, A. Investigating Nutrient Supply Effects on Plant Growth and Seed Nutrient Content in Common Bean. *Plants* **2022**, *11*, 737. [CrossRef]
39. Nicholls, J.W.F.; Chin, J.P.; Williams, T.A.; Lenton, T.M.; O'Flaherty, V.; McGrath, J.W. On the Potential Roles of Phosphorus in the Early Evolution of Energy Metabolism. *Front. Microbiol.* **2023**, *14*, 1239189. [CrossRef]
40. Wang, Y.; Wu, W.-H. Regulation of Potassium Transport and Signaling in Plants. *Curr. Opin. Plant Biol.* **2017**, *39*, 123–128. [CrossRef]
41. Barthole, G.; Lepiniec, L.; Rogowsky, P.M.; Baud, S. Controlling Lipid Accumulation in Cereal Grains. *Plant Sci.* **2012**, *185*, 33–39. [CrossRef] [PubMed]
42. Khalil, H.P.S.A.; Yusra, A.F.I.; Bhat, A.H.; Jawaid, M. Cell Wall Ultrastructure, Anatomy, Lignin Distribution, and Chemical Composition of Malaysian Cultivated Kenaf Fiber. *Ind. Crops Prod.* **2010**, *31*, 113–121. [CrossRef]
43. Podgórska, A.; Burian, M.; Gieczewska, K.; Ostaszewska-Bugajska, M.; Zebrowski, J.; Solecka, D.; Szal, B. Altered Cell Wall Plasticity Can Restrict Plant Growth under Ammonium Nutrition. *Front. Plant Sci.* **2017**, *8*, 1344. [CrossRef]
44. Zhang, W.; Qin, W.; Li, H.; Wu, A. Biosynthesis and Transport of Nucleotide Sugars for Plant Hemicellulose. *Front. Plant Sci.* **2021**, *12*, 723128. [CrossRef]
45. Kavanová, M.; Lattanzi, F.A.; Grimoldi, A.A.; Schnyder, H. Phosphorus Deficiency Decreases Cell Division and Elongation in Grass Leaves. *Plant Physiol.* **2006**, *141*, 766–775. [CrossRef] [PubMed]
46. Hasanuzzaman, M.; Bhuyan, M.H.M.B.; Nahar, K.; Hossain, M.S.; Mahmud, J.A.; Hossen, M.S.; Masud, A.A.C.; Moumita; Fujita, M. Potassium: A Vital Regulator of Plant Responses and Tolerance to Abiotic Stresses. *Agronomy* **2018**, *8*, 31. [CrossRef]
47. Peng, J.-S.; Gong, J.-M. Vacuolar Sequestration Capacity and Long-Distance Metal Transport in Plants. *Front. Plant Sci.* **2014**, *5*, 19. [CrossRef]
48. Sevanto, S. Drought Impacts on Phloem Transport. *Curr. Opin. Plant Biol.* **2018**, *43*, 76–81. [CrossRef]

MDPI AG
Grosspeteranlage 5
4052 Basel
Switzerland
Tel.: +41 61 683 77 34

Plants Editorial Office
E-mail: plants@mdpi.com
www.mdpi.com/journal/plants

www.ingramcontent.com/pod-product-compliance
Lightning Source LLC
Chambersburg PA
CBHW061505210926
51850CB00008B/1828